S0-BVP-390

DO NOT REMOVE
CARDS FROM POCKET

A volume in the series

Cornell Studies in Political Economy

EDITED BY PETER J. KATZENSTEIN

A full list of titles in the series appears at the end of the book.

Economic Containment

CoCom and the Politics of East-West Trade

Michael Mastanduno

Cornell University Press

Ithaca and London

First published 1992 by Cornell University Press.

International Standard Book Number 0-8014-2709-6 (cloth)
International Standard Book Number 0-8014-9996-8 (paper)
Library of Congress Catalog Card Number 92-52766

Printed in the United States of America

*Librarians: Library of Congress cataloging information appears on the
last page of the book.*

⊗ The paper in this book meets the minimum requirements of the
American National Standard for Information Sciences—Permanence of
Paper for Printed Library Materials, ANSI Z39.48-1984.

To my parents,

Salvatore and Elizabeth Mastanduno

Contents

List of Tables and Figure

Preface

As the cold war began, officials from the United States and Western Europe met to consider whether and how to prevent the Soviet Union and its allies from strengthening their military capabilities through East-West trade. It is unlikely any of them expected their informal "gentlemen's agreement" of 1949 that founded the Coordinating Committee that became known as CoCom to endure for over forty years and serve as an arena within which Western governments confronted the fundamental political problem of how to deal with the Soviet Union.

This book seeks to explain the cooperation and conflicts that have characterized the Western effort to use East-West trade as a strategic weapon. It steers between pure "power" and "regime" explanations and emphasizes leadership and the existence of common interests as key factors in the resolution of cooperation problems. It argues that to understand CoCom requires one to appreciate the interplay of international and domestic politics, and it attempts to show that the domestic structure of the United States, as reflected in export control policy, has profoundly affected the ability of the United States to exercise effective leadership in CoCom.

In writing this book I have incurred many debts, both personal and institutional. Faculty fellowships from Dartmouth and Hamilton Colleges, and research grants from Princeton University, the Brookings Institution, and the Institute for the Study of World Politics, provided the time and resources necessary to carry out the research and writing. Parts of the argument have appeared elsewhere, and I am grateful for permission to adapt or cite from those sources. An early version of Chapter 2 appeared as "Strategies of Economic Containment" in the July 1985 issue of *World Politics*. A version of Chapter 3 appeared as "Trade as a Strategic Weapon" in the Winter 1988 issue of *International*

Organization. Portions of what became Chapters 5 through 9 can be found in my contributions to *Controlling East-West Trade and Technology Transfer*, edited by Gary Bertsch (Duke University Press, 1988; reprinted with permission of the publisher); *East-West Trade and the Atlantic Alliance*, edited by David Baldwin and Helen Milner (London: Macmillan, and New York: St. Martin's Press, 1990; copyright © David A. Baldwin and Helen V. Milner, 1990); and "Searching for Security in a Global Economy," *Daedalus,* Journal of the American Academy of Arts and Sciences, Fall 1991, Vol. 120/4 (used by permission).

My greatest intellectual debt is to Robert Gilpin, who sparked my interest in the relationship between economics and security, and who provided the inspiration and example that sustained me in the initial stages of my career. I also owe a special thanks to John Ikenberry and David Lake, for the friendship, intellectual stimulation, and good humor that make the academic life so rewarding.

Many other scholars have been generous in their support. Kenneth Oye provided insight and direction to this project when both were lacking. David Baldwin, Peter Katzenstein, Lisa Martin, and Henry Nau read the manuscript in its penultimate draft and offered valuable suggestions. Others whose comments and reactions have been important include Robert Art, Gary Bertsch, Valerie Bunce, Beverly Crawford, Jeffry Frieden, Judith Goldstein, Joanne Gowa, Joseph Grieco, Stephan Haggard, Ed Hewett, Miles Kahler, Ethan Kapstein, Stephen Krasner, Charles Lipson, James McAdams, Helen Milner, John Odell, Robert Paarlberg, Louis Pauly, William Root, John Steinbruner, Angela Stent, Pete Suttmeier, and Raymond Vernon.

I am grateful to my students for engaging me on many of the ideas in this book, and especially to James Hulme, Michelle Sieff, Tyler Goldman, and Robert Martinage for research assistance. I also thank the many export control officials, on both sides of the Atlantic, for patiently answering my questions about an organization and activity they would have preferred not to discuss. I thank Roger Haydon for his skillful prodding on a subject I preferred not to discuss—the completion of this manuscript.

I have been privileged to know the value of a strong family. I thank my parents for making sacrifices in their lives so that I could have opportunities in mine, and I dedicate this book to them. My son, Michael Anthony, has taught me a great deal about keeping work in proper perspective. Finally, this book simply could not have been written without the love, support, and endurance of my wife, Melanie, who has always been there for me.

MICHAEL MASTANDUNO

Hanover, New Hampshire

Abbreviations

ASW	antisubmarine warfare
CCL	Commodity Control List
CG	Consultative Group
Chincom	China Committee
CoCom	Coordinating Committee
CTA	critical technologies approach
DoD	Department of Defense
DTSA	Defense Technology Security Administration
EAA	Export Administration Act
ECA	Economic Cooperation Administration
EDAC	Economic Defense Advisory Committee
EPCI	Enhanced Proliferation Control Initiative
ERP	Economic Recovery Program
FBIS	Foreign Broadcast Information Service
FRUS	Foreign Relations of the United States
GAO	General Accounting Office
GATT	General Agreement on Tariffs and Trade
HLM	high level meeting
IC/DV	Import Certificate/Delivery Verification
ICL	International Computers Limited
ICs	integrated circuits
IEA	International Energy Agency
IL	International List
ISA	International Security Affairs
LSI	large-scale integration
MCTL	Military Critical Technologies List

MFN	most favored nation
MIRV	multiple independently targetable reentry vehicle
MITI	Ministry of International Trade and Industry
MoD	Ministry of Defence
MOU	Memorandum of Understanding
NAS	National Academy of Sciences
NSA	National Security Agency
NSC	National Security Council
OECD	Organization of Economic Cooperation and Development
OEEC	Organization of European Economic Cooperation
OTA	Office of Technology Assessment
PDR	processing data rate
RSI	rationalization, standardization, interoperability
SGDN	Secrétariat Général de la Défense Nationale
SIOP	Single Integrated Operational Plan
STEM	Security and Technical Experts' Meeting
TACs	Technical Advisory Committees
TWEA	Trading with the Enemy Act
VLSI	very large-scale integration

Economic Containment

CHAPTER ONE

The Political Economy of CoCom

Should Western governments permit the sale of advanced telecommunications systems to the Soviet Union and Eastern Europe? That question proved to be highly contentious throughout the 1980s and even into the 1990s. For Western equipment producers, such as Alcatel of France and Siemens of West Germany, the answer was clear. Global competition was fierce, and every market counted. The rising costs of research and development forced firms to rely heavily on overseas sales to increase revenues, recoup investments, and maintain profits. The Soviet Union and Eastern Europe, with a stated commitment to modernize badly neglected telephone systems, were an obvious target. As recently as 1987, Moscow could receive simultaneously only sixteen long-distance calls, which was all the more remarkable because long-distance calls to other parts of the Soviet Union were channeled through Moscow.[1] Firms that established a foothold in these markets would reap immediate benefits and be well positioned to capture follow-on contracts for sales and services.

For Western governments the issue was more complex. On one hand, public officials shared an interest in the economic well-being of their firms and often sought to shape the global competition to their national advantage.[2] Success in the telecommunications sector could bring employment, income, and technological development to the national econ-

---

1. See Anthony Ramirez, "A.T.&T. in Link to Armenia," *New York Times*, September 10, 1991, p. D1.

2. When Italtel, Italy's public telephone authority, expressed interest in a joint-venture partner to modernize the Italian telecommunications network, both the U.S. and the West German governments lobbied aggressively at the highest levels on behalf of AT & T and Siemens, respectively. AT & T, which earlier had been outmaneuvered in the French market because of political intervention by Sweden and West Germany, was awarded the contract in 1989. See Thomas O'Boyle, "Once-Sleepy Siemens Turns Tiger in Fight for Telecom Business," *Wall Street Journal Europe*, March 15, 1989, p. 1.

omy. In the East-West context, these potential economic benefits were reinforced by political considerations. By the mid-1980s, Western governments were strongly committed to the success of Soviet and East European liberalizing reform efforts. Improved telecommunications infrastructures could accelerate those efforts by facilitating the links between individuals within those countries and between those countries and the Western world.[3] Access to information was a critical asset to both political democrats and economic entrepreneurs.

On the other hand, Western governments needed to weigh the military risks. Public telecommunications networks served the military as well as the civilian sector. Significant improvements in those networks, facilitated by Western technology, would mean a more efficient Soviet military with an improved capacity to direct and control its nuclear, conventional, and internal security forces. Moreover, the design and manufacturing know-how that would accompany such transfers might enhance Soviet military capabilities more broadly, by contributing to improvements in the development and production of missiles, radar, and other systems that relied intensively on microelectronics and computer networks.[4]

Conflict arose among Western governments because they perceived the trade-offs differently. The United States, locked throughout the cold war in global competition with the Soviet Union for military and political influence, emphasized the strategic risks of telecommunications transfers and sought strict controls.[5] West European governments, more dependent on exports for national economic well-being, were concerned that valuable economic and political opportunities might be lost and thus were wary of extensive controls.

As they had since 1949, Western governments sought to resolve these trade-offs between economics and security collectively, in a multilateral export control regime known as the Coordinating Committee, or CoCom. Conflict within CoCom over telecommunications exports was apparent by 1979, when the French government interpreted multilateral restrictions liberally and allowed the state-owned firm Thomson to sup-

3. Christine Westbrook and Alan B. Sherr, "U.S.-Soviet Joint Ventures and Export Control Policy," *Briefing Paper No. 3* (Providence, R.I.: Center for Foreign Policy Development, March 1990), pp. 8–10.

4. See, for example, "Alcatel's Soviet Joint Venture: A Sale the West Cannot Afford to Make," unpublished paper (Washington, D.C.: Center for Security Policy), March 21, 1989.

5. PD-53, entitled National Security Telecommunications Policy and signed by President Carter in 1979, and PD-59, a revision of U.S. nuclear doctrine signed in June 1980, singled out a survivable and reliable telecommunications infrastructure as a critical component of U.S. nuclear deterrence and war-fighting strategy. See Desmond Ball, "The Development of the SIOP, 1960–1983," in Ball and Jeffrey Richelson, eds., *Strategic Nuclear Targeting* (Ithaca: Cornell University Press, 1986), pp. 78–79. During the 1980s the United States sought as a national security priority to upgrade its own telecommunications network and was not about to see similar advantages conferred on its potential adversary.

ply the Soviet Union with digital switching equipment and technology worth $110 million. The United States protested, but to no avail. When the CoCom list came up for revision in 1982, however, U.S. officials made telecommunications a priority sector for stricter control. Other CoCom members acquiesced in 1984, but only after a bitter struggle and a U.S. concession to liberalize telecommunications exports automatically in September 1988. The compromise forced British, French, and West German firms to abandon major sales to the East. Sweden, though not a member of CoCom, suffered as well, as the Reagan administration coerced it to comply informally by threatening to restrict the access of Swedish firms to U.S. technology and components.[6]

Conflict reemerged in 1988. The Reagan administration sought to renege on its commitment to liberalize controls, contending that the Eastern bloc had not improved its indigenous capabilities to the extent CoCom members anticipated when they struck their compromise in 1984. Other CoCom governments refused to budge, asserting that controls were already outdated and that political developments in Eastern Europe suggested that, if anything, CoCom should be liberalizing more, rather than less, quickly. America's partners were also bothered by what they perceived as a fundamental inequity in the controls. In 1985, at U.S. initiative, CoCom liberalized restrictions on telecommunications exports to China. It thus appeared that the United States was supporting the maintenance of controls on trade with Eastern Europe and the Soviet Union, where West European firms were dominant, and supporting relaxation of controls on trade with China, where U.S. firms enjoyed some comparative advantage.

The United States grudgingly honored its commitment to liberalize in 1988, but controversy persisted nevertheless. Although more advanced equipment could now be sold to the East, manufacturing technology and know-how remained subject to CoCom review. The Belgian subsidiary of Alcatel tested the technology restriction in 1989 by proposing, in a deal with an estimated value of $1 billion, to sell to the Soviets its most sophisticated digital switching equipment and, more important, to construct a facility to manufacture that equipment in the Soviet Union.[7] Although the Belgian prime minister publicly endorsed the project, the United States vetoed the first phase of the deal in CoCom in February 1989 and made clear that it would continue to object.

The Alcatel joint venture and one proposed by Siemens were held in abeyance during 1989, as members of CoCom sought to absorb the impact of the collapse of Communist power in Eastern Europe, the

---

6. See *Export Control News* 2 (February 3, 1988), 2–6.

7. "Alcatel's Soviet Joint Venture," pp. 3–4, and Hugo Dixon and Terry Dodsworth, "Alcatel Close to Signing $1bn Soviet Contract," *Financial Times*, March 6, 1989, p. 1.

imminent collapse of the Warsaw Pact, and the coming unification of Germany. In June 1990 CoCom governments held a historic meeting and reacted to those changes by announcing that they had agreed to deep reductions in their export controls.

In the telecommunications sector, however, stalemate persisted. To the dismay and frustration of U.S. and West European firms, CoCom members could not agree to liberalize. The Bush administration demonstrated U.S. resolve by announcing in the context of the CoCom meeting that it had vetoed a proposed $500 million venture, led by a U.S. firm, to stretch a high-speed fiber-optic cable across the Soviet Union. Administration officials argued that the Soviet Union still represented a significant military threat to the West. High-speed fiber optics would enhance the command and control capabilities of the Soviet military and would frustrate U.S. efforts to monitor Soviet military communications.[8] The United States, supported by Great Britain, stood firm, and when CoCom's first "post–cold war" control list was unveiled in September 1991, strict controls on telecommunications exports remained.[9]

Why Study CoCom?

The coordination of export control policies has proved to be a politically difficult undertaking for the United States and its Western allies. The central purpose of this book is to explain the evolution of U.S. and Western export control policy, from the early postwar period to the beginning of the 1990s. What strategies did Western governments adopt in trade with the Soviet Union, and why? How did they resolve conflicts over the choice of a collective strategy? How effective were they in cooperating to restrict the release of sensitive technologies to the Soviet Union? Did their effectiveness vary over time, and if so, why?

I focus primarily on CoCom, whose members as of 1992 included the NATO allies (minus Iceland) plus Japan and Australia. CoCom's objective has been to monitor and control the flow of technologies and products of military significance to the Soviet Union, other members of the

8. *Export Control News* 4 (June 30, 1990), 3, 8–9. The proposed US West fiber-optic cable would transmit data at a rate of 565 megabits per second, well beyond the 140-megabits-per-second threshold observed at the time by CoCom.

9. See Bill Root, "The Core List: Expectations vs. Reality," ibid. 5 (July 29, 1991), 2–5. The conflict among Western governments continued into 1992, after it was revealed that the German company Carl Zeiss Jena, with the support of the German government, was planning to exploit a loophole in CoCom's rules to ship advanced fiber optic cable to the Soviet Union from the territory of former East Germany. Ibid. (December 18, 1991), p. 2.

Warsaw Pact, and the People's Republic of China.[10] The coordination of export controls in CoCom has constituted a regime situated squarely at the intersection of economics and national security; it has represented a collective effort by its members, who are military allies, to regulate their economic competition in order to respond more effectively to what they have perceived as common military threats.[11]

Why devote sustained attention, in a book-length study, to CoCom? One key reason is that policy coordination in CoCom has proved to be an important—yet relatively understudied—aspect of postwar international relations. CoCom has endured for over four decades, a striking example of sustained cooperation in economic statecraft. During the cold war, that cooperation extended beyond the membership of the organization and involved the United States in close, highly sensitive working relationships with neutral states, such as Sweden and Switzerland, and even those with special political ties to the Soviet Union, such as Finland and India. At the same time, CoCom was a source of considerable political controversy, reinforcing the East-West divide and also pitting the United States in conflict with other Western governments. CoCom played an obvious role in the military sphere, complementing and reinforcing U.S. and NATO deterrence strategies vis-à-vis the Soviet Union and Warsaw Pact. Though less obvious, it was and continues to be instrumental in helping to maintain a liberal international economic order among the non-Communist states. In short, export control policies and their coordination in CoCom have been an integral part of the postwar international system; to understand them is to understand more fully the dynamics of that system.

Nevertheless, students of international relations have devoted relatively little attention to CoCom. Although that attention has increased significantly over the past decade, we still know far less about export control coordination and CoCom than we do about other postwar international regimes and the institutions associated with them—trade and

10. CoCom was established during November 1949 and began to function on January 1, 1950. All current CoCom members joined by 1953, with the exception of Spain, which joined in 1985 after becoming a NATO member, and Australia, which joined in 1989. The United States, Britain, France, Italy, the Netherlands, Belgium, and Luxembourg struck the initial agreement. Norway, Denmark, Canada, and West Germany joined in early 1950, Portugal and Japan in 1952, and Greece and Turkey in 1953. See Gunnar Adler-Karlsson, *Western Economic Warfare, 1947–1967* (Stockholm: Almquist and Wiksell, 1968), p. 52. In addition to former Warsaw Pact members and the People's Republic of China, as of 1991 other proscribed destinations included Albania, Mongolia, Vietnam, and North Korea. In 1991, CoCom granted Poland, Hungary, and Czechoslovakia special status, making it possible for them to import most items on CoCom's Industrial List. The three pledged to cooperate with CoCom and to guard against unauthorized diversions of controlled technology. See *Export Control News* 5 (March 26, 1991), 9–10.

11. By "regime," I mean a set of rules, norms, and expectations around which actors' expectations converge in a given issue-area. This commonly accepted definition is provided by Stephen Krasner, "Structural Causes and Regime Consequences: Regimes as Intervening Variables," in Krasner, ed., *International Regimes* (Ithaca: Cornell University Press, 1983), p. 2.

the GATT, structural adjustment and the IMF, and development assistance and the World Bank.[12] It may be taken as one small sign of CoCom's relative obscurity that as late as 1974, the exhaustive *Yearbook of International Organizations* listed it as a defunct entity. As of 1989, the *Yearbook* recognized that CoCom existed but mistakenly asserted that it had originated as a part of NATO.[13]

This inattention to CoCom can be traced, in part, to its remarkably modest institutional presence and tradition of confidentiality. CoCom has operated discreetly from a small annex to the U.S. embassy in Paris, with a budget estimated during the mid-1980s at only $500,000.[14] Throughout the postwar era its deliberations and decisions have not, as a matter of course, been made public. Confidentiality was a necessary part of the early cold war compromise that created CoCom; for several West European states, participation in a system of economic discrimination targeted against Communist states was of dubious legality and potentially explosive politically. Even the nondescript appellation "CoCom" was considered classified information until 1953.[15] In part because of inertia and in part because of the continued political sensitivity of CoCom in some member states, the tradition of confidentiality endured beyond the early postwar period. CoCom's activities, to be sure, have become more transparent over time, and many useful sources of information have become available.[16] Nevertheless, the presumption and

12. There exists a voluminous literature, and at least several "core" books, on the workings of and political interaction within each institution. On the GATT, for example, see Kenneth W. Dam, *The GATT: Law and International Economic Organization* (Chicago: University of Chicago Press, 1970), and Gilbert Winham, *International Trade and the Tokyo Round Negotiation* (Princeton: Princeton University Press, 1986). On the IMF, see Margaret DeVries, *The International Monetary Fund, 1966–1971: System under Stress*, 2 vols., and *The International Monetary Fund, 1972–1978: Cooperation on Trial*, 2 vols. (Washington, D.C.: IMF, 1976 and 1985). On the World Bank, see Edward S. Mason and Robert E. Asher, *The World Bank since Bretton Woods* (Washington, D.C.: Brookings Institution, 1973), and Robert Ayres, *Banking on the Poor: The World Bank and World Poverty* (Cambridge: MIT Press, 1983). The sole "core" book on CoCom is Gunnar Adler-Karlsson's groundbreaking *Western Economic Warfare, 1947–1967*, which was published twenty-five years ago.

13. See *Yearbook of International Organizations, 1974*, 15th ed. (Brussels: Union of International Associations, 1974), entry 422d, p. 88, and *Yearbook of International Organizations, 1989–90*, 26th ed., vol. 1 (Munich: K.G. Saur, 1989), entry FF4582g, p. 1384. CoCom and NATO have had no formal relationship in the postwar era; the separation of the two institutionally was an explicit part of the compromise that created CoCom in 1949. See Chap. 3.

14. See Michael Mastanduno, "What Is CoCom and How Does It Work?" in Robert Cullen, ed., *The Post-Containment Handbook: Key Issues in U.S.-Soviet Economic Relations* (Boulder, Colo.: Westview Press, 1990), p. 75.

15. See Mutual Defense Assistance Control Act of 1951, *Reports to the Congress* (hereafter *Battle Act Reports*), no. 5 (November 23, 1954), p. 16.

16. The following sources have been most useful to me: (a) declassified U.S. executive branch documents and cables, particularly with regard to the origins and early development of CoCom; (b) U.S. congressional hearings and reports, spanning the entire postwar period; (c) reports from advisory groups (e.g., the Defense Science Board, the National Academy of Sciences) to the U.S. government; (d) interviews with U.S. and West European officials; (e) reporting in the business and technical press in the United States and Western Europe, usefully abstracted since 1983 in *Interflo: A Soviet Trade News Monitor*; and (f) the existing secondary literature, most of which is cited elsewhere in this chapter.

habit of confidentiality probably deterred some scholars, as well as being a source of frustration for journalists and other interested observers.[17]

The comparative inattention to CoCom can also be attributed to the fact that the issue of national security export controls does not fall neatly into conventional categories of analysis. During the postwar era, students of international political economy became accustomed to viewing the control of trade with military applications (as well as East-West economic relations more generally) as not being a part of "normal" political economy, with this term reserved for the bulk of economic transactions that take place among advanced industrial states and between those states and less developed ones.[18] For their part, specialists in national security affairs emphasized "high politics" issues such as deterrence, military strategy, arms races, and arms control, rather than "low politics" issues such as trade and technology transfer, even among potential military adversaries. Since the issue of strategic export control sits at the intersection of political economy and national security, the study of CoCom coordination has tended to fall into the cracks, as it were, that demarcate fields of scholarly investigation.[19]

Over the course of the 1980s, Western conflicts over Afghanistan, Poland, and the pipeline combined with Eastern efforts to reform their economies and integrate them more fully into the global economy to increase the political salience of Western policies on East-West trade and export control. There was, consequently, a significant increase in the amount of scholarly attention devoted to these issues and, by implication, to CoCom. Studies of Western policy that were relatively analytical or conceptual, however, tended not to focus directly on CoCom but to

17. See, for example, Susan Sachs, "Low Profile Group Affects Billions in World Trade," *Journal of Commerce* (September 22, 1987), p. 1. Sachs quotes Gerard Motel, the assistant director of the European Parliament's External Relations Committee, which sought in 1987 to produce a study of CoCom: "If you write to them, they don't write back. . . . If you phone them, they don't answer. If you have one of them in front of you, he will refuse to admit he works for CoCom. It is simply a club in which the members have a very private gentlemen's agreement."

18. In the preface to his authoritative survey of international political economy, Robert Gilpin notes that his focus is on "normal" economic activities—trade, monetary relations, and foreign investment—and neglects (presumably "abnormal") subjects such as East-West economic relations and the use of economic weapons for political ends. See *The Political Economy of International Relations* (Princeton: Princeton University Press, 1987), pp. xiv–xv. Students of economic statecraft (as Gilpin notes) do focus on such issues; yet, as David Baldwin has argued, the study of economic statecraft itself has been neglected relative to that of other foreign policy instruments. See David Baldwin, *Economic Statecraft* (Princeton: Princeton University Press, 1985), p. 4.

19. The sharp increase in interest among international relations specialists in "economics and security" that was evident by the late 1980s has begun to rectify the relative neglect of issues falling at the intersection of the two subfields. See, for example, Richard Rosecrance, *The Rise of the Trading State: Commerce and Conquest in the Modern World* (New York: Basic Books, 1986); Paul Kennedy, *The Rise and Fall of the Great Powers: Economic Change and Military Conflict from 1500 to 2000* (New York: Random House, 1987); Aaron Friedberg, *The Weary Titan: Britain and the Experience of Relative Decline, 1895–1905* (Princeton: Princeton University Press, 1988); and Ethan Kapstein, *The Insecure Alliance: Energy Crises and Western Politics since 1944* (New York: Oxford University Press, 1990), and *The Political Economy of National Security* (New York: McGraw-Hill, 1991).

treat it only to the extent that it bore on other problems.[20] The work devoted directly to CoCom tended to focus primarily on the current policies of, and conflicts among, its members, and on policy prescriptions.[21] What remain in short supply are studies of CoCom that integrate past patterns of Western interaction over export controls with those of the present, in a manner that is analytically or conceptually instructive.[22] This book is intended partly to fill that void and in so doing to enhance our understanding of several related issues of enduring concern in the international relations literature.

One such issue is economic statecraft, or the use of economic measures to achieve political objectives. Students of economic statecraft seek to understand why states employ it, the particular instruments or strategies available to them, and the conditions that facilitate or inhibit its effectiveness.[23] This book contributes to that effort by drawing upon existing literature to develop a framework for analyzing the relationship between international trade and military security. In Chapter 2, I outline and

20. The most important work includes Beverly Crawford and Stephanie Lenway, "Decision Modes and International Regime Change: Western Collaboration and East-West Trade," *World Politics* 37 (April 1985), 375–402; Angela Stent, *From Embargo to Ostpolitik: The Political Economy of West German-Soviet Relations, 1955–1980* (New York: Cambridge University Press, 1981); Bruce Jentleson, *Pipeline Politics: The Complex Political Economy of East-West Energy Trade* (Ithaca: Cornell University Press, 1986); Gary Bertsch, ed., *Controlling East-West Trade and Technology Transfer: Power, Politics, and Policies* (Durham: Duke University Press, 1988); Reinhard Rode and Hans-Dieter Jacobsen, eds., *Economic Warfare or Detente: An Assessment of East-West Relations in the 1980s* (Boulder, Colo.: Westview Press, 1985); Philip Hanson, *Trade and Technology in Soviet-Western Relations* (New York: Columbia University Press, 1982); William J. Long, *U.S. Export Control Policy: Executive Autonomy versus Congressional Reform* (New York: Columbia University Press, 1989); Richard J. Ellings, *Embargoes and World Power: Lessons from American Foreign Policy* (Boulder, Colo.: Westview Press, 1985); and David Baldwin and Helen Milner, eds., *East-West Trade and the Atlantic Alliance* (New York: St. Martin's, 1990).

21. See, for example, Angela Stent Yergin, *East-West Technology Transfer: European Perspectives* (Beverly Hills, Calif.: Sage, 1980); David Buchan, "Western Security and Economic Strategy toward the East," *Adelphi Papers*, no. 192 (London: International Institute for Strategic Studies, 1984); William A. Root, "Trade Controls That Work," *Foreign Policy* 56 (Fall 1984), 61–80; Robert Price, "CoCom after 35 Years: Reaffirmation or Reorganization," in Charles M. Perry and Robert L. Pfalzgraff, eds., *Selling the Rope to Hang Capitalism?* (Maclean, Va.: Pergamen-Brassey's International Defense Publishers, 1987), pp. 195–201; J. Michael Cleverly, "The Problem of Technology Transfer Controls," *Global Affairs* 4 (Summer 1989), 109–28; Henry R. Nau and Kevin Quigley, eds., *The Allies and East-West Economic Relations: Past Conflicts and Present Choices* (New York: Carnegie Council on Ethics and International Affairs, 1989); and Kevin Quigley and William J. Long, "Export Controls: Moving beyond Containment," *World Policy Journal* 7 (Winter 1989–90), 165–88.

22. In addition to Adler-Karlsson, *Western Economic Warfare, 1947–1967*, see R. J. Carrick, *East-West Technology Transfer in Perspective* (Berkeley, Calif.: Institute of International Studies, 1978); John R. McIntyre and Richard Cupitt, "East-West Strategic Trade Control: Crumbling Consensus?" *Survey* 25 (Spring 1980), 81–108; Gary Bertsch et al., "East-West Technology Transfer and Export Controls," *Osteuropa Wirtschaft* 26 (June 1981), 116–36; Gunnar Adler-Karlsson, "International Economic Power: The U.S. Strategic Embargo," *Journal of World Trade Law* 6 (September/October 1972), 501–17; Gary Bertsch, *East-West Strategic Trade, COCOM, and the Atlantic Alliance* (Paris: Atlantic Institute for International Affairs, 1983); and Timothy Aeppel, "The Evolution of Multilateral Export Controls: A Critical Study of the CoCom Regime," *Fletcher Forum* (Winter 1985), 105–24.

23. The seminal work in this literature is Baldwin, *Economic Statecraft*. See also Gary C. Hufbauer and Jeffrey J. Schott, *Economic Sanctions in Support of Foreign Policy Goals* (Washington,

clarify the logic of four export control (and promotion) strategies available to states seeking to influence the military capabilities of potential adversaries. I examine why states would (or would not) prefer to employ such strategies as instruments of economic statecraft. Subsequent chapters use the framework to explain the national policy preferences of the United States and other CoCom members in trade with the Soviet Union. The convergence or divergence of national strategies, in turn, sets up the problem of policy coordination.

A second issue concerns international cooperation. During the 1970s and 1980s, a vast literature emerged on the extent to which hegemony (i.e., a preponderance of power resources) is required to create international regimes and sustain effective cooperation in them.[24] That literature is also concerned with how hegemonic power is exercised and with the relationship between the power of hegemonic states and the interests of nonhegemonic states.[25] At a more concrete policy level, it considers whether and to what extent U.S. power declined during the 1970s and 1980s and the implications for the stability of the international economic regimes created after World War II.[26]

This examination of the Western export control regime finds no simple causal relationship between the relative power resources of the United States and the maintenance of effective cooperation in CoCom. Although relative power matters, other factors—the interests of nonhegemonic states and the leadership capacity of the United States—emerge as critical in determining the nature and extent of cooperation.

The evidence suggests, for example, that when the preferred export

D.C.: Institute for International Economics, 1983); Margaret P. Doxey, *Economic Sanctions and International Enforcement*, 2d ed. (New York: Oxford University Press, 1980); and Albert Hirschman, *National Power and the Structure of Foreign Trade* (Berkeley: University of California Press, 1945, reissued ed., 1980).

24. See, for example, Krasner, ed., *International Regimes*; Robert Keohane, *After Hegemony: Cooperation and Discord in the World Political Economy* (Princeton: Princeton University Press, 1984), and "The Theory of Hegemonic Stability and Changes in International Economic Regimes, 1967–1977," in Ole Holsti et al., *Change in the International System* (Boulder, Colo.: Westview Press, 1980), pp. 131–62; and Duncan Snidal, "The Limits of Hegemonic Stability Theory," *International Organization* 39 (Autumn 1985), 579–614.

25. See Charles Kindleberger, *The World in Depression* (Berkeley: University of California Press, 1973); Stephan Krasner, "State Power and the Structure of International Trade," *World Politics* 28 (April 1976), 317–43; David A. Lake, *Power, Protection, and Free Trade: International Sources of U.S. Commercial Policy, 1887–1939* (Ithaca: Cornell University Press, 1988); Arthur Stein, "The Hegemon's Dilemma: Great Britain, the United States, and the International Economic Order," *International Organization* 38 (Spring 1984), 355–86; and Scott James and David A. Lake, "The Second Face of Hegemony: Britain's Repeal of the Corn Laws and the American Walker Tariff of 1846," *International Organization* 43 (Winter 1989), 1–30.

26. In addition to Gilpin, *The Political Economy of International Relations*, and Kennedy, *The Rise and Fall of the Great Powers*, see Henry Nau, *The Myth of America's Decline: Leading the World Economy into the 1990s* (New York: Oxford University Press, 1990); Bruce Russett, "The Mysterious Case of Vanishing Hegemony," *International Organization* 39 (Spring 1985), 207–31; Susan Strange, "The Persistent Myth of Lost Hegemony," *International Organization* 41 (Autumn 1987), 551–74; and Joseph S. Nye, *Bound to Lead: The Changing Nature of American Power* (New York: Basic Books, 1990).

control strategies of the United States and its CoCom partners *diverged* fundamentally, the United States proved unable to prevail and determine the course of CoCom strategy, regardless of the extent to which a disparity in power resources existed. U.S. officials were forced to compromise and defer to the preferences of their allies both during the mid-1950s, a period of hegemonic ascendance, and during the early 1980s, a time of putative hegemonic decline.

When export control preferences *converged*, effective cooperation did not emerge spontaneously but required the exercise of U.S. leadership. I explain below why the United States must lead in CoCom; I also elaborate a specific set of responsibilities U.S. officials must undertake to maximize the potential for effective cooperation. For much of the postwar period, however, the United States has proved incapable of consistent, effective leadership. U.S. officials established a pattern of neglecting leadership responsibilities and of abusing their privileged position in CoCom. In explaining this behavior, I place less emphasis on power resources and more on U.S. domestic structure. Even during the 1970s and 1980s, the United States possessed adequate resources to lead fully and effectively in CoCom. One must look to the domestic level—in particular, to the existence of a powerful, yet divided Executive and to the absence of effective countervailing power from the U.S. private sector—to understand its inability to do so. Ironically, more so than its weakness, the considerable strength of the American state has constrained its leadership capacity.

Third, this book is aimed to enhance our understanding of postwar U.S. trade policy. The defining characteristic of that policy has been economic liberalism: a belief in free trade and a commitment on the part of the government to minimize its intervention in the marketplace. Scholars have traced the ideological and institutional roots of U.S. liberalism from the depression of the 1930s and the effort by U.S. officials to establish a liberal international economic order following World War II.[27]

By way of contrast, this book emphasizes that a powerful undercurrent of economic nationalism—a sense that the government must intervene in the market and regulate international economic transactions in the interest of national security—has coexisted uneasily with liberalism in postwar U.S. thinking and policy. Although one could trace the U.S. tradition of economic nationalism back to Alexander Hamilton, postwar export control policy, rooted in the "economic defense" experience of

27. See Charles Maier, "The Politics of Productivity: Foundations of American International Economic Policy after World War II," in Peter Katzenstein, ed., *Between Power and Plenty* (Ithaca: Cornell University Press, 1978), pp. 23–50; John Ruggie, "International Regimes, Transactions, and Change: Embedded Liberalism in the Postwar Economic Order," in Krasner, ed., *International Regimes*, pp. 195–233; and Judith Goldstein, "Ideas, Institutions, and American Trade Policy," in G. John Ikenberry, David A. Lake, and Michael Mastanduno, eds., *The State and American Foreign Economic Policy* (Ithaca: Cornell University Press, 1988), pp. 179–218.

World War II, offers the most prominent peacetime example of sustained economic intervention by U.S. officials in the interest of national security. In the chapters below I highlight both the accumulation of power by the U.S. Executive in pursuit of economic nationalism and, more important, the conflict between the dominant trait of liberalism and the recessive strain of nationalism in U.S. policy. Export control policy has had the potential to compromise the U.S. commitment to liberal trade policies. I trace the postwar effort by U.S. officials to manage the conflict between liberal and nationalist impulses, and emphasize the importance of an effective, multilateral export control regime in enabling them to do so effectively.

Finally, this study of CoCom should have policy relevance, despite the fact that the Soviet military threat and other key features of the postwar international system, including probably CoCom itself, will diminish in significance during the 1990s. As the cold war has ended, other issues highlighted in this book have risen to policy prominence. One is what Helmut Schmidt termed the "struggle for the world product," or economic and technological competition among the advanced industrial states.[28] That competition is likely to intensify as the Soviet threat recedes, as the pace of technological innovation continues to accelerate, and as defense capabilities continue to be driven by developments in the commercial sector.[29]

Another such issue concerns the proliferation of chemical, nuclear, and ballistic missile technologies to ambitious states in the developing world. On the impetus of the Persian Gulf War, the regulation of such technologies has already emerged as the key export control issue of the 1990s.[30] The experience of CoCom is instructive because many of the problems to be confronted are analogous to those faced by the West in its forty-year effort to deny sensitive technologies to the Soviets and their allies. Indeed, as discussed in Chapter 9, it is striking that patterns of U.S. behavior and multilateral coordination formed in the East-West context have been recreated in the nascent effort to build export control regimes to meet emerging threats in the North-South arena.

The remainder of this chapter outlines the relationship between U.S. policy and CoCom that is critical to the argument of the book. I elaborate in detail the conditions under which effective cooperation in CoCom might be achieved, the leadership responsibilities of the United

28. Helmut Schmidt, "The Struggle for the World Product," *Foreign Affairs* 52 (April 1974), 437–51.
29. See B. R. Inman and Daniel F. Burton, "Technology and Competitiveness: The New Policy Frontier," *Foreign Affairs* 69 (Spring 1990), 116–34, and Edward Luttwak, "From Geopolitics to Geo-Economics," *The National Interest* 20 (Summer 1990), 17–24.
30. See Janne E. Nolan, *Trappings of Power: Ballistic Missiles in the Third World* (Washington, D.C.: Brookings Institution, 1991), and National Academy of Sciences, *Finding Common Ground: U.S. Export Controls in a Changed Global Environment* (Washington, D.C.: National Academy Press, 1991).

States, and the factors that have constrained U.S. officials from carrying out those obligations fully and effectively. I then introduce the conflict between America's export control program and its postwar commitment to the maintenance of a liberal world economy. A final section briefly describes the chapters that follow.

EXPLAINING COOPERATION IN CoCom

The arguments and analysis of this book are grounded in the realist tradition in international relations.[31] I conceive of nation-states as the principal actors and assume that their interests are formulated and pursued by state (i.e., executive) officials. These officials may be either unified or divided in their approach to export control policy.[32] State officials have sought to use export control policies to promote a conception of the "national interest" defined in terms of some combination of military security, political opportunity, and economic welfare. In devising and pursuing export control strategies, executive officials have responded both internally, to considerations of public opinion and business interests, and externally, to the constraints and opportunities provided by the international economic and security environment.[33]

The coordination of national export control preferences in CoCom has represented a collective attempt by states to control markets or to limit some forms of economic interdependence in the interest of military security. Two aspects of cooperation in CoCom require explanation. CoCom's *strategy* refers to the broad criteria that have governed the construction and scope of CoCom's control list.[34] What general guidelines have CoCom members adopted in deciding what to control, and why? CoCom's *effectiveness* refers to the manner in which member states have implemented and administered the embargo. Given a choice of strategy, has the embargo been implemented stringently, or with laxity?

31. The best recent discussion is Robert Keohane, ed., *Neorealism and Its Critics* (New York: Columbia University Press, 1986), esp. the chapters by Keohane and Robert Gilpin, pp. 158–203 and 301–21.

32. That the state may be a divided actor deviates from a strict or pure realist approach, which would assume states to be *unitary* as well as rational actors. See the description of the realist research program in ibid., pp. 164–65.

33. An effort to develop a realist theory of the state as both internally and externally oriented is found in Michael Mastanduno, David A. Lake, and John Ikenberry, "Toward a Realist Theory of State Action," *International Studies Quarterly* 33 (December 1989), 457–74.

34. CoCom members have actually compiled three lists, which cover munitions (Group 1), atomic energy items (Group 2), and industrial items (Group 3). The third list has proved to be the most contentious historically, because it contains "dual-use" items of both commercial and military utility. Unless otherwise noted, when I refer to the "CoCom list" in this book, I am referring to the Industrial List (IL). Until 1991, that list was divided into eight segments, which cover metal-working machinery; chemical and petroleum equipment; electrical and power-generating equipment; general industrial equipment; transportation equipment; electronic equipment, including communications, radar, and computer hardware and software; metals, minerals, and their manufactures; and chemicals, metalloids, and petroleum products. The text of one CoCom

Conflicts over Strategy and U.S. Coercive Power

In principle, national security export controls could range from being very narrow and covering only weapons and other military items to being very broad and covering all exports to a target state. In practice, one set of Western conflicts has revolved around the choice of two strategies, which are based on different assumptions about the relationship between trade and national security. The two are analyzed in detail in Chapter 2, and the well-known distinction between them is critical to the argument of this book.

A *strategic embargo* seeks to prohibit only the trade that makes a direct and significant contribution to an adversary's military capabilities. In addition to purely military items, it focuses on the control of those commercial items that have a specific and important military use. Since the formation of CoCom in 1949, the United States and other member governments have agreed that multilateral controls should be maintained and should reflect, at least, the criteria of a strategic embargo.

During 1949–1958 and 1980–1984, however, the United States went further and sought to extend CoCom controls according to the criteria of *economic warfare*. Economic warfare is aimed to weaken the military capabilities of a target state by weakening that state's economy. The assumption is that because military power is ultimately dependent on an economic base, quantitatively and qualitatively, trade that significantly enhances the economy of an adversary indirectly enhances its military power and thus should be prohibited in the interest of national security.

U.S. officials found economic warfare most attractive when America's political relationship with the Soviet Union was confrontational and when U.S. officials viewed their strategic relationship with the Soviets as one of intensified military competition. Other CoCom members, in contrast, generally rejected the logic of economic warfare and proved reluctant to accept the extension of controls consonant with it. Their preference for confining controls to a more narrowly conceived strategic embargo can be traced to their relatively greater economic interest in East-West trade and their preference for a less confrontational political relationship with the Soviets. In addition, unlike the United States, they did not view themselves as mired in an arms race or a global struggle for influence with the Soviet Union.

Given this conflict over preferred export control strategies, to what

item—IL 1565 (computers)—taken from Great Britain's 1989 version of the CoCom list, is reprinted in Mastanduno, "What Is CoCom and How Does It Work?," pp. 79–105. In response to the end of the cold war, in September 1991, CoCom members formally adopted a more streamlined "core list" of industrial items, grouped into nine categories: advanced materials; materials processing; electronics; computers; telecommunications; sensors; navigation and avionics; marine technology; and propulsion. An abridged version of the core list is printed in *Export Control News* 5 (August 25, 1991), 2–4.

extent did the United States manage to obtain the compliance of other members in economic warfare and thereby determine CoCom strategy? The evidence put forth below suggests that the United States generally was unsuccessful. It failed to gain alliance support for economic warfare in the 1980–1984 period, despite a sustained effort, both within and outside CoCom, involving state officials at the highest levels. Similarly, the major alliance dispute of 1954 was settled in favor of the West European position, and U.S. officials eventually were forced to adjust their national export control strategy to an alliance consensus that reflected West European preferences.

The United States was unable to prevail for two reasons. First, unlike other issues (e.g., the removal of import barriers) in which a lack of cooperation would frustrate the interests of both sides, in export control strategy the United States required the support of other members, but not vice versa. The U.S. economic warfare strategy could not be viable in the absence of effective coordination. At the same time, failure to resolve the conflict of national preferences could actually work to the relative economic advantage of Western Europe and Japan. If the United States practiced a more restrictive control strategy than its allies, firms in the latter countries could exploit the differential and trade with the East, gaining economic benefits while frustrating U.S. controls.

Their lack of leverage within the export control issue area forced U.S. officials to consider cross-issue linkage. Domestic and international constraints, however, rendered the costs of such linkage prohibitive. In neither the 1950s nor the 1980s was the United States willing to open home markets sufficiently to compensate Western Europe for the economic costs of comprehensive economic denial in East-West trade. At the same time, the exercise of coercion was constrained by broader U.S. foreign policy interests. Both the Truman and the Eisenhower administrations resisted a coercive strategy (i.e., the denial of military or economic aid or of U.S. exports) in order to preserve the integrity of NATO and the nascent liberal transatlantic economic order. In the early 1980s the Reagan administration attempted intimidation but was forced to back down when it became apparent that the political costs of the subsequent transatlantic conflict outweighed whatever strategic gains might be realized from collective economic warfare. Overall, the nature of the East-West trade conflict, coupled with the constraint created by America's broader foreign policy commitments, served to provide other CoCom members with leverage over their more powerful alliance partner.

The only postwar circumstances in which America's allies supported economic warfare involved the outbreak of the Korean War. Their willingness to comply in that instance was not the result of the effective application of U.S. coercive power in the face of conflicting preferences, as the generally accepted interpretation contends. Rather, a shift in West

European preferences caused by the fear of imminent military conflict with the Soviet Union brought accord.[35] Intensified military competition and the fear of war facilitated U.S. leadership and were necessary conditions in making economic warfare attractive. When this fear subsided, so too did West European willingness to maintain the strategy.

CoCom's Strategic Embargo and U.S. Leadership

Since 1954, CoCom's control criteria have reflected that of a strategic embargo. The second aspect of cooperation in CoCom that requires explanation is the effectiveness, or strength, of that strategic embargo and its variations over time. When the export control preferences of the United States and its CoCom partners have been complementary, what determines the extent to which effective cooperation has been realized?

One must be careful to distinguish two meanings of "effectiveness."[36] The first concerns the impact of export controls on the Soviet defense sector or, to be more precise, the extent to which CoCom controls contributed to the preservation of U.S. and Western lead time over the Soviet Union and Warsaw Pact in the application of dual-use technology to military systems. To what extent did CoCom matter in East-West military competition?[37] The second directs attention to the internal political workings of CoCom and concerns the extent to which member states, given their commitment to a strategic embargo, faithfully formulate, implement, and administer their multilateral controls. Although I devote attention in this book to the first meaning, I am more concerned systematically with the second—with describing and explaining the extent to which member states cooperated effectively in CoCom. The two are, of course, related; an understanding of the nature and extent of

35. Lisa Martin usefully distinguishes "coercion games," situations of conflicting interests in which one player must extract the compliance of others, from "coadjustment games," situations of mutual interest in which states seek to coordinate policies. See *Coercive Cooperation: Explaining Multilateral Economic Sanctions* (Princeton: Princeton University Press, 1992). Although economic warfare generally involved a coercion game, I argue, in effect, that after the outbreak of the Korean War, it became a matter of coadjustment. Once preferences were compatible, U.S. leadership was required to sustain effective cooperation. See the discussion of leadership below.

36. See Bertsch et al., "East-West Technology Transfer and Export Controls," and John P. Hardt and Kate S. Tomlinson, "The Potential Role of Western Policy toward Eastern Europe in East-West Trade," in Abraham Becker, ed., *Economic Relations with the U.S.S.R.* (Lexington, Mass.: Lexington Books, 1983), pp. 105–25.

37. As I have argued elsewhere, to determine the precise impact of CoCom controls on the preservation of Western lead time in East-West military competition is a complex undertaking. Export controls are not the sole factor influencing the technology gap; of equal or greater importance are the efficiency of the U.S. weapons procurement system, the indigenous technological capabilities of the Soviet Union, and the ability of the Soviets to absorb and diffuse Western technology. And, even if one could isolate the effect of export controls, there are additional difficulties in ascertaining whether Western technology actually found its way into Soviet military systems and, if so, how significant a contribution it ultimately made. See Mastanduno, "The Management of Alliance Export Control Policy," in Bertsch, ed., *Controlling East-West Trade and Technology Transfer*, pp. 241–79.

cooperation in CoCom is a necessary step in understanding the impact of CoCom on Soviet capabilities and the preservation of Western lead time.

An assessment of CoCom's effectiveness—defined in the second sense of the term—can be made by examining such factors as enforcement, exceptions, and the coverage and interpretation of the control list. Although such indicators do not lend themselves to precise measurement, they can be used to provide a general picture of CoCom's effectiveness at a given point or across time.

The *enforcement* of agreed controls is a critical determinant of any embargo's strength or weakness.[38] Signs of CoCom's strength include evidence that member states have been able to prevent or hamper the export and illegal acquisition of controlled items, or have at least been willing to devote additional effort and resources to that task when evidence exists that it is required. Signs of weakness include proof of significant or routine diversions and inaction by member governments in response.

CoCom rules include an *exceptions* procedure that allows members to request and (if approved) undertake a one-time sale of a controlled item to a controlled destination on the grounds that it will be used for non-military purposes. The granting of exceptions, in and of itself, should not be taken as a sign of weakness, and in fact, the selective use of this procedure has helped the embargo to function more smoothly when economic pressures to export have been considerable and the risks of diversion minimal.[39] It should be taken as a sign of weakness, however, if exceptions granting becomes a matter of routine or if it takes place without adequate assurances that the item in question will not be diverted to military use. As a general rule, the less exceptions, the stronger the regime; the routine granting of exceptions has the potential to weaken CoCom by legitimizing (in the eyes of member governments and their firms) trade in strategic items and by increasing the likelihood that such items will be diverted to the military sector of a target state.

The *construction* and *interpretation* of the control list is similarly important. It is a sign of regime weakness if items of direct military significance are left off the list (consciously or inadvertently) or if member governments interpret controls differently, that is, some allow sales that others presume to be restricted. Conversely, the undertaking of list revisions that lead to the addition of items of military significance or of policies that lead to uniformity in interpretation can be taken as an indicator of regime strengthening.

The failure to decontrol items that have lost their strategic significance is not, by itself, a sign of embargo weakness. Such a practice is clearly a

---

38. See, for example, Doxey, *Economic Sanctions and International Enforcement.*
39. This point is made by Carrick, *East-West Technology Transfer,* and William Root, "CoCom: An Appraisal of Objectives and Needed Reforms," in Bertsch, ed., *Controlling East-West Trade and Technology Transfer,* pp. 417–41.

contributing factor, however, in that it strains the credibility of the regime in the eyes of member governments and their firms, increases the demand for exceptions, and exacerbates the burden of enforcement. Thus, the decontrol of less strategic items generally leads to a strengthening of the embargo effort.

An examination of CoCom's strategic embargo since 1954 reveals variation in its effectiveness, defined according to the above indicators.[40] A useful division of time is into four periods, covering 1955–1957 and, roughly, the 1960s, 1970s, and 1980s. CoCom was relatively ineffective in the 1955–1957 period. Member governments were locked in a policy conflict over China, which led some to demand routine exceptions, others to defy agreed controls openly, and most to question the continued viability of the regime. In contrast, between 1958 and 1968, cooperation in CoCom was comparatively effective. Exceptions were minimized, as were the problems in enforcement, list coverage, and interpretation that became evident in later periods. CoCom was again relatively ineffective between 1969 and 1979. Cynicism and discontent among members pervaded the embargo effort as exceptions mounted and enforcement was noticably lax. Finally, the 1980–1989 period presents something of a mixed picture. On one hand, significant improvements took place in list coverage and enforcement. On the other, the failure to decontrol less strategic items led to considerable friction among member governments, strained the credibility of the embargo, and hampered ongoing enforcement efforts. Exceptions proliferated rapidly, though largely for China rather than for the Soviet Union.[41]

The above pattern suggests that although CoCom members may deem an effective strategic embargo desirable, it does not emerge spontaneously. Member states have found themselves in a situation requiring cooperation—defined as conscious policy coordination—rather than in one of harmony, in which the independent pursuit of self-interest by states leads automatically to the realization of joint gains.[42] Although all members have preferred an effective strategic embargo to its absence, many also have preferred unilateral defection to unrequited cooperation.[43]

40. It should be evident that I refer here to changes within the CoCom regime, not to a change of regime. The general distinction between regime weakening and regime change is noted by Krasner, "Structural Causes and Regime Consequences."

41. An alternative characterization is offered by William Root, who suggests that CoCom was equally ineffective across the periods I have described. Root views the extent to which CoCom members have decontrolled less strategic items as the principal determinant of CoCom's effectiveness. See "CoCom: An Appraisal of Objectives and Needed Reforms," pp. 422–25.

42. See Keohane, *After Hegemony*, pp. 51–55, and Arthur Stein, "Coordination and Collaboration: Regimes in an Anarchic World," in Krasner, ed., *International Regimes*, pp. 115–40.

43. Kenneth Oye sets out the two conditions necessary to distinguish cooperation from harmony: mutual cooperation must be preferred to mutual defection, and unilateral defection preferred to unrequited cooperation. See "Explaining Cooperation under Anarchy: Hypotheses and Strategies," in Oye, ed., *Cooperation under Anarchy* (Princeton: Princeton University Press, 1986), p. 6.

Thus, the realization of effective cooperation in CoCom has required the visible hand of conscious political intervention. What determines the extent to which effective cooperation has been realized and, by implication, explains the variation noted above?

The critical factor has been the leadership role played by the United States. For the strategic embargo to be effective, the United States must play a role in CoCom structurally similar to that which "benevolent" hegemonic powers must play in order to maintain a liberal world economy.[44] The specific reasons why the United States must play such a role and the specific responsibilities it must undertake in CoCom are, of course, different.

The necessity of a U.S. leadership position can be traced to two factors: the structure of preferences among CoCom members and the institutional or regime features of CoCom itself. First, although both the United States and other members have accepted, in principle, the need for a strategic embargo, crucial differences have existed. As leader of the Western alliance, the United States always assumed the primary responsibility for ensuring the adequacy of alliance deterrence and defense. Moreover, throughout the cold war, it was engaged with the Soviet Union in an arms race and a struggle for geopolitical influence. Consequently, the United States has possessed the strongest interest among alliance members in assuring that Western technology does not contribute, purposely or inadvertently, to the military capabilities of target states. This is not to suggest that other CoCom members have been unconcerned but that the United States, by necessity, has been the *most* concerned. Over the course of the postwar era, it has invested the most in military systems and technology and has been the most responsive, in terms of defense expenditures and commitments, to improvements in target capabilities. U.S. officials have proved willing to err heavily on the side of caution and to risk overcontrolling exports to target states in the interest of national security.

Other members' interest in CoCom's strategic embargo can best be described as contingent.[45] They have proved willing to support it fully

44. Charles Kindleberger sets out the responsibilities (e.g., acting as a market for distress goods, and serving as crisis lender and lender of last resort) hegemonic states must undertake to foster a liberal world economy. See *The World in Depression*, chap. 14. Duncan Snidal distinguishes benevolent from coercive hegemonic states; the former bear the disproportionate costs of providing public goods that benefit all, whereas the latter structure international relations to their own particular advantage and not necessarily to the benefit of others. See "The Limits of Hegemonic Stability Theory." In relations with CoCom members, the United States has been forced to be a coercive hegemon in cases of economic warfare, where preferences conflict, and generally a benevolent one in the case of the strategic embargo, where preferences coincide. In securing the compliance of non-CoCom states with the strategic embargo, however, the United States most often has been compelled to act coercively. See below, Chaps. 4 and 8.

45. Throughout the postwar era, the difference in approach to the CoCom embargo *between* the United States and other CoCom members has far outweighed differences *among* America's CoCom allies, thereby making it reasonable to generalize about the positions of "other CoCom members" or about conflicts between the United States and "other members," as I do frequently

only if items under control are clearly "strategic," that is, shown to be directly useful to the military of target states and likely to enhance military capabilities meaningfully. They have considered controls aimed to influence the political behavior or to weaken the economy of target states as inappropriate for CoCom. Equally important, each government has been willing to control items deemed to be strategic only if assurances can be provided that all other suppliers—CoCom and non-CoCom—are similarly restricting the items in question.[46] Other member governments have placed considerable value on the economic benefits—both absolute and relative—of exporting to the East; thus, they have been reluctant to deny export opportunities to their firms when foreign availability has existed.

The combination of America's preponderant and other members' contingent interest in the strategic embargo has made it imperative that the United States exercise leadership in CoCom. For Western Europe and Japan, where economic interests have been a primary concern, effective participation has required that certain assurances be provided and conditions be met. For the United States, where military security interests have been paramount and have required multilateral coordination in order to be realized, there has been a strong incentive to provide such assurances and to satisfy such conditions.

Second, the peculiar institutional features of CoCom make the exercise of U.S. leadership all the more essential. In addition to confidentiality, the CoCom regime has been characterized by informality. CoCom does not enjoy treaty status; it was created as, and has remained, an informal "gentlemen's agreement." Its decisions are, formally speaking, nonbinding recommendations to member governments. And those governments, by conscious design, enjoy a considerable degree of national discretion in the implementation of export controls. Although CoCom members coordinate multilaterally on the composition of the control list, the administration and, most important, the enforcement of controls have been left fully in the hands of national governments. CoCom has no enforcement mechanism of its own and imposes no sanctions or penalties on individuals, firms, or governments that violate its directives.

---

in this book. This is not to suggest, however, that there exists a complete uniformity in the approaches and policies of other CoCom members. During the 1980s, for example, there was considerable disparity in the commitment of members to the enforcement of controls, with Great Britain and France performing more effectively than West Germany and Japan and far more effectively than, say, Greece, Portugal, Italy, or Denmark. See the survey in *Export Control News* 2 (December 22, 1988), 4–7. Similarly, some members (e.g., Great Britain, France, West Germany, Japan, Italy, and the Netherlands) have tended to participate far more actively in the routine business of CoCom (e.g., in list reviews and weekly meetings to consider exception requests) than have others, such as Greece, Portugal, and Turkey.

46. Robert Jervis emphasizes that states fear exploitation in international relations and that the prospects of cooperation can be enhanced if such fears are alleviated. See "Cooperation under the Security Dilemma," *World Politics* 30 (January 1978), 178–214.

CoCom's reliance on informality and national discretion suggests the potential for considerable variation in the application and enforcement of controls. Furthermore, it leaves open the opportunity for member governments to manipulate or tailor their national control practices in order to maximize the competitive advantages of their respective firms. Even if governments resist that temptation, resourceful firms or individuals seeking to evade CoCom controls can exploit variations in national procedures (e.g., in the resources devoted to enforcement or in the penalties applied to violators) and find, in effect, weak links in CoCom's chain. Since the CoCom regime, in and of itself, does little to protect against the exploitation of differences in control procedures, that task requires political intervention and, ultimately, the exercise of leadership.

What, in this context, does leadership entail? The structure of interests among member states and the character of the CoCom regime suggest that in order to maximize the effectiveness of cooperation in CoCom, the United States must undertake four responsibilities. First, it must take the lead in *maintaining the integrity of the control process*. Although in principle all CoCom members share this task, in practice the United States has borne primary responsibility for bringing control proposals to CoCom and establishing their strategic merit. More so than other members, the United States has possessed and employed the intelligence resources necessary to determine which civilian items are likely to be of military significance in target and NATO systems. In judging strategic merit, other CoCom members have tended to rely heavily on U.S. technical assessments; when U.S. officials have offered clear and specific evidence of an item's military relevance, other members (in the absence of evidence of non-CoCom supply or target production capability) have generally been receptive to adopting controls.[47] The credibility of the United States has been compromised, however, whenever other members have suspected that U.S. control proposals were driven by political, rather than narrowly military, considerations.[48] In that event, the incentives to cooperate have diminished, and even proposals justifiably guided by strategic considerations have been called into question.

To maintain the integrity of the control process has also required that the United States take steps to ensure that CoCom members interpret multilateral controls with some degree of uniformity and enforce them

47. See Root, "Trade Controls That Work."
48. By "political" I mean the use of export controls (or promotion) to influence the behavior or policies, as opposed to military capabilities, of target states. For systematic discussion, see Chap. 2. The distinction between influencing behavior and weakening military capabilities is embedded in U.S. export control legislation; the Export Administration Act (EAA) distinguishes "national security" from "foreign policy" controls and imposes special restrictions on the ability of the Executive to employ the latter. See James M. Montgomery, "The Cumbersome . . . Apparatus," in Cullen, ed., *The Post-Containment Handbook*, pp. 39–40, 48–54.

with some measure of vigilance. A lax commitment on the part of a major CoCom supplier weakens the embargo and can also have multiplier effects by providing other members with incentives to neglect controls. As enforcer of last resort, the United States has had the responsibility to ensure that other members meet, at least, a minimum standard of effective protection and, at best, harmonize their enforcement efforts at high levels of effectiveness.

Second, the United States must bear primary responsibility for *obtaining the compliance of non-CoCom suppliers* with CoCom controls. States that are not members of CoCom yet possess advanced industrial and technological capabilities (e.g., Sweden and Austria) have the potential to frustrate CoCom controls, to the extent they have been willing and able to export (or reexport) controlled items to target states. Most CoCom members have been strongly inclined to resist controls on strategic items when even modest sources of non-CoCom availability exist. At the same time, they have proved reluctant to take initiatives aimed at eliminating such sources. Similarly, CoCom itself has had no regular mechanism for ensuring the compliance of non-CoCom suppliers. Responsibility thus rests with the United States to establish bilateral agreements with key suppliers, assuring that they control both indigenous production and reexports in exchange for access to U.S. technology, and to act as a liaison between such suppliers and CoCom.[49]

Third, the United States must *set a domestic example*. The United States has been called the "conscience" of CoCom, which suggests that American behavior sets a standard for others.[50] In order to encourage other members to adopt and enforce controls faithfully, U.S. officials have had to ensure that U.S. firms are denied strategic trade with the targets of the embargo and that U.S. enforcement efforts are exemplary. As students of both military and economic statecraft have shown, self-imposed costs are an important means to demonstrate resolve and enhance the credibility of one's commitments.[51] Historically, considerable discontent has been generated in CoCom by the perception that U.S. policies enable U.S. firms to gain competitive advantages in East-West or intra-Western trade. When this perception has been widespread, the incentives for effective cooperation have diminished. The perception, however, has

49. Good discussions of U.S. efforts to extract "third-country" compliance with CoCom are found in Adler-Karlsson, "International Economic Power," and Hendrik Roodbeen, "Trading the Jewel of Great Value: The Participation of the Netherlands, Belgium, Switzerland, and Austria in the Western Strategic Embargo versus the Socialist Countries" (Rijksuniversiteit Leiden, the Netherlands: unpublished manuscript, 1991). During the 1980s, as non-CoCom suppliers proliferated, the United States sought to multilateralize the burden of securing their compliance through the "third-country initiative" in CoCom. See Chap. 8.
50. See GAO, *Export Controls: Need to Clarify Policy and Simplify Administration* (Washington, D.C.: GPO, March 1, 1979), p. 9.
51. See Thomas Schelling, *The Strategy of Conflict*, 2d ed. (Cambridge: Harvard University Press, 1980), pp. 123–24; Baldwin, *Economic Statecraft*, p. 107; and Martin, *Coercive Cooperation*.

not always been correct; in fact, U.S. firms have generally been more disadvantaged by export controls than their West European counterparts have been.

Finally, the United States can enhance CoCom cooperation by working to *minimize the administrative and economic burdens* of the control system. Specifically, this effort entails keeping the multilateral control list as streamlined as possible, by decontrolling relatively less strategic items and those readily available to target states either indigenously or through non-CoCom sources. It also calls for timely responses to the requests of other members for general exceptions or permission to reexport U.S.-origin equipment or technology. The extent to which participation in the control system poses an administrative or economic burden has been a critical issue to West European governments and firms. The more comprehensive the list, the more difficult it has been for governments to administer and enforce it and for firms to compete effectively for export markets. Although increasingly relevant in the U.S. context, these factors have weighed far more heavily for other members of CoCom, who are more trade-sensitive than is the United States.

The United States has occupied a pivotal position with regard to alleviation of the burdens of the control system. Its crucial role is accentuated by CoCom's unanimity principle. To add or subtract any item from the multilateral list requires the unanimous consent of all members. In principle, all members have a formal veto over the decontrol of any item and, more generally, list reduction. In practice, however, most CoCom members have not stood in the way of decontrol and, in fact, have encouraged it. The practical veto power, legitimized by the unanimity rule, has been held by the United States. Whether and to what extent the United States acquiesces has been the primary determinant of the pace and degree of decontrol.[52]

The above discussion suggests that in order to maximize effective cooperation in CoCom, it has been necessary for the United States to be simultaneously firm and accommodating. The United States has had to act as the conscience or guardian of the regime, the one state willing to stand and say no in the face of economic pressures, and assure that

52. The unanimity principle implies, of course, that other members have a veto over list additions and can exercise it as a bargaining chip to force the United States to acquiesce in the decontrol of less strategic items. This potential lever is difficult to employ in practice, however, because the military significance of a given item is likely to be most evident—and justifiable—when it is initially proposed for list addition. Over time, as technology and target capabilities progress, the military utility of that item is likely to decrease, yet still exist in absolute terms. Thus, there has been a built-in tendency for the list to become more comprehensive over time. The task for CoCom members has been to assure that less critical items are removed (or their performance or technological parameters upgraded) as new items are added. As R. J. Carrick and others have noted, CoCom ideally should administer a "rolling" embargo over time. See Carrick, *East-West Technology Transfer in Perspective*, p. 36. It has been primarily the United States, however, by virtue of its practical veto over decontrol, which determines whether the list "rolls" or simply becomes more comprehensive.

neither CoCom nor non-CoCom suppliers frustrate multilateral controls. Other CoCom members (individually or collectively) have simply lacked the combination of interest and capability necessary to undertake that task. At the same time, however, the United States has also been responsible for maintaining the credibility of the embargo by minimizing or alleviating administrative and economic burdens. The United States has been uniquely positioned to carry out that task as well, and to do so enhances the incentives of other members to participate faithfully. Faithful participation is crucial, given that CoCom places considerable discretion in the hands of member governments to implement and enforce the embargo.

The above discussion also emphasizes the fact that the policy behavior of the United States has constituted an important source of U.S. leverage in CoCom. Other members have commitments in principle to the strategic embargo, but strong economic interests and commitments as well. U.S. *behavior* has played a critical role either in providing them with incentives to implement the necessary controls and maintain them effectively or in serving as a convenient excuse for the evasion of such tasks. By maintaining the integrity of the control process, the United States can increase the collective security benefits of cooperation. By ensuring the compliance of non-CoCom suppliers, it can help to alleviate the fear of other members that cooperation will lead to their exploitation. By setting a domestic example, it can reduce its own potential to gain from exploitation and thus enhance the credibility of its leadership efforts. By minimizing administrative and economic burdens, it can decrease the costs of cooperation, defined in terms of foregone economic opportunities, for other members.[53]

The United States, of course, has possessed other key sources of leverage as well. Its military and intelligence assets have been crucial in determining the strategic significance of particular items and in allowing U.S. officials to make cogent arguments on military grounds for control (or decontrol) in CoCom. U.S. officials can also draw on the economic and technological resources of the U.S. government to compel other members to adopt, or compensate them for adopting, stricter adherence to CoCom.[54] Most important, U.S. officials can bring to bear the resources of private U.S. firms. The reliance of West European or Japanese firms on their U.S. counterparts as a source of supply or as a market has been a valuable instrument in shaping the behavior of other CoCom govern-

53. The general conditions under which one state can enhance the potential for cooperation through its behavior are elaborated in game theoretic terms by Jervis, "Cooperation under the Security Dilemma," pp. 170–86, and Oye, "Explaining Cooperation under Anarchy," pp. 4–11.
54. The Reagan administration, for example, explicitly linked the participation of West German firms in the Strategic Defense Initiative to a commitment from the West German government to improve export controls. See "Federal Republic of Germany-United States: Agreements on the Transfer of Technology and the Strategic Defense Initiative," reprinted in *International Legal Materials* 25 (July 1986), 957–77.

ments, particularly those most recalcitrant in the adoption and enforcement of controls. As discussed below, however, the disruption of intra-Western trade links can also be very costly to the United States, because it has the potential to breed discontent and resentment in CoCom and to earn U.S. firms reputations as unreliable suppliers in Western markets.

The Performance of the United States

Although the effectiveness of CoCom's strategic embargo has depended on U.S. leadership, the postwar performance of the United States has been, at best, mixed. At times, the United States has worked to preserve the integrity of the control process; at other times, its behavior has compromised it. At times, U.S. officials have undertaken systematic efforts to obtain the compliance of non-CoCom suppliers; at other times, they have been passive in the face of this problem. Sometimes the United States has set a strong domestic example; at other times it has not. Finally, until 1990 the United States proved unwilling or unable, with any degree of consistency, to minimize the economic and administrative burdens of the control system.

Although the overall record suggests inadequate and inconsistent leadership, there has been variation across time. U.S. performance was most inadequate during 1969–1979, as U.S. officials failed to set a domestic example, impose discipline on CoCom and non-CoCom suppliers, or minimize the burdens of the control system. During 1980–1989 the United States acted more effectively as the conscience of CoCom, yet it still managed to compromise the integrity of the control process and to exacerbate the economic and administrative burdens of the system. U.S. leadership was plagued in similar ways during 1955–1957, largely because of the obsession of U.S. officials with comprehensive controls on China. In relative terms, the United States was most effective during 1958–1968 at being simultaneously firm and accommodating in CoCom. Even then, however, signs of a willingness to compromise the integrity of the control process and increase economic burdens through unilateral action were apparent.

What accounts for the inability of the United States to lead fully and consistently in CoCom? Possible explanations can be located at the level of the international system, domestic society, or the state.[55] I find that international systemic factors, though important, are inadequate by themselves to explain U.S. behavior. A satisfactory explanation requires one to move to the domestic level and to focus in particular on the

55. For a discussion of the three approaches and a summary of the relevant literature concerning each, see G. John Ikenberry, David A. Lake, and Michael Mastanduno, "Introduction: Approaches to Explaining American Foreign Economic Policy," in Ikenberry et al., eds., *The State and American Foreign Economic Policy*, pp. 1–14.

distinctive structural features of the state, and of the relationship between state and society, in U.S. export control policy.

An explanation lodged at the level of the international system directs our attention to the relative power position of the United States. Defined in terms of control over economic resources, U.S. power clearly declined over the course of the postwar era. During the early 1960s, for example, the U.S. government initiated the production and controlled the release of many of the items deemed strategic in the context of the CoCom embargo. By the mid-1970s, initiative for the development of such technologies had passed from the U.S. government to the private sector, and control over their production and dissemination diffused from the United States to other CoCom members and eventually to non-CoCom states as well.

Consequently, during the 1960s it was much simpler for the United States to lead effectively in CoCom. As long as the United States dominated many of the relevant technologies, gaining acceptance for their control from other CoCom members, who had not yet developed a strong economic stake in their export, was not difficult. Since there were relatively few non-CoCom suppliers, obtaining their compliance was also a less complicated task. In the event of conflicts over particular items, the dependence of both sets of actors on the U.S. private sector as a source of equipment and technology provided a significant source of leverage to U.S. officials.

By the 1970s and 1980s, in contrast, leadership was more difficult. The economic stakes of other CoCom members in the export of dual-use technology increased, making it more difficult, as the telecommunications example demonstrated, to gain their acquiescence for controls. At the same time, the United States was forced to contend with a proliferation of non-CoCom suppliers, whose activities posed direct and indirect challenges to the integrity of CoCom.[56]

The relative distribution of economic and technological resources matters. It has helped to structure relationships in CoCom and to determine whether the leadership task is more or less challenging. By itself, however, it cannot explain the inadequacy of U.S. leadership. Even during the 1970s and 1980s, the period of *relative* decline, the United States still possessed sufficient economic and technological resources in *absolute* terms to lead effectively in CoCom. In areas critical to the embargo effort (e.g., computers and electronics), many West European firms remained heavily reliant on their U.S. counterparts as sources of supply. And, as the ambitious "third-country" effort demonstrated, such re-

56. As the number of potential suppliers increases, the prospect for effective cooperation decreases. See Oye, "Explaining Cooperation under Anarchy," pp. 19–22. Non-CoCom states could frustrate CoCom directly by selling to the Soviet Union and indirectly by encouraging CoCom members to defect in order to avoid the so-called sucker's payoff.

liance on U.S. technology continued to be an effective source of leverage for the United States in its efforts to obtain the compliance of non-CoCom suppliers, even during the 1980s. Moreover, bear in mind that throughout this period other CoCom states accepted in principle the need for a strategic embargo. Although it clearly faced resistance to certain proposals, the United States did not confront (as it did in the 1982 pipeline case or, more generally, in economic warfare) other governments that were so unalterably opposed to the controls in question that the only possible way to obtain their compliance was continually to coerce them into submission.

In addition, the relative decline of U.S. economic power should not have been expected to affect detrimentally the ability of the United States to carry out at least some of its leadership tasks. For example, the job of bringing control proposals to CoCom and establishing their strategic merit relies more heavily on military and intelligence assets—of which the United States has retained a preponderance—than on economic ones. Similarly, there is no reason to presume that U.S. officials should have been less able, in circumstances of relative decline, to minimize the economic burdens of the control system. Indeed, one could plausibly argue that they should have been more able, on the grounds that relative economic decline should have generated a greater sensitivity to the economic costs of controls and thus a greater willingness to support the liberalization of less strategic items.

A more satisfactory explanation requires attention to the domestic structure of the United States. Students of foreign economic policy typically characterize the United States as possessing a weak state.[57] Domestic "weakness" refers to the decentralization and fragmentation of the government structure, to the ease with which private actors penetrate and influence that structure, and to the dearth of resources and instruments available to state officials in pursuit of policy goals. The weakness of the state is viewed as a constraint on the ability of officials to articulate foreign economic policy objectives and attain them when confronted by domestic opposition.

In export control policy, however, the weak state characterization is not apt. To be sure, executive officials have been constrained by a fragmented and decentralized government structure. Yet, at the same time, they have possessed a wide array of powerful policy instruments to employ in pursuit of their preferred policies. And, executive officials have been well-insulated from the U.S. private sector, which has managed to muster remarkably little in the way of countervailing power and influence.

57. See Stephen Krasner, "United States Commercial and Monetary Policy: Unravelling the Paradox of External Strength and Internal Weakness," and Peter Katzenstein, "Domestic Structures and Strategies of Foreign Economic Policy," in Katzenstein, ed., *Between Power and Plenty*, pp. 51–88, and 295–336. See also Stephen Krasner, *Defending the National Interest* (Princeton: Princeton University Press, 1978).

These structural features set the United States apart from other advanced industrial states. In West European countries and Japan, the formulation of export control policies has been characterized more by cohesion and consensus than by division and fragmentation. Relations between industry and government have been close and collaborative rather than distant and confrontational. And, although other CoCom governments have retained the right to control exports for reasons of national security, in general they have neither sought nor exercised the far-reaching authority enjoyed by the U.S. Executive.[58]

That authority is grounded in U.S. export control law. Passed by Congress at the peak of cold war tension, the Export Control Act of 1949, later amended, has allowed the Executive to restrict U.S. exports in peacetime for reasons of national security or foreign policy or to prevent domestic economic shortages. Subsequent application has demonstrated that it is nearly impossible to conceive of an instance in which export controls could not be justified by at least one of these criteria. The law also authorizes the Executive to prohibit or curtail *all* exports, commercial as well as military, including intangible items such as technical data. And, although ostensibly directed at the Soviet Union and its allies, the Executive may restrict trade regardless of destination. As the cold war began, U.S. firms found that what had been a right to export in peacetime suddenly was transformed into a privilege to be granted by the state. The arbitrary nature of executive power was highlighted by the fact that export control decisions were generally made in secret and were exempt from judicial review, which left aggrieved exporters with no right to challenge decisions in federal court.[59]

A decentralized and divided Executive has wielded these powerful policy instruments. Responsibility for export control policy has been widely dispersed within the U.S. government, creating the potential for an ongoing struggle for influence among agencies whose policy preferences are at times complementary and at times conflictual. After World War II, Commerce was granted the interagency lead in national export control policy (a legacy of its wartime role as administrator of short supply controls), whereas State assumed lead responsibility for multilateral coordination. Defense's formal role was less well defined, although its influence has been considerable. Other agencies (e.g., the National Security Council staff, Treasury, the CIA, and the Department of Energy) have also participated extensively at certain times or on certain issues. And, as in other areas of trade policy, the Executive as a

58. See Stent Yergin, *East-West Technology Transfer: European Perspectives*, and the country-specific essays in Bertsch, ed., *Controlling East-West Trade and Technology Transfer*, and in Baldwin and Milner, eds., *East-West Trade and the Atlantic Alliance*.

59. Franklin D. Cordell and John L. Ellicott, "Judicial Review under the Export Administration Act of 1979: Is It Time to Open the Courthouse Doors to U.S. Exporters?" in National Academy of Sciences, *Finding Common Ground*, appendix H, pp. 321–35.

whole shares responsibility with the Congress. Although Congress has proved willing to delegate extensive authority to the Executive, it has also seen fit to reassert itself routinely in the formulation and implementation of export control policy.[60]

Finally, this powerful, yet fragmented state has been only minimally constrained by the U.S. private sector. In overall terms, the history of U.S. export control policy has been one of the subordination of business interests to the pursuit of national security and foreign policy goals by the state. American firms have been consistently frustrated by the byzantine nature of the U.S. control system, their variable access to it, and their inability to influence decisively the substance of policy.

The tradition of business deference was established by the early 1950s, because U.S. firms were either uninterested in trade with the Communist world or suppressed their interest for fear of being stigmatized as "traders with the enemy." By the 1960s, as firms in Western Europe and Japan entered Eastern markets, American industry began to lobby for the removal of overly restrictive export controls. They have lobbied ever since, with only modest success. Major triumphs for industry have been short-lived and usually followed by reassertions or extensions of executive authority.[61] In general, U.S. firms have been on the defensive, seeking to temper what they have considered to be the most destructive aspects of U.S. export control policy.

These distinctive structural features of U.S. export control policy have constrained U.S. leadership in CoCom. The combination of a powerful and insulated, yet fragmented state has produced or facilitated patterns of policy behavior that have, in turn, complicated the ability of U.S. officials to carry out the leadership tasks necessary for effective cooperation in CoCom.

For example, the combination of a powerful Executive with an ambitious foreign policy agenda, and the lack of countervailing influence from the private sector, has facilitated the "sanctions habit," or the routine reliance of U.S. officials on export controls to influence the political behavior of other governments. Other CoCom governments have sought to avoid using foreign policy controls, largely because they are especially costly economically and have the potential to earn firms reputations as

60. See, for example, the discussion of the Battle Act in Chap. 3 and of the Toshiba incident in Chap. 8. In the latter case, Congress mandated formal sanctions against a foreign firm for a CoCom violation; this action defied CoCom's norm, which leaves enforcement, and the imposition of penalties against violators, to the discretion of the violating firm's government.

61. For example, the initial opening of U.S. trade with the Soviet Union during the détente of the early 1970s was followed by the Jackson-Vanik Amendment, restrictions on export credits, and only marginal progress on the relaxation of export controls. Similarly, the liberalizing Export Administration Act of 1979 was followed almost immediately by severe sanctions against the Soviet Union following the invasion of Afghanistan and the imposition of martial law in Poland. Even the eagerly anticipated "core list" of 1991 left U.S. industry with major disappointments. See Joanne Connelly, "Relaxed CoCom Core List Draws Mixed Reactions," *Electronic News*, June 3, 1991, p. 1.

unreliable suppliers. The United States, in contrast, has dominated the postwar employment of economic sanctions by an overwhelming margin. According to one survey of 107 cases covering 1945 to 1982, the United States participated in 75 of these, far ahead of the United Kingdom (14), France (12), and the Soviet Union (11). In 63 cases, the United States acted unilaterally. The targets of U.S. sanctions have spanned the globe, and the instruments have included restrictions on or the promotion of arms, economic aid, technology, grain, and energy supplies.[62]

Although economic sanctions per se are not detrimental to U.S. performance in CoCom, they become entangled with the multilateral regime all too easily when the targets are CoCom destinations and the sources of leverage include CoCom-controlled items. The entanglement may be inadvertent, as when export promotion became a key unilateral instrument of U.S. policy toward the Soviet Union during the early 1970s and toward China during the early 1980s. It may also be purposive, as when U.S. officials sought to use CoCom as a convenient forum for the coordination of punitive sanctions against China in the mid-1950s and against Poland and the Soviet Union in the early 1980s.

In either case, U.S. policy has worked at cross-purposes with its leadership role and the effectiveness of cooperation in CoCom. Export promotion in CoCom-controlled items has compromised the ability of the United States to set a domestic example and has encouraged other members to be lax in enforcement and in their interpretations of CoCom directives. Efforts to manipulate CoCom controls as a means to punish particular targets have compromised the integrity of the control process and have led other members to demand exceptions and defy CoCom regulations. And, the detrimental consequences for U.S. leadership have been magnified when U.S. foreign policy initiatives have had differential economic effects, appearing to favor American firms. This differential has led other members to suspect a mercantilist conspiracy between U.S. government and industry to dominate certain export markets, a situation that further decreases their willingness to cooperate.

The weak position of industry and fragmentation within the Executive have reinforced a second distinctive U.S. policy trait: an insensitivity to the economic costs of East-West controls. The United States has consistently struck the balance between economic and strategic or foreign policy interests in East-West trade overwhelmingly in favor of the latter.[63] Although the Commerce Department and at times other agencies have championed economic concerns, the Executive as a whole has never

62. See Ellings, *Embargoes and World Power*, pp. 123–24 and app. C, pp. 161–63. A chronology of sanctions cases is also provided by Hufbauer and Schott, *Economic Sanctions in Support of Foreign Policy Goals.*

63. See National Academy of Sciences, *Balancing the National Interest.* The one significant exception has been grain, although even in that case foreign policy considerations have loomed large, as evidenced by the controversial embargo of 1980.

reached a clear consensus that the potential economic benefits of exporting to the East, as a national interest, are worth defending.

That attitude is a legacy of early cold war policy, in which the combination of national security imperative, ideological crusade, and overall economic dominance led the United States to downplay almost entirely the economic stakes in East-West trade. It has persisted, even as the U.S. economy has become more reliant on foreign markets to maintain domestic economic growth and prosperity.[64] Neither private industry nor its allies in Commerce and the Congress have managed to precipitate a fundamental shift in how the balance between economic and security objectives is struck in U.S. export control policy. As it increased its influence over that policy during the 1970s and 1980s, the Defense Department assured the maintenance of a conservative bias in the formulation and administration of U.S. controls, despite the changing role of the United States in the world economy.[65]

The relative insensitivity of the United States to the economic costs of East-West restrictions has constrained its ability to relieve the economic and administrative burdens of the multilateral control system. A willingness to tolerate an overly comprehensive control list that errs heavily on the side of caution nationally has translated into tolerance of the same multilaterally. A willingness to accept long delays in the processing of export license requests nationally has spilled over and affected the manner in which the U.S. control system handles the CoCom exception requests of other governments and the reexport requests of foreign firms. Such considerations, judged to be less critical by the U.S. government, have been judged to be more crucial by America's trade-sensitive allies, whose incentives to cooperate fully in CoCom have been influenced by the extent to which they have perceived the control system as posing an unwarranted economic burden.

Yet another distinctive feature of U.S. export control policy has been its inconsistency and unpredictability. Although the policies of West European states and Japan have been generally uniform over time, that of the United States has lurched from the comprehensive denial of the 1960s, to the linkage strategies and export promotion of the 1970s, and back to comprehensive denial from 1980 to 1984.[66] Since 1985, U.S. policy has shifted again, in the direction of export liberalization. The pattern of inconsistent policy has been evident not only across time but

64. For an analysis of the response (or lack thereof) of export control policy to globalization trends in the world economy, see Michael Mastanduno, "The United States Defiant: Export Controls in the Postwar Era," *Daedalus* 120 (Fall 1991), 91–112.

65. James M. Montgomery has recently noted in an analysis of the Export Administration Act that "the ethos of economic warfare lingers, faintly in the law itself, quite markedly in the way it is administered." See "The Cumbersome . . . Apparatus," p. 39.

66. See Philip Hanson, "Soviet Responses to Western Trade Policies," in Baldwin and Milner, eds., *East-West Trade and the Atlantic Alliance*, p. 50.

also on particular export control issues (e.g., computers, energy equipment, and technology) within a given time frame.

Inconsistency and unpredictability have been primarily a function of the fragmentation of the U.S. government. As the initial cold war consensus collapsed during the 1970s and 1980s, conflict intensified within the U.S. Executive and between the Executive and Congress.[67] The State, Commerce, and Defense Departments engaged in an ongoing struggle and sought to defend both their institutional interests and their conceptions of the appropriate balance between economic, diplomatic, and security objectives in export control policy. Their battles frequently ended in stalemate or in the temporary triumph of one agency, whose views were translated into policies subsequently sabotaged by other agencies. These problems were exacerbated, particularly during the Carter and Reagan administrations, by the lack of intervention by and guidance from the White House, which often allowed bureaucratic conflicts to be aired publicly and to fester without resolution.

The detrimental impact on U.S. leadership and CoCom has been profound. Especially during the 1970s and 1980s, U.S. officials found it difficult to preserve the integrity of the control process multilaterally, given that they were deeply divided internally over the substance of policy and often spoke with different voices to CoCom allies, leaving the latter to wonder who was in charge.[68] Similarly, they found it hard to relieve the administrative and economic burdens of control, as interagency conflicts led to delays in licensing and the inability to reach a consensus on the decontrol of less strategic items. The rise of the Defense Department, which by the early 1980s had become the dominant actor domestically and began to assert itself multilaterally as well, was a key factor in this regard. Defense officials posed the major obstacle to the liberalization of less strategic items; they also created profound discontent among CoCom allies by seeking to dictate policy in a context in which compromise and consensus building had been the historical norm.[69]

---

67. See Bruce Jentleson, "From Consensus to Conflict: The Domestic Political Economy of East-West Energy Trade," *International Organization* 38 (Autumn 1984), 625–60.

68. As Robert Putnam has noted, "Involuntary defection, and the fear of it, can be just as fatal to prospects for cooperation as voluntary defection." See "Diplomacy and Domestic Politics: The Logic of Two-Level Games," *International Organization* 42 (Summer 1988), 439. During the early 1980s, some CoCom members were reluctant to tighten controls as urged by the Reagan administration, in part because they doubted the sincerity of the abrupt shift in U.S. policy. "Will the United States continue a restrictive policy, or will it shift again next year, leaving us vulnerable economically?" French officials wondered. Interviews, Foreign Ministry, Paris, June 1982.

69. Defense carried its struggle into the post–cold war era, resisting the core list liberalizations of 1990. One high official of the Defense Technology Security Administration, when reminded that President Bush personally approved liberalization, reportedly responded that Bush "didn't speak for DTSA." See Robert Kuttner, "How National Security Hurts National Competitiveness," *Harvard Business Review*, January-February 1991, 148.

A fourth and final critical feature of U.S. policy involves the extraterritorial reach of export controls. Throughout the postwar era, the U.S. Executive, as an extension of its far-reaching domestic authority, has reserved the right to control the movement of U.S.-origin products, components, and technologies, even after they have left U.S. territory.[70] If U.S.-origin items are involved, American officials have presumed that the trade of other CoCom members with the Soviet Union or any other destination, including trade among CoCom members, falls under the control of the U.S. government. This distinctive exercise of authority applies to CoCom-controlled items as well as to items controlled unilaterally by the United States for either national security or foreign policy reasons. U.S. officials have administered it through a set of regulations that require foreign-based firms to obtain written permission from the Commerce Department prior to reexporting U.S.-origin items from one foreign destination to another. The failure to comply exposes foreign firms to the same set of legal and administrative sanctions that apply to U.S.-based violators of U.S. export control laws.[71]

Reexport controls have been attractive to U.S. officials, by providing an additional means to stem the flow of U.S. technology to the former Soviet Union and other controlled destinations. Their routine use, however, has compromised U.S. leadership in CoCom. Other CoCom governments have consistently viewed U.S. reexport controls as an infringement on their sovereignty and as a U.S. vote of no confidence in the multilateral regime. The exercise of such controls, because they are so contentious, has diverted attention in CoCom away from the inadequate performance (e.g., lax enforcement) of other members; attention has instead focused on the unacceptable unilateral behavior of the United States. America's extraterritorial assertion has also increased the administrative burden of the control system, by controlling some intra-CoCom transactions that might otherwise not be regulated and by requiring firms in some cases to acquire double permission—from their home government and then separately from the United States—in order to export to the East.[72] Finally, reexport controls have reinforced the repu-

70. See John Ellicott, "Extraterritorial Trade Controls—Law, Policy, and Business," in Martha L. Landwehr, ed., *Private Investors Abroad—Problems and Solutions in International Business in 1983* (New York: Matthew Bender, 1983), pp. 3–35; Kenneth W. Abbott, "Defining the Extraterritorial Reach of U.S. Export Controls: Congress as Catalyst," *Cornell International Law Journal* 17 (Winter 1984), 79–158; and Janet Lunine, "High Technology Warfare: The Export Administration Act Amendments of 1985 and the Problem of Foreign Reexport," *New York University Journal of International Law and Politics* 18 (Winter 1986), 663–702.

71. U.S. Department of Commerce, International Trade Administration, *Export Administration Regulations* 15 CFR (October 1, 1985), part 374.

72. Until the late 1970s, foreign firms sometimes required double permission even from the United States to export to the East: a CoCom exception approval, granted by the State Department, and reexport approval, granted by the Commerce Department. See GAO, *Export Controls: Need to Clarify Policy and Simplify Administration*, p. 14. Commerce eventually waived its reexport license requirement for items reviewed in CoCom.

tation of the United States as a "technological imperialist," a state that seeks to use its privileged position in CoCom to hamper the performance of foreign firms competing with those based in the United States.

To summarize, distinctive features of U.S. export control policy—the sanctions habit, insensitivity to economic costs, inconsistency and unpredictability, and the routine assertion of extraterritoriality—have been the proximate constraint on the ability of the United States to lead fully in CoCom. Those features, in turn, can be traced primarily to domestic structural attributes (i.e., the existence of a powerful, yet fragmented Executive, largely unconstrained by the private sector) which emerged in the crisis circumstances of the early cold war and which endured, despite changes in U.S.-Soviet relations and in the position of the United States in the world economy.

The general inadequacy of U.S. leadership rendered CoCom more controversial and less effective than it otherwise would have been, particularly during the 1970s and 1980s. It did not, however, lead to the complete collapse of cooperation in CoCom. Even in the absence of fully effective U.S. leadership, continued participation in the multilateral regime was assured by the core of common interests in national security export controls shared by the United States and its principal allies.

Put differently, until 1989 the postwar international security environment was an accommodating one for cooperation in CoCom. CoCom members perceived a well-defined military threat from the Soviet Union, participated in formal alliance structures designed to meet that threat, and found their perceptions reinforced by a clear ideological conflict between East and West. The post-1989 security environment, characterized by a dramatically diminished military threat, the transformation or dissolution of alliance structures, and the disappearance of the ideological conflict, promises to be far less forgiving. In the context of the 1990s, inadequate U.S. leadership will probably result not only in a less effective CoCom but also in the collapse of CoCom altogether. Indeed, as I argue in Chapter 9, even with effective leadership, CoCom may not endure. In a radically altered security environment, U.S. leadership will be a necessary condition for CoCom's survival, but it may not be a sufficient one.

THE DUAL SIGNIFICANCE OF CoCom

CoCom played its principal postwar role in the *East-West* context, as an adjunct to U.S. (and NATO) defense policy. U.S. officials perceived it as a means to enhance deterrence, by denying or delaying the Soviet Union's access to the West's most advanced dual-use technologies and thereby helping to preserve the lead time of the United States and NATO members in the application of those technologies to military systems. Less well recognized, yet equally significant, however, has been CoCom's role in the context of *intra-Western* trade relations. CoCom has been a key element in support of U.S. efforts to pursue a liberal trade policy nationally and to take the lead in maintaining a liberal economic order internationally. It has done so by serving, in effect, as a firebreak, with the potential to insulate the liberal orientation of U.S. trade with other Western states from the economic nationalism inherent in America's export control program.

A natural tension has existed in postwar U.S. thinking between a belief in minimal government intervention and the primacy of the market-place, on one hand, and a desire by government to intervene systematically in the market and regulate transactions in the interest of national security, on the other. That tension, as it manifests itself in trade policy, can be minimized to the extent national security export controls are confined to direct trade with potential military adversaries (i.e., the Soviet Union and its allies during the cold war and, perhaps, developing states bent on the acquisition of mass destruction capabilities during the 1990s). It can become more acute, however, if controls are extended beyond direct trade with potential adversaries and affect U.S. trade with the rest of the world, especially that with other advanced industrial states. Such an extension becomes considerably more likely if other states refuse to participate in an export control program or if their participation (e.g., coverage or enforcement of controls) is judged by U.S. officials as less than adequate. In such circumstances U.S. officials confront a difficult dilemma: they can maximize the free flow of goods and technologies within the nontargeted world, thereby compromising military security by risking the leakage of critical technology through noncooperators to controlled destinations, or they can maximize military security by extending controls to trade with noncooperators, thereby compromising their ability to pursue liberal trade policies.

The dilemma is not merely hypothetical, given the formidable array of policy instruments executive officials have accumulated in the postwar era with the potential to disrupt intra-Western trade in the name of national security. As discussed above, executive officials can control tangible and intangible trade to all destinations and have claimed authority over U.S.-origin items beyond the borders of the United States. Even where U.S.-origin items are not involved, U.S. officials have interpreted U.S. law as providing authority to regulate the trade of foreign-based U.S. multinationals.[73] Violators of U.S. control laws—whether they be

73. See Ellicott, "Extraterritorial Trade Controls," and Robert B. Thompson, "United States Jurisdiction over Foreign Subsidiaries: Corporate and International Law Aspects," *Law and Policy in International Business* 15 (1983), 319–400.

American or foreign individuals or firms—may be blacklisted by the U.S. government and denied the right to export (or receive exports) from U.S. territory. Recent legislation has extended the Executive's punitive authority further, allowing it to deny foreign violators of U.S. or CoCom regulations the right to export into the U.S. market.[74]

For most of the postwar period, U.S. officials chose not to employ these trade weapons to the full extent; they instead sought to minimize the potential conflict between their commitment to liberalism in the West and their export controls directed at the East. The key factor in managing that conflict has been the existence and effectiveness of CoCom. The creation of the regime in 1950 institutionalized export control cooperation and provided the initial justification for the United States to insulate intra-Western trade from East-West restrictions. When the effectiveness of other members' enforcement began to lag in the early 1950s, U.S. officials felt compelled to tighten intra-Western controls. In 1954, however, the United States extracted a commitment from its CoCom allies to tighten their enforcement, and in exchange it scaled back restrictions on both East-West and intra-Western trade. That arrangement held together fairly well for two decades. It began to unravel during the 1970s as the effectiveness of CoCom waned and as it became evident that the control of dual-use technology was increasingly critical to the maintenance of U.S. qualitative military superiority over the Soviet Union.

In light of these factors, in 1976 an influential Defense Department task force recommended that the United States expand intra-Western restrictions in trade with both CoCom and neutral states as part of an effort to protect more effectively the most critical dual-use technologies.[75] The Carter administration rejected that advice, judging that such a policy would be disproportionately costly to the United States in its relations with other Western states. The Reagan administration, however, did expand intra-Western restrictions, both to protect U.S. technology and to coerce other governments, within CoCom and outside it, to improve their national control systems. Although that effort bore some fruit, it also proved contentious diplomatically and costly to the United States economically, because it left many U.S. firms with reputations as unreliable suppliers in the highly competitive context of technology-intensive trade within the industrial West. The high costs drove the United States in 1988 to seek and achieve a compromise in CoCom, similar to the one reached in 1954, that involved commitments to streamline the control list, improve enforcement, and facilitate intra-Western trade.

74. Public Law 100-418, *Omnibus Trade and Competitiveness Act of 1988*, August 23, 1988, sec. 2444.
75. Defense Science Board Task Force, *An Analysis of Export Control of U.S. Technology: A DoD Perspective* (Washington, D.C.: Office of the Director of Defense Research and Engineering, 1976). This document is known as the Bucy Report.

The overall lesson is clear: the stronger the cooperation in CoCom has been, the less the incentive there has been for the United States to expand intra-Western restrictions. When the effectiveness of CoCom weakened, U.S. officials were tempted to blur the distinction between East-West trade and that between the United States and the rest of the world. This is somewhat ironic, because U.S. leadership has been the key factor in determining the effectiveness of CoCom. The United States has held, within its own hands, the means to minimize the conflict between its preferred policies in what has been, until very recently, a bifurcated world economy.

The radically altered security environment that emerged in 1989–1990 has mixed implications for the resolution of the conflict between liberalism and nationalism in U.S. trade policy. On one hand, the new realities finally forced the United States to accept drastic reductions in the multilateral control list. A more focused CoCom embargo facilitates enforcement and obviates the need for extensive intra-Western restrictions. CoCom members have committed to harmonize their enforcement practices in preparation for the elimination by 1992 of all licensing restrictions in trade among CoCom countries. A more streamlined list also facilitates the task of obtaining cooperation from non-CoCom suppliers, removing another incentive for the extension of controls. CoCom members are even making it possible for former target states (e.g., Poland, Hungary, Czechoslovakia, and, perhaps, the newly independent Baltic states) to be integrated into intra-Western trade networks and to be removed from the target list altogether, to the extent they provide adequate safeguards against the leakage of critical Western technology to the former Soviet military.

On the other hand, the fundamental conflict between liberalism and nationalism in U.S. export control policy may reemerge in a different context during the 1990s. As the United States takes the lead in extending restrictions in trade with developing states in order to frustrate the development of nuclear, chemical, and biological weapons, it is likely to confront other suppliers less willing or able to replicate U.S. controls to the full extent. U.S. officials will once again face the dilemma of either tolerating gaps in controls in the interest of free trade or obstructing trade with noncooperators in an effort to extract their compliance. The conflict may even prove to be more severe than it has been in the East-West context, given the considerable number of potential target states, the widespread availability of many of the items useful in the development of weapons of mass destruction, and the probable difficulty of maintaining an effective international consensus.[76]

76. National Academy of Sciences, *Finding Common Ground*, p. 13.

U.S. POLICY AND CoCom: AN OUTLINE OF THE BOOK

Chapter 2 extends the analytical framework of this book by exploring the logic of four export control strategies: economic warfare, strategic embargo, tactical linkage, and structural linkage. The four may be viewed as strategies of economic containment, in that they represent attempts to contain the expansion of an adversary's military power and political influence using economic, rather than (or in addition to) political and military, means. Each strategy is based on a different conception of the relationship between trade and national security. Taken together, they serve as a guide to the varying East-West trade preferences of the United States and other Western states and to the collective policy that results from the attempt to coordinate these national preferences.

The remaining chapters examine the interaction of U.S. and CoCom export control policy as it has evolved over time. Chapter 3 discusses the origins of CoCom, the rise and demise of economic warfare as alliance strategy, and the initial attempt by the United States to resolve the conflict between East-West and intra-Western trade. Chapter 4 takes up the consolidation of the multilateral strategic embargo during the 1958–1968 period, and the domestic struggle within the U.S. government and between government and industry to adjust U.S. national controls to the alliance consensus.

Chapter 5 examines the shift in U.S. export control strategy during the 1970s to tactical linkage and its implications for Western economic competition and for U.S. leadership and cooperation in CoCom. Chapter 6 takes as its point of departure the Bucy Report and its recommendation for a fundamental reorientation of export control policy from an emphasis on products to one on militarily critical technologies. The Bucy Report is worthy of detailed consideration because its analysis and recommendations structured the U.S. and alliance debates over export control policy during both the Carter and the Reagan administrations.

Chapters 7 and 8 take up different aspects of that debate during the 1980s. The former examines U.S. efforts to reimpose economic warfare as alliance strategy. The Carter administration used the Soviet invasion of Afghanistan to reopen the question, and the Reagan administration pursued it with vigor, privately in CoCom and more publicly in efforts to gain alliance support for export credit restrictions and the termination of the gas pipeline project. Each of these initiatives met with failure. By 1985 the United States abandoned its multilateral efforts and reconsidered its domestic policy.

Chapter 8 shifts from the question of CoCom's strategy to that of its effectiveness, and it examines U.S. efforts to revitalize the strategic embargo. I demonstrate that the United States undertook some of its leadership responsibilities effectively, but neglected or abused others. Consequently, CoCom was somewhat strengthened, yet it also proved to be a source of considerable controversy among its members. U.S. dissatisfaction with the pace of progress in CoCom led it to tighten restrictions on intra-Western technology transfer, exacerbating further the dispute in CoCom and triggering a business-government conflict in the United States. A compromise solution finally emerged in CoCom in 1988, the implementation of which was overtaken by the dramatic events of 1989 and 1990.

Chapter 9 summarizes the argument, discusses its conceptual implications, and analyzes the prospects for export control policy during the 1990s. I examine CoCom's reaction to the revolutionary events of 1989–1991 and the possibility of CoCom's endurance in a post–cold war world. I also examine briefly the shift in export control policy to an emphasis on North-South proliferation and the extent to which the experience of the cold war helps us to understand U.S. policy and multilateral coordination in that context.

CHAPTER TWO

Strategies for Trade with an Adversary

Economic statecraft can take a variety of forms, including not-so-obvious ones such as free trade.[1] Particularly when employed by economically dominant states, free trade or, more generally, economic liberalism, can be a powerful foreign policy instrument.[2] Following World War II, liberalism became the main foreign economic strategy of the United States and was used to strengthen the domestic economy, promote the economic recovery of allies, foster the development of third world states, and generally bolster U.S. political and strategic influence.

For economically dominant and technologically advanced states, however, trade with geopolitical competitors and potential military adversaries poses special problems. Although free trade may bring economic and possibly political benefits, it also carries national security risks in that it may contribute significantly to an adversary's military capabilities. For this reason, throughout the cold war U.S. officials treated trade with the Soviet Union and its allies differently from U.S. trade with other destinations. Although the general thrust of postwar U.S. foreign economic policy has been to minimize state intervention or extricate the state from the market, with regard to the Soviet Union and other Communist states, U.S. officials were unwilling to leave trade in private hands. They sought to regulate such transactions in order to minimize their strategic risk and maximize their potential political benefit.

A general commitment to regulate trade leaves open the specific issues of what should be controlled, for what purposes, and under what conditions. This chapter argues that there are four broad strategies available

1. David Baldwin, *Economic Statecraft* (Princeton: Princeton University Press, 1985), pp. 44–46.
2. E. H. Carr, *The Twenty Years' Crisis, 1919–1939*, 2d ed. (London: Macmillan, 1946), and Robert Gilpin, *U.S. Power and the Multinational Corporation* (New York: Basic Books, 1975).

to government officials seeking to use trade as an instrument of state-craft in dealing with potential military adversaries.[3] Each is based on a different conception of the relationship between trade and national security, and they differ in the manner and extent to which they affect trade. Preference for one strategy is driven by a particular combination of economic, political, and strategic considerations (see Table 1). The likelihood that a given strategy will be effective is influenced by the ability of government officials to overcome domestic constraints and problems of international coordination and by the extent to which the target state is vulnerable or responsive to the economic instruments employed.

Both within the United States and between the United States and its major allies, much of the postwar conflict over East-West trade involved the issue of which strategies should be employed and how they should be implemented. Thus, a systematic examination of the strategies provides a framework for analyzing the evolution of U.S. and Western policy. The next three sections of this chapter explore the logic and assumptions of each strategy. A final section examines the interests and circumstances that underlie a government's preference for a particular strategy or strategies.

ECONOMIC WARFARE

Economic warfare may be considered the waging of hostilities against an adversary by economic rather than military instruments. Its immediate or proximate objective is to weaken the economy or economic potential of a target state by denying it the benefits of international economic exchange. Although it may be an end in itself, inflicting economic damage may also be considered as a means to achieve a more particular strategic objective. A state or states may use economic warfare in an attempt to weaken an adversary's economic potential in order, ultimately, to weaken its *military* capabilities or potential.

The use of economic warfare as a strategic instrument became viable and attractive during the twentieth century. The most salient attempts of the prior century—the Continental System of Napoleon and the British counterblockade—failed because of the economic self-sufficiency of the targets involved. Neither household consumption nor the war production efforts of the British and French governments were more than marginally dependent on external sources of supply or markets.[4] By

3. An earlier discussion of these strategies is found in Michael Mastanduno, "Strategies of Economic Containment: U.S. Trade Relations with the Soviet Union," *World Politics* 37 (July 1985), 506–16. See also Philip Hanson, *Western Economic Statecraft in East-West Relations*, Chatham House Paper no. 40 (London: Routledge and Kegan Paul for the Royal Institute of International Affairs, 1988).

4. See Eli Heckscher, *The Continental System* (Oxford: Clarendon Press, 1922), pp. 364–65.

1914, however, economic interdependence had increased significantly. In addition, with the advent of mechanized fighting forces, the economic infrastructures of the leading powers became a critical component of their ability to wage war effectively and thus an attractive target for economic warfare.[5] In both world wars, economic warfare was used to complement the military strategies of the principal belligerents.[6] The case of postwar U.S. policy toward the Soviet bloc is distinctive in that it represents the most systematic attempt to employ economic warfare as a strategic instrument during peacetime.

According to the logic of this strategy, the relationship between the trade of the sanctioning state and the military capabilities of the target state is *indirect*. The target economy can be viewed as a mediating variable between the two. The strength of the argument for economic warfare as a strategic weapon rests on the strength of two links: that between trade and the target economy and that between the target economy and the military capabilities of the target state.

The economic and military capabilities of the modern industrial state are closely related. As both classical liberal and mercantilist thinkers understood, military power is ultimately based on a foundation of economic power, both qualitatively and quantitatively.[7] The level of technology, labor, and management skills available to the economy as a whole is generally that which is available to the military sector. The greater the level of its economic development, the more probable it is that a state will be able to produce a wide array of technologically sophisticated weapons systems. Similarly, the greater the magnitude of overall productive capacity and the higher the rate of economic growth, the greater the amount of resources available for military use and for other uses, such as consumption and investment. Historically, there has been a direct relationship across countries between the level of gross national product (GNP) and the level of military expenditures.[8]

Second, in accordance with the law of comparative advantage, a state that engages in trade earns aggregate economic benefits. By exporting what it produces efficiently and by importing those items needed for production and consumption but produced domestically only at relatively high cost, that state "saves" economic resources. Those "extra" resources can, in turn, be devoted to military or civilian spending, invest-

5. See Bernard Brodie, "Technological Change, Strategic Doctrine, and Political Outcomes," in Klaus Knorr, ed., *Historical Dimensions of National Security Problems* (Lawrence: University of Kansas Press, 1976), pp. 263–306, and Margaret P. Doxey, *Economic Sanctions and International Enforcement* (London: Oxford University Press, 1980), chap. 2.
6. The classic work on the latter effort is W. M. Medlicott, *The Economic Blockade*, 2 vols. (London: HMSO, 1952 and 1959).
7. See Edward Mead Earle, "Adam Smith, Alexander Hamilton, Friedrich List: The Economic Foundations of Military Power," in Earle, ed., *Makers of Modern Strategy* (Princeton: Princeton University Press, 1943), pp. 117–54.
8. Klaus Knorr, *The Power of Nations* (New York: Basic Books, 1975), p. 50.

ment, or some combination of the three. *The critical assumption made on behalf of economic warfare is that the additional resources released by trade with an adversary are ultimately devoted, wholly or in part, to military pursuits, which results in a significant improvement in the adversary's military capabilities.* Or, what is logically the same, the resources provided by trade allow the target to maintain a level of military expenditures and capabilities that it would not be able to, given the competing demands on its aggregate resources, in the absence of trade.

Given the two critical links, the most vulnerable target is the state that is both highly dependent on trade with the sanctioning state[s] and, at the same time, unable to provide the level of resources demanded by its military sector (or able to provide those resources only if it engages in trade). The least vulnerable target is the state that is largely self-sufficient (i.e., trade with the sanctioning state[s] allows only a very modest release of resources) and whose military sector's demand is easily satisfied at the pre-trade level of aggregate resources. In this case, the resources released by trade are likely to be channeled to the civilian rather than the military sector. Between these two extremes, an infinite number of target situations can be contemplated, the result of varying both the marginal contribution of trade to the economy and the propensity of the state to allocate resources at the margin between civilian and military pursuits.

The extent of the target's trade dependence is determined by a variety of factors, which include the size of its economy and population, its level of economic development, and the quality and composition of its natural resource endowment. Government policy is also a critical factor: a state may consciously choose, by pursuing a strategy of autarchy, to sacrifice the economic benefits of trade in exchange for the expected political benefits of autonomy and self-sufficiency. More ambitiously, it may attempt to maintain both sets of benefits by expanding the economic base under its control through conquest or political domination of its neighbors. Japan's desire for a "Co-Prosperity Sphere" and Germany's for *Lebensraum* (living space) represented attempts to implement this strategy prior to World War II.[9] In general, small, less developed states tend to be more trade-dependent, and hence potentially more vulnerable to economic warfare, than are large, economically advanced states.

The second key relationship, that between the economy and the military, will be influenced primarily by the international security predicament of the target state. For example, a state engaged in or mobilizing for protracted war will devote as many of its economic resources as possible to the war effort. In such circumstances, military production and consumption constitute the overwhelming priority; any resources

9. See Alan Milward, *War, Economy, and Society, 1939–1945* (Berkeley: University of California Press, 1977), chap. 1.

created through trade will be used either to increase the war effort or to sustain it by helping to maintain civilian consumption at an acceptable level. In either case, the state at war or preparing for war is an attractive target for economic warfare.

Different types of war exact military demands of different magnitudes. A "total" war, fought by conventional means against an adversary of roughly equivalent economic and military strength, demands a greater commitment of resources than does a "limited" war, fought against a smaller and weaker adversary. During World War II, the major belligerents devoted 50 percent of their economic resources to the war effort. Yet, U.S. military spending was only approximately 10 percent of GNP at the height of the Vietnam War.[10] All other things being equal, the state engaged in or preparing for a long war against a major adversary is more vulnerable to economic warfare than is a state involved in a limited war against a weaker opponent.

Alternatively, the prospect of a total nuclear war, such as that envisioned by U.S. strategic planners during the early 1960s, renders economic warfare strategically irrelevant. In the event of war, America's Single Integrated Operational Plan (SIOP) called for the immediate and simultaneous destruction of all important Eastern military and industrial targets.[11] If such a threat were to be carried out, economic warfare would be of little strategic value in either the mobilization or war-fighting phases. Indeed, during the 1950s, British officials were prompted by NATO's adoption of a massive retaliation strategy to argue that economic warfare against the Communist bloc should be abandoned, on the grounds that it had become strategically obsolete.[12] In the nuclear age, the rationale for economic warfare as a complement to military warfare must rest on the assumption that a protracted conventional war is worth preparing for and can be conducted below the level of all-out nuclear exchanges.[13]

Finally, a state need not be fighting a war or preparing for one that is imminent to be a potentially attractive target for economic warfare. The sustained increases in military spending needed to maintain one's position in an arms race may become enough of a burden on the allocation of economic resources that the marginal contribution of trade begins to take on strategic significance. The same point applies in the case of

10. Knorr, *Power of Nations*, p. 84.

11. On the SIOP of 1960, see Thomas Powers, "Planning a Strategy for World War III," *Atlantic Monthly*, June 1982, p. 92, and David Rosenberg, "The Origins of Overkill: Nuclear Weapons and American Strategy, 1945–60," *International Security* 7 (Spring 1983), 6–7.

12. The British contention is discussed in a memorandum from the U.S. deputy director for intelligence, The Joint Staff, to the chairman of the Joint Chiefs of Staff, December 7, 1955, p. 1. Reprinted in *Declassified Documents Quarterly* (1976), item 358B.

13. In the face of strategic parity and the decline of détente during the late 1970s, U.S. officials became more attentive to issues of conventional mobilization. See Paul Bracken, "Mobilization in the Nuclear Age," *International Security* 3 (Winter 1978–79), 74–93.

target states that maintain extensive and costly foreign commitments. During 1980–1984, the Reagan administration's preference for economic warfare was based in part on the premise that the Soviet Union could not sustain the accelerated pace of the arms race, or maintain its foreign commitments, without the assistance of Western technology and credits.

The above discussion implies that the state which perceives little threat to its national security and which is not engaged in war, war mobilization, or an arms race is a relatively unattractive target for economic warfare as a strategic weapon. The resources released in trade with such a state are more likely to be channeled into civilian consumption and investment than into military pursuits.

In addition to its international security predicament, the potential vulnerability of a state to economic warfare may be affected by domestic characteristics. In general, it is plausible to expect governments of authoritarian states, in control of centralized, command economies, to be better able to determine the absolute and marginal allocation of economic resources than are governments of democracies, in charge of decentralized, market economies.[14] All else being equal, the former are better equipped to impose economic hardship on the consumer sector in the interest of the military. As the recent changes in the Soviet Union and Eastern Europe dramatically demonstrate, however, this is not to imply that the discretion of such regimes in resource allocation is absolute or that inflicting hardship on the consumer is without political consequence. It does suggest, however, that stable Communist governments may be better able than their democratic counterparts to frustrate attempts at economic warfare directed against them.

Assuming a state decides to pursue economic warfare as a strategic instrument, which items should it deny to the target state? Applied most rigorously, the logic of economic warfare calls for a complete trade embargo. Unless the target state trades foolishly, any commercial transactions will contribute, however marginally, to its economy and thus will release resources that could be put to military use.

The logical extreme notwithstanding, the fact that certain items in trade contribute more significantly than do others to the target economy suggests that economic warfare can rationally be practiced selectively. In light of the economic cost of an embargo and the need for multilateral coordination, it is sensible for the sanctioning state to focus its efforts on those items that, if unavailable, would bring the most harm to the target economy. The Allied powers in World War II reasoned in this fashion; while attempting to maintain a comprehensive embargo, they devoted

14. See Stephen D. Krasner, "Domestic Constraints on International Economic Leverage," in Klaus Knorr and Frank Trager, eds., *Economic Issues and National Security* (Lawrence: University of Kansas Press, 1977), pp. 160–81.

special attention to denying the German economy such items as precision ball bearings, chrome ore, and wolfram, often by purchasing them preclusively from neutral suppliers.[15]

Precisely which items will be most significant will vary across target economies and within a single economy over time. In general, the most important can be termed "bottleneck" items—those for which target demand is highly inelastic but for which the cost of domestic production is either prohibitive or very high, at least in the short run.[16] The impact of denying bottleneck items effectively will be to reduce the overall output of the target economy, particularly if the items in question are intermediate goods, essential to the production process, rather than finished goods.

In general, one can expect technology to be a bottleneck item.[17] Given the same amount of inputs, an economy that introduces improved technology into a production process will have a greater output (possibly of higher quality) than it had previously. Moreover, unlike raw materials or finished products, technology is not "used up" in the production process. Rather, it provides the recipient with an enduring improvement in capabilities.[18] Through the diffusion of know-how, the recipient can improve productivity and efficiency at production facilities other than that at which the technology was initially introduced.[19] Thus an entire sector, rather than a single factory, may enjoy the benefits of modernization. Such modernization will also indirectly benefit the economy as a whole, particularly if it takes place in key infrastructural sectors such as communications, transportation, or energy.

Over time, the transfer of advanced technology tends to bring other indirect benefits to the recipient, such as improvements in the quality and skill level of the labor force and in the management and organization of production. More abstractly, along with the specific know-how involved may be transferred the ethos of the donor society or the general character of its approach to economic problem solving. Along these lines, in 1976 the Bucy Report suggested that the transfer of computers would enhance the "cultural preparedness" of the Soviet Union to exploit advanced technology.[20] The more "active" the mechanism by which technology is transferred (i.e., the more informal and prolonged the

15. See Medlicott, *Economic Blockade*, vol. II, and Milward, *War, Economy, and Society*, pp. 294–328.
16. Peter Wiles, *Communist International Economics* (Oxford: Basil Blackwell, 1968), p. 462.
17. Ibid., p. 465.
18. See Defense Science Board Task Force, *An Analysis of Export Control of U.S. Technology: A DoD Perspective* (Washington, D.C.: Office of the Director of Defense Research and Engineering, 1976), p. 25.
19. The effective diffusion of technology, of course, should not be presumed. For a good discussion of the problems faced by the former Soviet Union in absorbing and diffusing Western technology, see Thane Gufstafson, *Selling the Russians the Rope? Soviet Technology Policy and U.S. Export Controls* (Santa Monica: RAND Corporation, 1981), and Philip Hanson, *Trade and Technology in Western-Soviet Relations* (New York: Columbia University Press, 1982).
20. *Analysis of Export Control*, p. 25.

contact between donor and recipient), the more likely these intangible effects are to be realized.

In summary, large-scale technology transfers not only increase the level of output but, over the longer run, may also enhance the technical sophistication of the economy as a whole. They bring the recipient both economic growth and development and provide a base upon which further technological advances—either indigenous or transferred—can be built. As noted above, the level of technological sophistication of the economy as a whole is generally that which is available to the military, or as one analyst has noted, "military developments have come to depend more and more intimately on the combined technological skills of the nation's entire industry."[21]

The above discussion has focused on economic warfare as a strategic instrument, that is, one designed to weaken the military capabilities of an adversary. It is important to recognize that the strategy may be employed for other purposes as well, such as to overthrow a regime or to signal strong disapproval of a regime's behavior or its very existence.[22] Economic warfare is often used to convey publicly that the sanctioning state considers the target government to be illegitimate, unfit to conduct relations with, and unworthy of the respect and recognition generally accorded to sovereign governments. In these circumstances economic warfare tends to be comprehensive, involving the cutoff of all trade and other economic interaction with the target.

Signaling has been an important objective in many postwar cases of economic warfare, including the Arab embargo of Israel, the UN sanctions against Rhodesia, and the U.S. embargoes of Cuba, North Korea, Vietnam, and Kampuchea.[23] It also figured prominently in U.S. trade policy toward the Soviet Union. During the 1960s, for example, the United States maintained economic warfare long after it had been abandoned by other Western states and ceased to have any significant impact on Soviet economic or military capabilities.[24] As I discuss below, this fact

21. Gufstafson, *Selling the Russians the Rope?*, p. 66. See also Alexander H. Flax, "The Influence of the Civilian Sector on Military R&D," in Franklin A. Long and Judith Reppy, eds., *The Genesis of New Weapons Systems* (Elmsford, N.Y.: Pergamon, 1980), pp. 113–36.

22. On the importance of signaling in economic statecraft, see Baldwin, *Economic Statecraft*, especially chap. 6. On the utility of economic warfare to undermine a target government, see Johan Galtung, "On the Effects of International Economic Sanctions, with Examples from the Case of Rhodesia," *World Politics* 19 (April 1967), 378–416, and Donald Losman, *International Economic Sanctions: The Cases of Cuba, Rhodesia, and Israel* (Albuquerque: University of New Mexico Press, 1979).

23. See the statement of William A. Root, Department of State, in U.S. Congress, House Committee on Foreign Affairs, Subcommittee on International Economic Policy and Trade, *Extension and Revision of the Export Administration Act of 1969*, hearings, 96th Cong., 1st sess., February 15–May 9, 1979, p. 132; Richard Stuart Olson, "Economic Coercion in World Politics, with a Focus on North-South Relations," *World Politics* 31 (July 1979), 441–75; and Losman, *International Economic Sanctions*, pp. 47, 131.

24. See Chap. 4, below, and Baldwin's discussion of the East-West trade embargo in *Economic Statecraft*, pp. 235–50.

suggests that broad political considerations may be as important as narrowly defined strategic calculations in determining the preference of a government for this strategy.

STRATEGIC EMBARGO

The overriding purpose of a strategic embargo is to deny or delay improvements in the military capabilities of an adversary. In pursuit of this goal, government officials do not attempt to weaken the economy of the target state. Instead, they attempt to prohibit the export of only those items that could make a *direct* and *specific* contribution to target military capabilities, regardless of their economic impact. Furthermore, a strategic embargo is designed to be highly selective in the scope and extent of its controls. It is intended both to protect national security and to interfere with trade to the minimum extent possible.

The system of rules governing the seizure of contraband during wartime is a classic example of a strategic embargo. Those rules, codified in the 1909 Declaration of London, separated traded goods into three categories, according to their purpose and destination.[25] "Absolute contraband" referred to items suited exclusively for war purposes, such as arms and ammunition. Such items, carried by a neutral ship, could be seized by a belligerent if they were found to be destined to territory either belonging to or occupied by the enemy or its armed forces. Goods that could serve both peaceful and military purposes were considered "conditional contraband" and could be seized only if destined for the government or armed forces of the enemy. Finally, "free goods" were those unlikely to be considered of use in war; they could not be seized by a belligerent, regardless of destination.

The assumption underlying the contraband rules was that only goods used directly by the target government or military to carry out its war effort should be subject to trade denial. Because it served only military purposes, absolute contraband was certain to be used in war, and thus, its particular destination in the target state was deemed irrelevant. Conditional contraband, it was assumed, would only be used by the military if destined for the government or the army; if destined for the civilian population of the enemy, it would serve peaceful purposes. Free goods were considered to be inherently peaceful and could therefore be sold even to the target government. In this fashion, the rules purported to protect both the security interests of belligerents and the trading interests of neutrals.

25. See Robert W. Tucker, *The Law of War and Neutrality at Sea*, International Studies of the Naval War College (Washington, D.C.: GPO, 1955), chap. 9.

Although the distinctions were plausible on paper, the contraband rules proved to be impractical when tested in the circumstances of total war. They were abandoned by the belligerents immediately at the outbreak of both world wars. Their demise points, more generally, to the two major problems that must be resolved in any attempt to fashion a rational strategic embargo, whether in wartime or peacetime.

First, the criteria of "military use" and the desire for controls to be highly selective may be incompatible. All items in international trade, even ostensibly "peaceful" goods, can conceivably serve military purposes. Food can be used to feed soldiers as well as civilians. Wool can be used to produce clothing for civilians or uniforms for soldiers. Trucks can carry troops to war or produce to market. And the same computers useful in weather forecasting may be valuable in the design of nuclear weapons. In short, virtually every material, product, or technology can be defined as having a "dual use," that is, both military and civilian applications. Even in peacetime, the state wishing to control any trade that *could* be put to military use by an adversary may find itself exercising comprehensive controls, more befitting of economic warfare than of a strategic embargo.

The second problem is potentially more troublesome. Even if a selective embargo can be maintained, controlling items most likely to be of military *use* may not adequately reflect the contribution of trade to target military *capabilities*. That is, items not directly used by the military may nevertheless be of greater military *significance* to the target state— because of their economic significance—than are items actually put to military use. This assertion follows simply from the argument that the target military is but one part of an integrated target economy.

For example, a strategic embargo focusing solely on the likelihood of military use might be led to prohibit the sale of ballistic missiles to an adversary while permitting the sale of wheat.[26] If the target economy can produce ballistic missiles far more efficiently than wheat (and assuming a well-fed target military), however, the sanctioning state had better embargo the latter rather than the former. The sale of wheat relieves a more significant burden in the target economy and releases resources which, theoretically at least, can be used to produce the missiles that have been embargoed. Practically speaking, the missile embargo would be meaningless; the sale of wheat permits the target economy to have both products.

This argument, based on the law of comparative advantage, is a powerful one. It has led most students of economic sanctions to reject categorically the idea of a strategic embargo based on the denial of trade that is of direct military utility. For example, Theodore Osgood argued in 1957 that much of the confusion over the U.S. export control pro-

26. Similar examples are developed in Thomas Schelling, *International Economics* (Boston: Allyn and Bacon, 1958), chap. 30.

gram resulted from the attempt to categorize commodities as either "strategic" or "peaceful."[27] Thomas Schelling similarly noted that "there is little difference between retarding the Soviet economy as a whole and retarding military production capacity . . . one must not confine his search to the munitions production field if he is looking for the shortages that might most retard Soviet war potential."[28] As this statement suggests, students of sanctions have generally advocated economic warfare—based on the bottleneck effect—rather than a strategic embargo, if trade controls are to be used for strategic purposes.[29]

Despite the power of this argument, it is possible to devise an economically sensible, selective strategic embargo that still concentrates on items of direct military utility. To do so, the control criteria must move beyond an exclusive concern with military use to a consideration of both military use and military *significance*. Rather than merely asking whether a dual-use export will be put to military use, one must ask, If this item is put to military use, will it make a significant contribution to target military capabilities? A "significant" contribution should be defined as one that, within a specified time horizon, enables the target to achieve an otherwise unattainable military capability given the character of its resource base. Somewhat more technically, an export makes a significant contribution if it allows the target to attain a level of military output that lies beyond its current production-possibility frontier.[30]

According to this revised conception of a strategic embargo, the export of ballistic missiles should be prohibited if (1) the target state is unable to produce them at all; (2) the missiles it can produce are markedly inferior in quality to those it wishes to purchase; or (3) it can produce missiles of comparable quality but cannot produce them in the quantity it desires for its military purposes. In such cases, along with the missiles themselves, the target should also be denied the technology (either technical data or specialized equipment) and the specialized materials essential to the production of the missiles. At the same time, if the target can produce them in sufficient quantity and of comparable quality, missiles need not be embargoed, even if they may be used militarily.[31]

27. See Theodore Osgood, "East-West Trade Controls and Economic Warfare" (Ph.D. diss., Yale University, 1957), p. 101.
28. Schelling is quoted in ibid., p. 101. See also Schelling, *International Economics*, pp. 498–504.
29. In addition to Schelling and Osgood, see Peter Wiles, *Communist International Economics*, p. 464. See also Baldwin, *Economic Statecraft*, pp. 214–24.
30. A good discussion of the difference between trade that is "capability-enhancing" and "resource-releasing" is Robert Klitgaard, "Limiting Exports on National Security Grounds," in *Commission on the Organization of the Government for the Conduct of Foreign Policy*, vol. 4, pt. VII (Washington, D.C.: GPO, 1975), pp. 455–56.
31. It would be difficult *politically*, of course, to permit the export to an adversary of items likely to be put to military use, even if those items were inconsequential with regard to military significance. The postwar U.S. government did not allow the sale of any weapons to the Soviet Union, despite the fact that in some cases Soviet weapons may have been superior to those produced by the United States.

In this manner, a strategic embargo can satisfy the requirements of both selectivity and economic rationality. First, even though any item could be put to military use, only a select number of exports could be expected to make a significant contribution to military capabilities if put to such use. The more developed the military capabilities of the target (alternatively, the narrower the gap between sanctioner and target military capabilities), the more selective the strategic embargo can be.

Second, if the (strategic) exports embargoed are beyond the target's production possibilities, then the resources released by nonstrategic trade will not, in the short run, enable the target to produce strategic items. The sale of wheat may increase the aggregate economic resources of the target (and thus the potential share of the military), but it will not release the *kind* of resources (e.g., technical expertise and specialized equipment) that are unavailable domestically and required to produce ballistic missiles. Thus one could arguably prohibit the sale of missiles, but not of wheat. Over time, of course, the resources released by nonstrategic trade can contribute indirectly to qualitative military advances, for example, if they were devoted to a research and development effort that resulted in a strategic military breakthrough. But this point simply reiterates one made in the economic warfare discussion: technology transfer (or additional resources devoted to technological advancement) contributes to economic development, and *in the long run*, a more developed economic base can accommodate a more sophisticated array of military capabilities.

To say that a strategic embargo prohibits the export of dual-use products and technologies that are beyond the production-possibility frontier of the target is simply to say that the sanctioning state seeks to exploit the bottleneck effect. It is important to emphasize, however, that a strategic embargo aims to embargo only items that are military bottlenecks in the target. A crucial distinction between a strategic embargo and economic warfare is that the former seeks to exploit *military* bottlenecks, whereas the latter is concerned primarily with *economic* bottlenecks. A strategic embargo is designed to permit the export even of items that relieve economic bottlenecks in the target, so long as these items are not simultaneously military bottlenecks. An important implication is that for a strategic embargo, strengthening the economy of the target state, even significantly, is not in and of itself detrimental to the security of the sanctioning state.

Many items, for example, telecommunications networks in the former Soviet Union, will simultaneously be economic and military bottlenecks. Alternatively, an item may be an economic, though not a military, bottleneck if the target can produce only a limited quantity of it and if fulfilling the requirements of the military is considered the first priority. In the early 1980s, the Soviet Union may have been able to produce enough large-scale integrated circuits (ICs) to satisfy its military needs, but not the demands of its economy as a whole. Such was the assessment of the CIA in 1982.[32] Therefore, a strategic embargo would have permitted the sale of the ICs in question; economic warfare, which is aimed to stifle the modernization of the economy as a whole, would not have. A strategic embargo, however, would deny the technology needed to produce the ICs if that know-how could be used to produce more sophisticated ICs, with more advanced military applications, which were currently beyond the Soviets' production-possibility frontier.

As is the case with economic warfare, technology is the key to the military bottleneck effect of concern to a strategic embargo. In modern weaponry, technology has replaced raw materials as the limiting production factor of primary importance. Military strength has become increasingly dependent on the direct application of dual-use technologies to weapons and military support systems. Indeed, America's postwar extended deterrence strategy, both conventional and nuclear, relied explicitly on the ability of the United States to balance the quantitative military advantages of the Soviet Union with qualitatively superior weapons systems based on superior technology.[33] The strategic embargo, with the potential to contribute to the ability of the United States to maintain its technological lead, thus was and continues to be an important element of U.S. defense strategy.

This section has sought to establish that, at least in principle, it is possible to devise a strategic embargo which is both selective and economically rational. It may be difficult, however, to implement such an embargo in practice. First, the precise determination of military bottlenecks, particularly in a closed society that effectively guards its military activities, is a difficult intelligence task. It requires detailed knowledge of military systems, production methods and capabilities, and resource allocation at the margin. The vast resources of the U.S. intelligence community notwithstanding, such information has often been difficult to obtain. As a consequence, the designers of America's—and CoCom's—strategic embargo often resorted to a practice known as "mirror imaging" in their construction of control lists. That is, U.S. military use became a surrogate for Soviet military significance; what is of military utility in the U.S. defense sector was judged to be of probable military significance to the Soviets. The obvious problems are that U.S.

32. U.S. Central Intelligence Agency, "Soviet Acquisition of Western Technology" (Washington, D.C., April 1982), p. 9. The CIA estimated that by combining their legal and illegal acquisitions from the West, the Soviets could meet 100 percent of their military requirements or 50 percent of all their microelectronic needs.

33. See, for example, the statement of William Perry, director of Defense Research and Engineering, in U.S. Congress, Senate, Committee on Banking, Housing, and Urban Affairs, Subcommittee on International Finance, *Trade and Technology, Part II: East-West Trade and Technology Transfer*, hearings, 96th Cong., 1st sess., November 28, 1979, pp. 25–26.

weapons systems have incorporated a vast array of dual-use components and technologies, and that the design and production of Soviet military systems may not always have mirrored the process as it takes place in the United States. Mirror imaging may thus lead to an embargo that, from the perspective of target military significance, simultaneously over-controls and undercontrols.

A second problem concerns the rapid proliferation of dual-use technologies that has taken place within the past two decades, outside the direct control of the U.S. government. One result is that U.S. officials may be unaware of commercial technologies, developed in the private sector, that have potentially important military applications.[34] Moreover, it is often difficult to determine whether, and to what extent, commercially useful technology will have *future* military applications. In light of these circumstances, failure to control could result in technologies slipping away to an adversary before their military significance is appreciated at home. At the same time, the danger of controlling all technologies that *might* have current or future military application is in compromising the selective nature of the embargo.[35]

LINKAGE: TACTICAL AND STRUCTURAL

Whereas economic warfare and a strategic embargo aim to weaken the capabilities (economic and/or military) of a target state and rely on either comprehensive or selective trade denial, linkage strategies are characterized by a reliance on some degree of trade *expansion* as a means to influence the *behavior* or *policies* of a target government.

Two variants may be distinguished. *Tactical* linkage is intended to condition or calibrate trade according to changes in target behavior. It rewards (or promises to reward) good behavior by permitting or promoting trade, and it punishes (or threatens to punish) bad behavior by restricting trade. *Structural* linkage relies exclusively and unconditionally on trade expansion. It is based on the notion that the development or intensification of a positive economic relationship can be used to induce or reinforce desirable changes in the domestic or foreign policy of a target, by influencing that government's allocation of resources.

The underlying assumption of tactical linkage is that trade relations are sufficiently important that the target government will be prepared to change its policies, in a manner desired by the sanctioner, in order to

---

34. For an elaboration of this point, see Gufstafson, *Selling the Russians the Rope?*, p. 5.

35. During the Reagan administration, disagreement existed between the State and Defense Departments over the extent to which technologies that *might* have future military applications should be controlled. As one State official put it, the DoD position was that technologies should be "born controlled" rather than "born free." Interview, Department of State, Office of East-West Trade, Washington, D.C., May 12, 1982, p. 5.

maintain them. Unlike economic warfare, in pursuing tactical linkage a sanctioner does not strive to isolate a target government. It accepts the legitimacy of that government, at least implicitly, and tries to shape its conduct. Rather than political polarization, tactical linkage requires the establishment and continuance of political communication between sanctioner and target.

The League of Nations sanctions against Italy in 1935–1936 are a classic example of tactical linkage.[36] The primary objectives of these sanctions were to demonstrate resolve to Germany and to force Mussolini to stop his war against Ethiopia, "not necessarily to topple him as an authority or end the fascist regime in Italy."[37] Weakening Italy's military or economic capabilities appears to have been a secondary goal at best given that oil, the critical bottleneck item, was excluded from the embargo.

Other examples of tactical linkage include the Arab oil embargo against the Western states in 1973–1974 and the Jackson-Vanik Amendment to the U.S. Trade Act of 1974. By withholding the export of a single key commodity, the Arab states attempted to demonstrate their collective determination to confront Israel and to force Western Europe and Japan to change their political stance in the Arab-Israeli conflict. With the Jackson-Vanik Amendment, the United States sought to alter the emigration policies of Communist regimes by either offering to grant or threatening to withhold most-favored-nation (MFN) status.[38]

As the above examples suggest, the economic measures associated with tactical linkage vary considerably. They may be positive or negative, depending on whether a reward or punishment is considered politically more effective by the sanctioner.[39] They may involve a range of commodities, including economic bottleneck items, as in the case of the Arab oil embargo, or military bottlenecks, as when an industrial nation sells a developing nation advanced fighter aircraft or other defense equipment in exchange for political concessions.

In addition to specific commodities, an economic relationship itself may constitute a significant source of potential leverage. This has been particularly relevant in the U.S.-Soviet context. Soviet officials indicated during the 1960s and early 1970s that their willingness to cooperate with

---

36. See George W. Baer, "Sanctions and Security: The League of Nations and the Italian-Ethiopian War, 1935–1936," *International Organization* 27 (Spring 1973), 165–79, and Doxey, *Economic Sanctions and International Enforcement*, pp. 42–55.

37. The quote is from Olson, "Economic Coercion in World Politics," p. 475. See also Baldwin, *Economic Statecraft*, pp. 154–65.

38. On Jackson-Vanik, see Paula Stern, *Water's Edge: Domestic Politics and the Making of U.S. Foreign Policy* (Westport: Greenwood Press, 1979). On the Arab oil embargo, see Roy Licklider, "The Utility of the Arab Oil Weapon, 1973–74," in David Leyton-Brown, ed., *The Utility of International Economic Sanctions* (London: Croom Helm, 1987), pp. 167–81.

39. See David A. Baldwin, "The Power of Positive Sanctions," *World Politics* 24 (October 1971), 19–38, and *Economic Statecraft*.

the United States diplomatically would be significantly enhanced by the normalization of bilateral economic relations, which they took to be an important symbol of political status or prestige. Thus, the granting of MFN status—accepted by the Soviets as the cornerstone of a normal economic relationship—became an important political instrument for U.S. officials as they sought to initiate the détente process, despite the fact that it was of relatively little importance economically.

In general, however, for tactical linkage to be effective, the sanctioning state must control a good portion of the target's trade or financial relations, either absolutely or in terms of a few key commodities. Beginning with little or no bilateral trade, a sanctioning state will need to "invest" in future leverage by building up its trade relations with the target to the level at which the threat of denial is effective politically. In his classic study of the 1930s, Albert Hirschman demonstrates how the Germans consciously structured their trade with smaller East European nations to the point where economic dependence translated into political influence.[40]

When tactical linkage is practiced as a *strategy* (i.e., repeatedly against a single target over time), the sanctioning state must be prepared to employ both positive and negative measures in order to maintain credibility. Trade denial will not force a target to change its behavior unless the target believes trade will be restored if and when it complies. Similarly, trade rewards can bring concessions only if the target is convinced that they will be removed if the concessions are not forthcoming. In recognition of this fact, Samuel Huntington argued in 1978 that an effective tactical linkage strategy must be one of "conditioned flexibility"; the sanctioning state must be willing and able to "open and close the economic door."[41] Using a different metaphor, George Shultz referred less approvingly to tactical linkage as "lightswitch diplomacy."[42]

The need for conditioned flexibility suggests that over the long term a consistent or systematic tactical linkage strategy may be difficult to implement, particularly for democratic governments in charge of decentralized economies. First, as Huntington and others have noted, a tactical linkage strategy requires a significant degree of centralization in the foreign policy process. Decision-making authority must be concentrated, rather than dispersed, if trade relations are to be calibrated in a timely and precise manner with changes in a target's political behavior. Second, the strategy requires both the mobilization of private economic interests and their long-term subordination to political considerations.

40. Albert Hirschman, *National Power and the Structure of Foreign Trade* (Berkeley: University of California Press, 1980).
41. See Samuel Huntington, "Trade, Technology and Leverage: Economic Diplomacy," *Foreign Policy*, no. 32 (Fall 1978), 63–80.
42. Shultz's comment, made in October 1978, is noted in Bruce Jentleson, *Pipeline Politics: The Complex Political Economy of East-West Energy Trade* (Ithaca: Cornell University Press, 1986), p. 159.

Firms must be willing to engage in trade when it is politically opportune, yet they must also retreat from the market when political circumstances change. Not surprisingly, government employment of tactical linkage has tended to generate greater discontent in the U.S. business community than has either economic warfare or a strategic embargo, both of which, for better or worse, tend to allow greater predictability in long-term trade relationships.

Structural linkage is based on the premise that trade expansion in and of itself can induce or reinforce desirable changes in the domestic or foreign policy of a target state. Proponents of the strategy argue that to produce such effects, trade need not, and should not, be conditional. The unconditional expansion of trade in certain commodities, even with a potential adversary, can enhance a sanctioning state's security by restructuring the choices, the incentives, and, ultimately, the behavior of a target government.

Several examples help to illustrate the link between trade expansion and desirable target behavior. During the late 1970s, some Western observers expressed the concern that the impending decline in Soviet oil output would tempt the Soviets to intervene, perhaps even militarily, in the Persian Gulf in order to secure their long-term access to energy resources.Some argued that in response the West should make available to the Soviets technologically sophisticated energy production and exploration equipment, a bottleneck item in the Soviet economy.[43] A Soviet regime confident of sustained domestic output would have less incentive to pursue an adventuristic course in the Persian Gulf. It would be unnecessary for Western governments to link tactically the provision of technology to a Soviet promise not to intervene, because the very desire to do so would be diminished by the technology transfer. Indeed, one could further argue that a tactical link would actually be counterproductive; the Soviets would be more likely to refuse access to technology than to accept formal or informal constraints on their foreign policy autonomy.

A second example concerns the promotion of trade in those civilian items that are "resource-absorbing" rather than "resource-releasing" in an effort to induce a target government to devote more resources to the civilian and less to the military sector. One way to exploit the resource absorption effect might be to offer the export of "luxury" items that the target would not otherwise produce or consume. John Maynard Keynes called this a "policy of temptation"; he suggested in 1939 that the British try it with regard to Germany as an alternative to economic warfare in

---

43. See, for example, the editorial in *Washington Post*, October 11, 1978. The discussion that took place during the Carter administration is described in Jentleson, ibid., pp. 151–58. A Soviet military thrust to solve domestic energy problems provides the opening setting for Tom Clancy's popular novel *Red Storm Rising* (New York: G. P. Putnam's Sons, 1986).

the attempt to diminish the German military buildup.[44] Similarly, the sanctioning state might export civilian bottleneck products and technologies that require an additional commitment of target resources before they may be fully utilized. In the early 1970s, U.S. proponents of détente argued that if automobile technology were sold to the Soviet Union, it would require that the Soviets construct roads, service stations, and repair facilities, thus drawing resources away from potential use by the military.[45]

Structural linkage might appropriately be considered the pursuit of economic warfare by other means. Supporters of economic warfare seek to reduce the overall level of the target's economic *resources* without necessarily attempting to influence the target's preferences (i.e., how those resources are to be allocated). They assume that because it has fewer aggregate resources, the target will be forced to devote less to both military and nonmilitary uses. Advocates of structural linkage, however, seek to influence the target's preferences, even if in the process they bring about an increase in its economic capabilities. Such an increase implies that more resources are available for possible use by the military; but the underlying premise is that "additional" resources will be devoted to nonmilitary use and that, because of the nature of the trade and its political impact, even resources previously allocated to the military will be drained away to the civilian sector. The outcome ultimately sought, as in the case of successful economic warfare, is a decrease in the target's military capabilities.

That the expansion of trade could be conceived as a means to reduce an adversary's military capabilities highlights the complexity of the relationship between trade and national security. Indeed, it is possible to consider technology transfers of even direct military significance as potentially enhancing the security of a sanctioning state. The transfer of advanced telecommunications technology, for example, could improve an adversary's command and control systems and thus its ability to avoid or terminate an accidental nuclear war.[46] President Reagan's by-now-infamous offer to share "star wars" technology with the Soviets, on the grounds that such a transfer would encourage Soviet officials to appreci-

---

44. See Medlicott, *Economic Blockade*, vol. I, p. 39.

45. See, for example, John P. Hardt and George D. Holliday, *U.S.-Soviet Commercial Relations: The Interplay of Economics, Technology Transfer, and Diplomacy* (Washington, D.C.: GPO, June 10, 1973). The more general claim that interdependence leads to the erosion of state autonomy is found in Robert O. Keohane and Joseph Nye, *Power and Interdependence: World Politics in Transition* (Boston: Little, Brown, 1977). A comprehensive review of East-West economic relations from the perspective of interdependence is Franklin Holtzman and Robert Legvold, "The Economics and Politics of East-West Relations," *International Organization* 29 (Winter 1975), 275–322.

46. See William Root, "CoCom: An Appraisal of Objectives and Needed Reforms," in Gary K. Bertsch, ed., *Controlling East-West Trade and Technology Transfer: Power, Politics, and Policies* (Durham: Duke University Press, 1988), pp. 430–33.

ate the advantages of strategic defenses and would reduce their incentives to initiate a nuclear exchange, is based on similar reasoning. Such initiatives, intended to induce "stabilizing" behavior on the part of an adversary through the mechanism of capability-enhancing technology transfers, are compatible with the logic of structural linkage.[47]

The notion of structural linkage and the economic policies associated with it gained increasing political prominence during the late 1980s and early 1990s, as the Soviet Union and new governments in Eastern Europe embarked on ambitious economic and political reform programs. To proponents of the strategy, Eastern efforts to decentralize the economy, reduce military expenditures, convert military industries to civilian ones, and liberalize the political system provided unprecedented opportunities for the West to reinforce and render irreversible such desirable changes by deepening East-West trade and financial links.[48]

EXPORT CONTROL STRATEGIES: DETERMINANTS AND PREFERENCES

Given a commitment to use trade with an adversary as an instrument of statecraft, the choice of strategy is driven by economic, political, and strategic considerations. Specifically, those factors likely to be most significant to a sanctioning government are the degree of its trade dependence on the target, its overall political orientation to the target government, and its military relationship with the target, including such elements as the probability of war and the nature of peacetime military competition (see Table 1).[49] Differences in these factors account for dif-

47. Yet another example would be the sale of sophisticated milling equipment to improve the stealth of Soviet submarines. The more invulnerable are Soviet submarines, the less likely the Soviet government would be to resort to nuclear weapons in a crisis. The illicit sale of advanced milling equipment by the Toshiba Machine Company created an uproar in the U.S. defense community when it was revealed in 1987. See Chap. 8.

48. Mark Palmer, at the time U.S. ambassador to Hungary, argued in 1989 that U.S. policy in Eastern Europe should seek to "build institutions, programs, and extensive connections of such deepening mutual interest that they will not be shaken by the inevitable setbacks in individual countries or in the overall East-West climate. The West should be designing a comprehensive network of relationships that makes the borders irrelevant." See "U.S. and Western Policy—New Opportunities for Action," in William E. Griffith, ed., *Central and Eastern Europe: The Opening Curtain?* (Boulder, Colo.: Westview Press, 1989), p. 389.

49. The table depicts the "ideal" economic, political, and strategic circumstances for the adoption of each strategy. There are, of course, other possible combinations. A sanctioning state may be dependent on a target economically, yet engaged in intensified military competition. Examples include the relationship between frontline states and South Africa during the 1980s and between Western Europe and the Soviet Union at the outbreak of the Korean War. A sanctioner may also have a relationship with a target that is confrontational politically, yet stable in terms of military competition, such as that between the United States and North Korea since the 1960s. In these and other cases of crosscutting economic, political, and strategic circumstances, it is difficult to predict in general the preferred strategy of the sanctioning state.

Table 1. Choice of export control strategies

Preferred strategy	Sanctioner's relationship to target		
	Economic	Political	Strategic
Economic warfare	less dependent	confrontational	intensified military competition
Strategic embargo	more dependent	competitive	stable military competition
Tactical linkage	less dependent	competitive	stable military competition
Structural linkage	more dependent	cooperative	declining military competition

ferences in the preferred strategies of the United States and its Western allies and of the U.S. and West European governments over time.[50]

The degree of trade dependence is significant because the use of export controls involves economic costs for the sanctioning state. The potential political or strategic benefits of trade controls must be weighed against the economic costs of applying them.[51] The more dependent the sanctioning state is on trade with the target, either in the aggregate or with respect to specific sectors or products, the more costly it will be to implement strategies of trade denial and thus the less attractive those strategies become. Along with the direct economic costs of foregoing trade that is prohibited, indirect costs must also be considered, such as those associated with the administration of controls, the possible retaliation by the target government, and the potentially discouraging impact of controls on the long-run calculations of importers and exporters. An indirect cost of critical importance in the East-West trade context is that of so-called second-order sanctions, or those applied against third countries that refuse to comply with the controls in question.

The direct economic costs of economic warfare will be greater than those of a strategic embargo, simply because the former controls will be broader in scope. The attempt to deny trade that makes a direct or indirect contribution to target military capability will result in a more

50. The export control preferences of a state are also affected, of course, by the nature and extent of multilateral coordination. For purposes of analytical simplification, in this section I assume that export control preferences are derived nationally and are not affected by the preferences or coordination attempts of other sanctioning states. The question of how export control preferences interact in the East-West trade area is discussed conceptually in Chap. 1 and receives sustained attention in those that follow.

51. Because trade denial hurts the sanctioner as well as the target, the objectives of strategic trade controls are often cast in terms of "net advantage" or "relative strategic gain." The ideal situation for the sanctioner is one of extreme asymmetry—to be invulnerable or insensitive to the denial of bilateral trade while the target is extremely sensitive or vulnerable to it. For systematic discussion, see Schelling, *International Economics*, pp. 496–98, and Hirschman, *National Power and the Structure of Foreign Trade*.

comprehensive control list than will the attempt to deny only items of direct military significance. Since the government that practices economic warfare would probably not permit trade of direct military significance, the items controlled under a strategic embargo are likely to be a *subset* of those controlled for the purposes of economic warfare.

Economic warfare will also have greater indirect costs. The political hostility associated with it will probably affect even trade not explicitly covered by the embargo, either because the target government retaliates by imposing a counterembargo or because some firms in the sanctioning state are reluctant to engage in what may be perceived publicly as trading with the enemy. Given its arguably defensive nature, a strategic embargo is less likely either to inspire retaliation or to discourage private traders from initiating or maintaining trade in nonstrategic items.

The economic costs of tactical linkage are similarly likely to be greater than those of a strategic embargo, in the long run. In the short run, tactical linkage may actually bring economic benefits if the sanctioning state uses an increase in trade as a reward for past or future desirable behavior. When practiced against a single target over time, however, the strategy will probably be extremely costly. By routinely subordinating trade to political considerations, the sanctioning government creates an environment of unpredictability and uncertainty in trade relations. Over time, its firms may gain reputations as unreliable suppliers and lose markets to competitors who, unburdened by such political constraints, can be counted on to honor their commitments and meet delivery schedules. The political requirements of an effective tactical linkage strategy are in direct conflict with the economic requirements of a successful export promotion strategy, because export promotion requires stability and predictability, whereas tactical linkage demands that governments be willing and able to intervene and manipulate trade at a moment's notice, in accordance with changes in the target's political behavior. A strategic embargo, by contrast, is less burdensome to exporters because its trade controls are more predictable and stable over time.

The economic costs of a strategic embargo, practiced against an adversary of roughly equivalent military capabilities, are likely to be minimal to the extent that the control criteria remain narrowly focused on civilian items of military use and significance. As I argue above, however, the selectivity of the embargo may be compromised if military significance is equated with military use and the probability of the latter is judged conservatively, or if controls are applied to any technology that *might* have a current or future significant military application. An embargo governed by such criteria is likely to become quite comprehensive and may come to carry the direct economic costs generally associated with economic warfare. These costs will be compounded if second-order sanctions are applied, as U.S. officials learned during the 1980s.

Structural linkage is the least costly strategy economically. Because it relies on unconditional trade expansion, it produces a situation in which the strategic interests of sanctioning governments and the economic interests of export-oriented firms coincide. The coincidence will be strongest for those items selected and promoted by the sanctioning government, because of their potential to generate resource absorption or more general stabilizing effects.

The choice of strategy is also influenced by the overall political orientation of the sanctioning government toward the target. Export control strategies, like foreign economic policies more generally, are framed within and tend to complement the broader foreign policy relationship between states.

Economic warfare is compatible with and tends to reinforce a confrontational foreign policy approach to dealing with an adversary. A government that has taken such a stance politically will be less likely to tolerate, and will find it more difficult to justify, allowing trade in even seemingly innocuous items. The criticism directed at President Reagan, largely from his own supporters, for lifting the grain embargo while simultaneously depicting the Soviet Union as an "evil empire" that could not be trusted or bargained with, demonstrates this point. Economic warfare enhances the credibility of a confrontational foreign policy strategy by signaling the willingness of the sanctioning government to forego business as usual and to accept economic sacrifices if necessary in an effort to isolate and damage the target's economy.[52]

A strategic embargo complements a political orientation best described as competitive, that is, one guided by neither confrontation nor close cooperation but by a recognition of the potential for significant conflicts of interest between sanctioner and target. By permitting a good amount of "peaceful" trade, a strategic embargo can accommodate and reinforce a less confrontational political approach. At the same time, by closely monitoring a narrow range of items deemed strategic, this strategy reflects a concern over the capabilities and intentions of a potential adversary and a reluctance on the part of the sanctioning government to lower its guard completely.

Tactical linkage is similarly compatible with a competitive political orientation. Its underlying assumption is that the target government is a worthy negotiating partner whose policies are subject to influence through economic interaction. Premised on the belief that relative economic advantages should be exploited to bring political concessions, tactical linkage is a natural complement in the economic realm to a political approach that seeks to avoid confrontation and yet at the same time recognizes the existence of or potential for significant conflicts of interest.

52. Baldwin emphasizes the importance of accepting costs as a means to establish the credibility of sanctions. See *Economic Statecraft*, chap. 6.

Structural linkage complements and is reinforced by a cooperative political orientation. Sanctioning governments will be relatively most receptive to endorsing or promoting trade that will enhance the economic and/or military capabilities of an adversary, especially as a sustained policy over time, when political relations with that state are characterized by either the absence or diminution of significant conflict. Not surprisingly, the argument that the West should help Gorbachev economically gained momentum in the late 1980s as the Soviet leader relaxed political repression domestically and as the United States and Soviet Union moved to diffuse long-standing regional conflicts and made meaningful progress in arms control. By the same logic, the existence or reemergence of significant conflict would make it politically difficult to justify unconditional trade expansion and would increase the incentives for Western governments to resort to tactical linkage.

The third important factor influencing the choice of export control strategy is the nature of military competition between sanctioner and target. As suggested earlier, economic warfare is most appropriate in circumstances of intensified military competition, for example, the acceleration of an arms race or the mobilization for an imminent war. As the Reagan administration sped up U.S.-Soviet arms competition during its first term, it attempted economic warfare in order to force the Soviet Union, in the words of National Security Adviser Clark, "to bear the brunt of its economic shortcomings."[53] NATO leaders similarly opted for economic warfare at the outbreak of the Korean War, on the assumption that the Soviets might be preparing to initiate a more general conventional war. In both instances, Western officials accepted the idea that the primary function of the Soviet economy was to support defense-related activities.

A strategic embargo is more appropriate when military competition between potential adversaries follows a stable, regular pattern and is expected to continue to do so over a foreseeable time period. In such an environment it is logical for a sanctioning government to seek to deny an adversary strategic advantage from trade with significant military applications while it allows nonstrategic trade on the assumption that the resources released by it are unlikely to be translated into major improvements in military capability. Put somewhat differently, a strategic embargo is apt when government officials are prepared for a military competition of indefinite duration and do not expect either to eliminate it altogether or to gain a decisive advantage in it.

Because tactical linkage is usually employed to influence foreign policy behavior rather than military capabilities, the nature of the bilateral military relationship is relatively less important in determining its appro-

53. *New York Times*, May 22, 1982, p. A1.

priateness as an export control instrument. Nonetheless, an environment of stable military competition is most conducive to its employment as a systematic strategy. The existence of intensified military competition makes it politically difficult to justify "rewarding" a potential adversary with economic benefits; similarly, declining military competition renders it more difficult to justify "punishing" an adversary through trade denial.

Structural linkage, particularly when employed to influence military capabilities by seeking to exploit the "resource absorption" effect, is most appropriate in circumstances of declining military competition. Such an environment is most accommodating for a sanctioning government to expand trade unconditionally and for a target government to react to an increase in aggregate economic capabilities by devoting less, rather than more, to defense activities. Alternatively, in an atmosphere of intensified military competition, few sanctioning governments would risk contributing to a significant expansion of target economic capabilities.

Different interests and circumstances—economic, political, and strategic—have led the United States and other Western governments to differ in their preferred export control strategies. In general, economic considerations have weighed more heavily for Western Europe and Japan, whereas political and strategic factors have been more prominent in the calculations of U.S. officials. Specifically, throughout most of the postwar era, West European and Japanese governments have rejected economic warfare as being economically costly and politically and strategically counterproductive. Economic costs and skepticism regarding the potential political benefits have similarly deterred them from preferring tactical linkage. To the extent linkage strategies have been advocated or adopted, they have tended to follow the logic of structural linkage. Finally, Western Europe and Japan have been willing to choose a strategic embargo and have preferred controls to be narrowly focused on items of direct military significance.

U.S. officials, in contrast, have found both economic warfare and tactical linkage to be attractive export control strategies during different postwar periods. They have also favored a strategic embargo, though consistently one broader and more comprehensive in scope than that preferred by other Western states. That the U.S. has been less trade-sensitive, that it placed itself in a global struggle with the Soviet Union for political influence, and that it accepted primary responsibility within the Western alliance for assuring the adequacy of Western deterrence and defense—overall, these facts have prompted U.S. officials to resort more readily to export controls as an instrument of statecraft.

In light of their divergent preferences and the crucial fact that coordination is required for any export control strategy to be pursued effectively, it is not surprising that East-West trade policy has persisted as an issue of controversy among Western governments. The initial struggle over export control strategy in CoCom, and its resolution, is the subject of Chapter 3.

CHAPTER THREE

CoCom's First Decade: The Rise and Demise of Economic Warfare

Following World War II, the United States possessed the influence and interests necessary for global economic leadership. Driven by lessons derived from the experience of the 1930s, American officials sought to restructure international economic relations according to the principles of liberalism. Such an arrangement was presumed to satisfy both U.S. economic interests, by providing access to overseas markets for U.S. firms, and U.S. security interests, by ensuring that the competitive political and economic bloc system that had led the world to war would not reemerge.[1]

The commitment of U.S. officials to the primacy of the market in foreign economic policy was tempered, however, by the emergence of geopolitical competition with the Soviet Union. American officials quickly came to believe that U.S. security interests would be best served if the Soviet Union, Eastern Europe, and the People's Republic of China were isolated from, rather than integrated into, the liberal world economy. In this regard they were influenced by the economic lessons not of the 1930s but of World War II, namely, that comprehensive economic denial was a necessary and attractive strategic weapon in a struggle with hostile and implacable adversaries. While economic policy toward the West was influenced by the ghost of Hoover, that toward the East was shaped by the ghost of Hitler.[2]

1. See, for example, Charles Maier, "The Politics of Productivity: Foundations of American International Economic Policy after World War II," and Stephen Krasner, "U.S. Commercial and Monetary Policy: Unravelling the Paradox of External Strength and Internal Weakness," *International Organization* 31 (Autumn 1977), 607–71; Robert Gilpin, *U.S. Power and the Multinational Corporation* (New York: Basic Books, 1975); Fred Block, *The Origins of International Economic Disorder* (Berkeley: University of California Press, 1977); and Robert Pollard, *Economic Security and the Origins of the Cold War, 1945–1950* (New York: Columbia University Press, 1985).

2. I am grateful to James Kurth for the "Hoover-Hitler" image. On the importance of economic warfare to the United States during World War II, see David L. Gordon and Royden

The need to bifurcate U.S. foreign economic policy posed two related problems for U.S. policy-makers. First, for comprehensive restrictions on economic relations with the East to be effective, the United States required the support of other Western governments. Those governments, however, tended not to share the enthusiasm of the United States for the strategy. Second, assuming that effective coordination could not be realized, the need to maintain the integrity of U.S. controls threatened to compromise the ability of the United States to pursue liberal foreign economic policies in the intra-Western context. East-West export controls could "spill over" and disrupt intra-Western trade. During the first postwar decade, the management of these twin problems occupied a position of prominence on the U.S. foreign policy agenda.

I discuss three main topics in this chapter. First, I establish and account for the divergent export control preferences of the United States and Western Europe and the bargaining process that led to the formation of CoCom in 1949. For the United States, strategic considerations, in particular the reality of a geopolitical and expected military conflict with the Soviet Union, were primary and led to a preferred strategy of economic warfare. For West European governments, recognition of the strategic risks of Eastern trade combined with the need to engage in it for the purposes of economic recovery. The result was a desire to limit controls to those of a more narrow strategic embargo. Given preferences that were partly complementary and partly in conflict, the United States and its West European allies were willing to coordinate controls, yet they possessed different expectations regarding the structure of the new regime and the collective strategy that would be pursued within it. Compromises struck in these early years, particularly with respect to CoCom's institutional character, continued to influence the operation of the multilateral regime for four decades.

Second, the chapter explains the rise and relatively rapid demise of economic warfare as an alliance strategy. The dominant explanation of this pattern in the East-West trade literature focuses on the utility of U.S. coercive power. I challenge that interpretation and provide an alternative that emphasizes U.S. leadership in the context of a shift in West European export control preferences triggered by the outbreak of the Korean War.[3] Only in truly extraordinary circumstances, and for a relatively brief duration, were America's allies prepared to reconceptualize the relationship between economic benefits and security risks in East-West trade and join the United States in economic warfare. I argue that

---

Dangerfield, *The Hidden Weapon: The Story of Economic Warfare* (New York: Harper and Brothers, 1947), and W. M. Medlicott, *The Economic Blockade*, vol. II (London: HMSO, 1959).

3. For a version of this argument, see Michael Mastanduno, "Trade as a Strategic Weapon: American and Alliance Export Control Policy in the Early Postwar Period," *International Organization* 42 (Winter 1988), 121–50.

because of its prohibitive costs, the coercion of U.S. allies was not a viable strategy for U.S. officials. Moreover, to focus on it leads one to underestimate the significance of U.S. multilateral commitments and of the constraints those commitments placed on executive officials seeking alliance support for their preferred export control strategy.

Third, I examine the link between East-West export controls and the ability of the United States to promote liberalism in the intra-Western context. During the early 1950s, a lag in West European enforcement efforts left U.S. officials with a dilemma: either tolerate the leakage of U.S.-origin items to the East or restrict U.S. exports to Western Europe in an effort to prevent that leakage. By 1954, the United States managed to resolve this problem. In exchange for CoCom list reductions, West European governments agreed to improve their enforcement of controls, which in turn enabled the U.S. government to avoid interference in intra-Western trade. This tacit arrangement, in which the allies agreed to protect indigenous and U.S.-origin technology and equipment in exchange for easy access to the United States as a source of supply, served the broad interests of the United States and other CoCom governments, and their manufacturing industries, until its breakdown in the late 1970s.

DIVERGENT EXPORT CONTROL STRATEGIES

For the United States, the division of the postwar international economy into polarized blocs was neither planned nor inevitable. During and immediately following the war, U.S. officials hoped the Soviet Union and Eastern Europe could be induced to participate in the multilateral trade and financial institutions that were to govern global economic relations. American officials encouraged the Soviets to join the Bretton Woods agreement and considered offering them a sizable postwar reconstruction loan.[4] In 1945, Secretary of State Cordell Hull tried personally to convince Stalin to cooperate in the contemplated expansion of free trade.[5] Such initiatives were consistent with the postwar political strategy toward the Soviet Union planned by Roosevelt, which has been called "containment by integration."[6] By offering the Soviets a prominent stake in the postwar order, rather than isolating them from it, U.S. officials hoped to enlist their cooperation in its construction and maintenance.

4. Joan Spero, *The Politics of International Economic Relations*, 4th ed. (New York: St. Martin's Press, 1990), p. 305.

5. Herbert Feis, *Churchill, Roosevelt, Stalin: The War They Waged and the Peace They Sought* (Princeton: Princeton University Press, 1967), p. 642.

6. John Lewis Gaddis, *Strategies of Containment: A Critical Appraisal of Postwar American National Security Policy* (New York: Oxford University Press, 1982), p. 9.

With regard to economic relations, containment by integration called for a strategy of tactical linkage. Soviet behavior could be shaped by the creative use of U.S. economic instruments, or as Ambassador Averell Harriman suggested in 1945, the Soviets "should be given to understand that our willingness to cooperate wholeheartedly with them in their vast reconstruction problems will depend on their behavior in international matters."[7] Harriman advised that the State Department gain control of all the activities of agencies dealing economically with the Soviet Union, "so that pressure could be put on or taken off, as required."[8] The economic incentives the United States could offer included a continuation of lend-lease, a postwar loan, and reparations demanded by the Soviets from the U.S. zone of Germany. President Truman was receptive to the idea of linkage, because he believed the Soviet Union "was susceptible to pressure, especially economic pressure, which could be used to control, discipline, and punish it."[9]

Despite its presumed attractiveness, the exercise of tactical linkage proved to be short-lived and failed to achieve meaningful results. In response to what it perceived as Soviet political intransigence (e.g., on the question of Poland's future), the Truman administration abruptly terminated lend-lease aid to the Soviets in May 1945 and held in abeyance the postwar reconstruction loan. The Soviets refused to change their position, which suggests that the administration may have overestimated the extent to which its wartime ally would be dependent on American assistance for reconstruction and, more important, the extent to which it would be willing to make political concessions in order to obtain it.[10] Free elections and the relinquishing of political control over Eastern Europe—the demands frequently voiced by members of Congress—were clearly out of the question. Indeed, Soviet officials may even have believed that leverage ultimately rested with them, given their assumption that the end of the war would bring a capitalist crisis of unemployment and that the United States would thus be compelled to seek export markets on generous credit terms in the East.

More important, the nascent tactical linkage strategy was gradually overtaken by the emergence of a new political approach to the Soviet Union, articulated most clearly by George Kennan in the infamous "long telegram" of February 1946.[11] Kennan's message was that the Soviet state was driven by an unrelenting hostility toward the United States and

7. Feis, *Churchill, Roosevelt, Stalin*, p. 646.
8. Harriman is quoted in Daniel Yergin, *Shattered Peace: The Origins of the Cold War and the National Security State* (Boston: Houghton Mifflin, 1977), p. 93.
9. Ibid., pp. 92–93. See also Gaddis, *Strategies of Containment*, pp. 11, 16.
10. Feis, *Churchill, Roosevelt, Stalin*, p. 645, and Gaddis, *Strategies of Containment*, p. 17.
11. Telegram, the chargé in the Soviet Union (Kennan) to the secretary of state, February 22, 1946, reprinted in *Foreign Relations of the United States* (hereafter *FRUS*), 1946, 6:696–709.

"being impervious to the logic of reason," could not be influenced by U.S. inducements, either political or economic.[12] His analysis implied that tactical linkage, the effectiveness of which rested on the assumption that the target would respond rationally and predictably to rewards and punishments, could not be effective. Instead, an attempt to isolate the Soviets economically and weaken their military-industrial capabilities without attempting to influence their behavior would be more compatible with the emerging U.S. political approach. Economic warfare was the logical economic counterpart to the more confrontational political and military strategy of containment.

By 1948, political events had conspired to polarize relations between East and West and to reinforce to U.S. officials the analysis of the long telegram. The pronouncement of the Truman Doctrine in 1947 signaled U.S. willingness to intervene against Communist infiltration in the free world. The Communist coup in Czechoslovakia in February 1948 demonstrated that the Soviets would also maintain a sphere of influence. By the summer of that year, the United States, Britain, and France had decided to support in principle the formation of a West German state and had already begun to plan for a military alliance. In the economic realm, Congress approved the $17 billion aid package called for by the Marshall Plan. Judging it as another U.S. attempt to extract political concessions, the Soviet Union rejected the (half-hearted) U.S. offer to participate and pressured Poland and Czechoslovakia to do the same.[13] As the Economic Recovery Program (ERP) for Western Europe got under way, the State Department was convinced that the Soviets were making, and would continue to make, every effort to sabotage it.[14] The Soviets' counterresponse to the ERP was to foster the economic integration and self-sufficiency of Eastern Europe, in order to consolidate their political and economic control over the region. By mid-1948, most observers considered the division of Europe to be an accomplished fact.[15]

Economic Warfare as American Strategy

By early in 1948, Truman administration officials had concluded that the strategic risks of East-West trade outweighed any potential economic or political benefits; thus, they sought to restrict U.S. exports in order to

12. Ibid., pp. 706–7.
13. Given that the antagonism between the United States and Soviet Union was already evident, some analysts have questioned the sincerity of the U.S. offer. See, for example, Gabriel Kolko and Joyce Kolko, *The Limits of Power: The World and U.S. Foreign Policy, 1945–1954* (New York: Harper and Row, 1972), p. 363.
14. See "United States Exports to the U.S.S.R. and the Satellite States," paper prepared by the Policy Planning Staff, Department of State, November 26, 1947, reprinted in *FRUS*, 1948, 4:490.
15. Gaddis, *Strategies of Containment*, p. 73.

"prevent further increase in the war potential of the East European economies."[16]

Controls for that purpose required special authority in order to be exercised during peacetime. The United States had relaxed its wartime controls in 1945, with the exception of a "Positive List" of commodities retained under control because they were in short supply domestically. In March 1948, the Commerce Department creatively expanded the notion of short supply and determined that for the purpose of European recovery, *all* exports to both Western and Eastern Europe must be placed on the Positive List, under licensing control.[17] By regulating exports to both Western and Eastern Europe, U.S. officials hoped to maintain the appearance of nondiscrimination, as well as to assure that U.S. exports to the West had the maximum possible impact on economic recovery.

Once all exports were placed under control, the problem was to determine which were actually to be *prohibited* from trade with the East. U.S. officials resolved this in 1948–1949 by developing a series of categories, within which products, technologies, and equipment could be placed.

Items considered to be of the highest strategic significance were designated class 1A. The criteria of 1A were:

(a) Materials or equipment that are designated or used principally for the production and/or development of arms, ammunition, and implements of war.
(b) Materials or equipment that could contribute significantly to the war potential of the Soviet bloc where the items incorporate advanced technology or unique technological know-how. It applies only to goods sufficiently important to the war potential that the absence of an embargo would permit a significant advance in Soviet bloc technology over its present level of development.
(c) Materials or equipment that would contribute significantly to the war potential of the Soviet bloc in that the items, if embargoed, would maintain or create a critical deficiency in the war potential of the Soviet bloc.

Class 1A was highly selective and as of January 1950 the 1A list contained 167 items, which comprised mainly specialized machine tools (40 items), petroleum equipment (15 items), chemicals and chemical

16. See Report by the Ad Hoc Subcommittee of the Advisory Committee of the Secretary of Commerce, May 4, 1948, reprinted in *FRUS*, 1948, 4:536.
17. See Gunnar Adler-Karlsson, *Western Economic Warfare, 1947–1967* (Stockholm: Almquist and Wiksell, 1968), p. 22.

equipment (31 items), precision scientific and electronic equipment (42 items), and certain nonferrous metals (12 items).[18]

The criteria of class 1A were those of a strategic embargo. They included items of direct and specific military utility (a) and both bottleneck materials (c) and technologies (b). Though U.S. officials did not deem it necessary to distinguish explicitly between military and purely economic bottlenecks, the fact that 1A items were considered to be of the highest strategic significance in vital war-supporting industries suggests that they were viewed as military or, more likely, as simultaneously military and economic bottlenecks. Because the Soviets were rapidly rebuilding their military capabilities along with (and perhaps at the expense of) their civilian economy, shortages of equipment or materials were believed by U.S. officials to be felt in both areas.[19]

Items of secondary strategic significance were designated class 1B. Since their main contribution was to the general development of Soviet industrial potential, they were of indirect military significance. Materials and equipment on the 1B list were considered strategic only if shipped to the East in substantial quantities. In early 1950 approximately three hundred items comprised the 1B list, including primary commodities such as lead, copper, and zinc, and common industrial and transportation items, such as trucks, steel rails, and freight cars.[20] The embargo of such items was clearly directed at the maintenance and development of the Soviet economy; to prohibit 1B exports, along with those of 1A, was to engage in economic warfare.

In the actual practice of U.S. licensing policy, the distinction between the 1A and 1B lists was relatively insignificant. The United States maintained an unconditional embargo on 1A items, and its formal policy with regard to 1B items was to license limited quantities liberally. In reality, however, the 1B restrictions were virtually identical to the 1A total embargo. In 1949, for example, only $770,000 out of $22 million in requests to export 1B items was approved.[21]

The preference of Truman administration officials for economic warfare was primarily driven by strategic considerations. Administration officials believed that the United States would be locked in a long-term geopolitical struggle with the Soviet Union. Events such as the Czechoslovakian coup, the Berlin blockade, and the Communist takeover in

18. Report by the Ad Hoc Subcommittee, May 4, 1948, pp. 539–40, and telegram from the secretary of state to Certain Diplomatic Offices, April 26, 1950, reprinted in *FRUS*, 1950, 4:87–93.

19. See NSC-68: Report to the National Security Council by the executive secretary, April 14, 1950, reprinted in *FRUS*, 1950, 1:248, 257.

20. Telegram from the secretary of state to Certain Diplomatic Offices, April 26, 1950, p. 89; telegram from the secretary of state to Certain Diplomatic Offices, January 12, 1950, reprinted in *FRUS*, 1950, 4:65–66; and memorandum by the secretary of commerce (Sawyer) to the National Security Council, June 8, 1950, reprinted in *FRUS*, 1950, 4:141–43.

21. Telegram from the secretary of state to the Office of the United States High Commissioner for Germany, March 31, 1950, reprinted in *FRUS*, 1950, 4:80.

China reinforced this prevailing conception. Yet although the Soviet Union was a formidable adversary ideologically and militarily, it was also considerably weaker than the United States economically. Free trade, particularly in industrial equipment and technology, would provide greater benefits to the Soviets than to the United States. Economic warfare thus presented a national security opportunity in that it could help to perpetuate the backwardness of the Soviet economy, hinder reconstruction efforts, and delay the expansion of Soviet military capabilities and commitments.

Three related perceptions, shared by administration officials, provided the more specific strategic rationale for economic warfare. First, U.S. officials were firmly convinced that the Soviet economy was a "war economy," completely subservient to the demands of military production. They estimated that in 1950, 40 percent of Soviet resources were allocated to military purposes and heavy industrial investment, the great majority of which was war-supporting.[22] NSC-68, the most important statement justifying U.S. cold war policies, stated:

> The Kremlin has no economic intentions unrelated to its overall policies. Economics in the Soviet world is not an end in itself. The Kremlin's policy, in so far as it has to do with economics, is to utilize economic processes to contribute to the overall strength, particularly the war-making capacity of the Soviet system. The material welfare of the totalitariat is severely subordinated to the interest of the system.[23]

U.S. officials contended that Soviet industrial potential was military potential and that attempts to distinguish between the two were fruitless and dangerous. Export controls, they believed, "must be broad and deep enough to affect the entire production complex of the Soviet state."[24]

Second, U.S. export control policy was premised on the expectation that a protracted war with the Soviet Union was inevitable, either in the short or longer run.[25] U.S. officials believed that if war were to break out

22. See NSC-68, April 14, 1950, p. 257.
23. Ibid., p. 248.
24. Telegram from the secretary of state to the Embassy in the United Kingdom, August 22, 1950, reprinted in *FRUS*, 1950, 4:174–76. Acheson states (p. 175): "We do not consider it possible to draw [a] distinction in certain basic industrial categories between strategic fields or uses and normal peace-time industry or peace-time uses, when we are dealing with controlled economies of [the] Soviet bloc which have kept to a bare minimum their "peace-time" production in order to divert a large proportion [of] national production to direct military preparations and to [the] development [of] military potential."
25. See the memorandum by the associate chief of the Economic Resources and Security Staff (Armstrong), undated, reprinted in *FRUS*, 1950, 4:117–18. A good discussion of American thinking in the early postwar period regarding the strong likelihood of war and the possibility of even an American preventive war is Marc Trachtenberg, "A 'Wasting Asset': American Strategy and the Shifting Nuclear Balance, 1949–1954," *International Security* 13 (Winter 1988–89), 5–49. Trachtenberg notes (p. 22) that American officials took for granted that the likely third world war would be a protracted one; as I argued in Chap. 2, the attraction of economic warfare is enhanced when officials prepare for a long war against an adversary of roughly equivalent military strength.

imminently, it would be most important to have denied the Soviets 1A items. If war were to come in five or ten years, however, it would at present be "more important to keep basic production equipment from the Soviets than to deprive them of strategic military equipment."[26] Much of the latter would become obsolete over a span of ten years. Because they intended to prepare simultaneously for both contingencies, U.S. officials deemed it necessary to embargo both 1A and 1B items.

A third factor concerned the issue of Soviet vulnerability, which was treated in detailed reports by the State Department's Office of Intelligence Research and by the CIA in 1951.[27] Both reports recognized that the Soviet economy, especially when integrated with those of Eastern Europe, was relatively self-sufficient. Since bloc trade with the West was but 1 percent of GNP in 1950, a general embargo would be of only limited effectiveness. What was demanded from the West, however, consisted mainly of items "essential for military preparedness and the economic basis of military preparedness," including vital raw materials and semi-manufactured goods such as metals, abrasives, ball-bearings and industrial diamonds, and capital goods and equipment ranging across transportation, energy, and heavy industrial sectors.[28] An effective embargo would impair current levels of production, future production capacities, and the ability of the Soviets to wage a protracted war. Importantly, both reports recognized that to carry out such an embargo effectively would require the full support of the Western allies.

During the first postwar decade the great majority in the U.S. Congress supported the Executive's economic warfare strategy. Yet, although the preferences of executive officials were driven primarily by strategic considerations (i.e., by a conception of the close relationship between trade, the Soviet economy, and Soviet military power), those in Congress were motivated more by political or ideological concerns. As one analyst of this period has noted, for most of the U.S. public and Congress, "the general view tended to view [sic] East-West trade as an evil which should be eliminated."[29] As early as 1948, seven out of ten Americans believed East-West trade should be severed completely.[30] Most congressional dis-

26. Memorandum by the associate chief of the Economic Resources and Security Staff, undated, p. 117. See also "Trade of the Free World with the Soviet Bloc," report prepared by the Economic Cooperation Administration, February 1951, reprinted in *FRUS*, 1951, 1:1042–45.
27. "Vulnerability of the Soviet Bloc to Existing and Tightened Western Export Controls," report prepared by the Office of Intelligence Research of the Department of State, January 20, 1951, reprinted in *FRUS*, 1951, 1:1035–45, and "Vulnerability of the Soviet Bloc to Economic Warfare," National Intelligence Estimate No. 22, prepared by the CIA, February 19, 1951, reprinted in *Declassified Documents Quarterly* (1976), item 17C.
28. "Vulnerability of the Soviet Bloc to Existing and Tightened Western Export Controls," January 20, 1951, p. 1037.
29. See Suchati Chuthasmit, "The Experience of the United States in Controlling Trade with the Red Bloc, 1948–1960" (Ph.D. diss., Fletcher School of Law and Diplomacy, 1961), p. 226.
30. Adler-Karlsson, *Western Economic Warfare*, p. 33.

cussions of the issue took on the character of a moral crusade, reflecting the fervor of anti-Communist sentiment that had developed in the American public during this period.

Congress not only supported the Executive's economic warfare strategy; more significantly, it also delegated to the Executive remarkably powerful policy instruments in order to implement it. Historically, the Executive could restrict U.S. exports only in times of war or special emergencies; as noted above, the Truman administration was forced to resort to the manipulation of short-supply controls to initiate its strategy in 1948. With the passage of the Export Control Act of 1949, however, Congress gave the Executive the right to employ export controls for national security purposes in peacetime. Although the delegation of authority was initially intended to be a temporary response to the exigencies of the cold war, it was renewed periodically by Congress and eventually became recognized as permanent. Moreover, as two legal scholars have noted:

> Probably no single piece of legislation gives more power to the President to control American commerce. Subject only to the vaguest standards of "foreign policy" and "national security and welfare," he has the authority to cut off the entire export trade of the United States, or any part of it, or to deny "export privileges" to any or all persons. Moreover, the procedures for implementing this power are left almost entirely to his discretion, and at the same time heavy administrative and criminal sanctions may be imposed for violation of any export regulation he may introduce.[31]

In the circumstances of the cold war, executive officials, while committing the United States and other states to the creation of a liberal world economy marked by minimal government intervention, were simultaneously empowered by Congress to interfere drastically with global trade, in the interest of national security.

Although the potential strategic and political benefits of economic warfare were judged by U.S. policy-makers to be substantial, the direct economic costs of the strategy were minimal. U.S. firms, unlike their West European counterparts, had not developed a significant stake in Eastern markets. U.S. corporate officials tended to share both a concern for the risks and a healthy skepticism regarding the potential benefits of commerce with a state-trading nation.[32] And, the U.S. political climate served as an additional inhibiting factor. Firms that expressed even a passing interest in East-West commerce risked exposure to the charge of

31. Harold J. Berman and John R. Garson, "United States Export Controls—Past, Present, and Future," *Columbia Law Review* 67 (May 1967), 792.

32. This point is made well by Bruce W. Jentleson, "From Consensus to Conflict: The Domestic Political Economy of East-West Energy Trade Policy," *International Organization* 38 (Autumn 1984), 635–36.

Figure 1. U.S. trade with Eastern Europe, 1947–1956

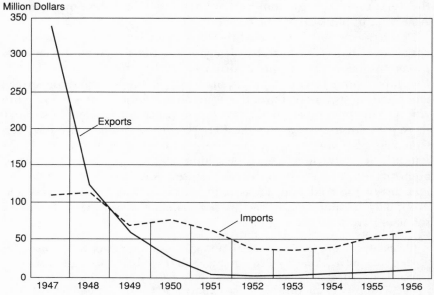

Source: Department of Commerce, *Export Control,* Thirty-Seventh Quarterly Report (Third Quarter, 1956).

"trading with the enemy." Such a reputation could do great harm to a firm's position in domestic markets. Consequently, U.S. firms tended to avoid Eastern trade, even that not considered strategic by the state, and when corporate officials spoke publicly on the subject, they often expressed their strong opposition to it.[33]

Figure 1 depicts U.S. trade with the Soviet Union and Eastern Europe from 1949 to 1956. The combination of strategic controls, public pressure, and business reluctance resulted by 1951 in virtually the complete absence of U.S. exports.

Western Europe: Trade and a Strategic Embargo

Unlike that of the United States, the primary objective of the West European states in East-West trade was economic. Their smaller economies were more dependent than that of the United States on trade in general and on East-West trade in particular. In 1951, for example, while U.S. exports comprised only 5 percent of its GNP, for the Netherlands, the comparable figure was 50 percent; for Belgium, 34 percent;

33. See, for example, the statements of corporate officials in U.S. Congress, Senate, Committee on Governmental Operations, Permanent Subcommittee on Investigations, *East-West Trade*, hearings, 84th Cong., 2d sess., February 15–17, March 6, 1956.

for Britain, 26 percent; and for France and West Germany, 16 percent.[34] In 1937, trade with Eastern Europe had accounted for 10 percent of West European exports and 11 percent of its imports.[35] After the war, West European states anticipated the restoration of their traditional trade with the East.

Imports from the East traditionally concentrated on a few critical commodities. Given the supply scarcity of the immediate postwar world and the imperatives of reconstruction, these commodities took on even greater significance. Western Europe required large quantities of timber from the Soviet Union, coal from Poland, potash from East Germany, and foodstuffs from the Ukraine. To obtain such imports, Western Europe needed to export to the East. Trade with Eastern states was negotiated annually and conducted bilaterally, and what a Western nation could acquire depended on what it had to offer. Prior to the war, Eastern Europe generally demanded agricultural products and simple manufactures, such as textiles. After the war, it began to require capital goods, machinery, and equipment in order to fulfill rapid industrialization plans. Initially, Western Europe exported such items primarily as a means to obtain vital supplies from the East. As their economies recovered by the end of the Korean War and they began to seek additional markets, the weight of East-West trade shifted from the import to the export side.

Despite the primacy of economic considerations, by 1948 the West European states, behind the initiatives of the British and the French, displayed a strong willingness to embargo exports to the East selectively for reasons of national security. The French considered a strategic embargo a "matter of legitimate defense" and an appropriate response to the aggressive policies of the Soviet Union. Similarly, the British determined that security considerations outweighed those of trade in items properly classified as strategic.[36] The two nations independently compiled export control lists in 1948 but by early 1949 collaborated on what became known as the Anglo-French list. Other West European states were willing to abide by it, and Italy and Germany, under pressure from the United States, accepted even greater restrictions and observed the U.S. control list. The Anglo-French list was somewhat shorter than the U.S. 1A list and was composed of civilian items having direct military significance. Secretary of State Acheson, confidentially commenting on their efforts, noted that by 1950, West European states had made "considerable progress in formulating a joint program of control over exports to Eastern Europe of goods that have a clear-cut strategic and

---

34. Chuthasmit, "The Experience of the United States in Controlling Trade with the Red Bloc," p. 282.

35. Figures are provided in U.S. Congress, Senate, Committee on Foreign Relations, *East-West Trade*, 83d Cong., 2d sess., April 9, 1954, p. 7.

36. British and French perspectives are discussed in the memorandum by the Associate Chief of the Economic Resources and Security Staff, undated, pp. 119–20.

military significance, even when the imposition of such controls involves to some extent the sacrifice of national trade interests."[37]

West European export control efforts were subject to two significant constraints. First, each state was willing to embargo strategic exports only if all other suppliers were willing to do the same. Given the primacy of economic interests, each required assurances that it would not be losing business to competitors by controlling trade.[38] Early in their discussions with the United States, French officials asserted unequivocally that they would not control any item not agreed to by *all* other Western countries. All must abide by a similar list, so that "complying countries are not put at a severe economic disadvantage compared to those who do not."[39] The British, who drew up their list first, reserved the right to reconsider it if others couldn't be persuaded to participate. Belgium informed the United States that it would not implement any controls until its government was absolutely certain that all OEEC members (Organization of European Economic Cooperation, which included neutral Sweden and Switzerland) maintained an equally rigorous embargo. The fact that West Germany's early control list, having been dictated by U.S. occupation officials, was more restrictive than those of other West European nations brought a strong formal complaint in early 1950 from Chancellor Adenauer regarding the unfair burden placed on German industry.[40]

Second, West European governments were strongly averse to applying any controls that suggested, either in appearance or in actuality, a strategy of economic warfare. In terms of the U.S. lists, they readily accepted most of 1A but had strong reservations about even considering 1B. The extension of controls to cover the basic industrial infrastructure of the Soviet bloc, they believed, took the West beyond purely defensive means and would ultimately lead to a total blockade. West Europeans feared economic warfare would result in the loss both of export markets, and of essential imports if the East chose to retaliate.[41] Even if it did not retaliate, extending the controls to basic industrial equipment and raw materi-

---

37. Telegram from the secretary of state to Certain Diplomatic Offices, April 26, 1950, p. 87.
38. West European governments preferred unilateral defection to unrequited cooperation. See Kenneth A. Oye, "Explaining Cooperation under Anarchy: Hypotheses and Strategies," in Oye, ed., *Cooperation under Anarchy* (Princeton: Princeton University Press, 1986), p. 6.
39. The statement, attributed to Hervé Alphand, director general for financial and economic affairs, French Ministry for Foreign Affairs, is reported in a telegram from the U.S. ambassador in France (Caffery) to the acting secretary of state, January 19, 1949, reprinted in *FRUS*, 1949, 5:69.
40. On the British, see ibid., p. 71; on Belgium, see the telegram from the chargé in Belgium (Millard) to the secretary of state, October 13, 1949, reprinted in *FRUS*, 1949, 5:148–49; on the Germans, see the letter from Chancellor Adenauer to the chairman of the Allied High Commission for Germany (McCloy), February 2, 1950, reprinted in *FRUS*, 1950, 4:73–74.
41. See, for example, the discussion of the concerns of Denmark, Norway, and Italy in the telegram from the U.S. deputy representative on the North Atlantic Council (Spofford) to the secretary of state, reprinted in *FRUS*, 1951, 1:1006–8.

als would leave the West Europeans with relatively little to bargain with in order to obtain what they needed. This problem confronted the British, who traditionally traded rubber (which the United States wanted to prohibit) to the Soviet Union for timber and grains. Similarly, Denmark's postwar negotiations for Polish coal were jeopardized because the Danes, under pressure from U.S. officials, could not assure the supply of items usually required by Poland, such as cargo ships and automotive parts.[42] If Eastern supplies were cut off, Western Europe could not easily obtain replacements from the United States, given their shortage of dollars. Expressing their opposition to the U.S. 1B concept, French officials noted that "security comes first, but economics can't be ignored."[43]

The perceived strategic risks of economic warfare were equally important. West European governments did not share the enthusiasm of the United States for using economic warfare as a peacetime weapon, believing it to be inextricably and inevitably linked to military warfare, as it had been during World War II. Engaging in economic warfare, or even the appearance of doing so, significantly increased the possibilities of an East-West military confrontation. Although the United States— geographically isolated and militarily capable—may have been willing to accept such a risk, in the early postwar period West European states were especially wary of provoking their powerful Eastern neighbor, given the inadequacy of their own defenses and the uncertainty of the U.S. commitment to assist them. Especially sensitive to this dimension of the problem were those states (Norway, Denmark, and Belgium, in particular) which were directly exposed to Soviet aggression and which had been overrun by the Germans only a short time earlier.[44] Swiss officials, expressing their reluctance to cooperate informally with U.S. 1B initiatives, frankly noted that Switzerland did not want to incur the hostility of the Soviet Union and then be left "holding the bag" if U.S. foreign policy interests shifted from Europe to Asia.[45]

To summarize, before 1950, West European governments, suspicious of Soviet intentions yet not convinced of the inevitability of war, were prepared under certain conditions to deny the Soviet Union trade judged to be of direct military significance. In contrast with the United

42. On Britain, see the minutes of the Second Meeting of the United States–United Kingdom Foreign Ministers' Meeting in Washington, September 11, 1951, reprinted in *FRUS*, 1951, 1:1183; on Denmark, see "Export of Automotive Parts via Denmark for Poland," report prepared by the Department of State and the Economic Cooperation Administration, August 27, 1951, reprinted in *FRUS*, 1951, 1:1169–74.

43. The French statement was made at a tripartite discussion of U.S. 1B proposals at Paris, May 8–9, 1950, and is reported in the memorandum by the associate chief of the Economic Resources and Security Staff, undated, pp. 119–20.

44. West European concerns are reported in "Trade of the Free World with the Soviet Bloc," February 1951, p. 1043.

45. The Swiss reminded U.S. officials of the striking reversal of U.S. policy toward the League of Nations after World War I. Memorandum by Roswell D. McClelland of the Office of West European Affairs, Department of State, June 11, 1951, reprinted in *FRUS*, 1951, 1:1103–4.

States, however, their economic and strategic predicament led them to consider the costs and risks of economic warfare to be excessive. Internal instability, rather than external aggression, was the primary national security challenge, to which the best response was to foster economic recovery. The restoration of East-West trade could contribute to that objective and help to stabilize East-West political relations as well.[46]

Coordinating Controls: The Formation of CoCom

Bilateral negotiations between the U.S. and West European governments began early in 1948 and culminated in the formation of a multilateral control system in November 1949. In October 1949, some West European members of the OEEC held a series of informal meetings in which they decided to seek agreement on a common list of commodities to control.[47] The United States supported this initiative and was later invited by the West Europeans to join the group. The multilateral mechanism that resulted served the interests of both the United States and Western Europe, in different ways.

For Western Europe, what became known as CoCom resolved the dilemma of uncertainty that had plagued each state from the time it had accepted, in principle, the necessity of a strategic embargo. Before its formation, each was faced with diplomatic pressure from the United States to implement or extend controls unilaterally. The choice for each was either to acquiesce and risk losing trade in areas others ultimately proved unwilling to control, or to resist and risk incurring the political and possibly economic costs of defying the United States. By bringing the West European states together in a multilateral forum, CoCom provided what each government required: immediate and up-to-date information regarding what all other participating states were willing or unwilling to embargo and thus the knowledge of whether and to what extent economic sacrifices were strategically justifiable.[48] CoCom enhanced the incentives for each government to cooperate by reducing transaction costs and creating a shared sense of obligation among the members not to exploit opportunities in trade collectively deemed strategic.

46. U.S. officials also recognized the important role East-West trade could play in West European recovery efforts. See William Diebold, "East-West Trade and the Marshall Plan," *Foreign Affairs* 26 (July 1948), 709–22. U.S. officials hoped, of course, that essential imports could be obtained from the East without the West having to export "strategic" items in return (see below).

47. See the telegram from the special representative in Europe for the Economic Cooperation Administration (ECA) (Harriman) to the administrator of the Economic Cooperation Administration (Hoffman), October 15, 1949, reprinted in *FRUS*, 1949, 5:150, and the telegram from the secretary of state to Certain Diplomatic Offices, April 26, 1950, p. 89.

48. Robert Keohane emphasizes the importance of regimes in facilitating cooperation by reducing transaction costs and conveying information. See Keohane, *After Hegemony* (Princeton: Princeton University Press, 1984), pp. 97–103.

As its willingness to control was not contingent on the behavior of others, uncertainty and the fear of losing export opportunities did not pose a similar problem for the United States. Indeed, upon agreement of multilateral controls, U.S. officials explicitly reserved the right to maintain national controls that extended beyond the consensus of the group. For the United States, the chief value of the multilateral regime was that it increased the scope and effectiveness of controls.[49] As Robert Wright, a participant in the early negotiations, noted in 1976: "If we were to attempt to accomplish the same purpose with a series of bilateral arrangements, we feel the result would be a lower level of control, absence of uniformity, and a complicated process of negotiation. In fact, it was substantially for this reason the decision was made in 1949–50 to shift the approach to a multilateral organization."[50]

The process of bilateral negotiations had indeed proved frustrating to U.S. officials. It resulted in a stalemate, as most states delayed their acceptance and implementation of controls until others took action. U.S. efforts to assure each state that others were cooperating were generally met with skepticism.[51] In addition, the longer it took U.S. officials to convince others to comply, the more in jeopardy was the continued cooperation of those states which had already implemented their controls (e.g., Italy) and which were de facto being penalized.

By March 1949, the State Department's attempt to obtain agreement on its 1A list had reached an impasse. Though willing to move somewhat beyond the Anglo-French list, the British were prepared to accept only 121 out of 163 1A items. France, Belgium, and the Netherlands, which considered the British total as marking the upper limit of their cooperation, had agreed to control significantly less. By July, despite tedious and sustained negotiations, virtually no further progress had been made. More important, until the negotiations were completed, the West Europeans were unwilling to implement the controls that had been agreed to, choosing instead to use the tentative lists solely for the purposes of information and guidance.[52] As the allies demurred, U.S. anxiety increased over the contribution of Western trade to the mobilization of Soviet economic and military power.

The agreement to form a multilateral control system paved the way for the deadlock to be broken. At the November meetings, the United States

49. See Oye, "Explaining Cooperation under Anarchy," p. 20.

50. Wright's statement can be found in U.S. Congress, House, Committee on International Relations, Subcommittee on International Trade and Commerce, *Export Licensing of Advanced Technology: A Review*, 94th Cong., 2d sess., March 11–30, 1976, p. 54.

51. The problem of assuring the continued support of the early cooperators is discussed in the telegram from Harriman to Hoffman, October 15, 1949, p. 151.

52. This was particularly the case for Belgium and the Netherlands. See the telegram from the secretary of state to the Embassy in the Netherlands, October 28, 1949, reprinted in *FRUS*, 1949, 5:162–63.

and its West European allies agreed to form two permanent bodies: the Consultative Group (CG) and the Coordinating Committee (CoCom).[53] The former was composed of high-level representatives from the participating states and was expected to meet relatively infrequently to resolve major policy disputes and set broad guidelines for the export control program. CoCom, composed of technicians and lower-level officials, was to meet more frequently to manage the task of implementing the agreed controls.

The participants consented to construct three international lists. List I was composed of items to be embargoed unconditionally. List II contained items for which members agreed they would restrict their exports to Eastern Europe to "reasonable" quantities and exchange information on what was actually exported. List III contained items still under consideration, for which agreement to control had not been reached.[54] For items to be placed on or removed from the control lists required the unanimous consent of the members.

With the multilateral consensus behind them, each state was willing to put CoCom controls into effect immediately. By January 1950, 144 items on the U.S. 1A list (which had grown to 177) were accepted for CoCom List I, 6 were retained on List II, and 27 on List III. Those that the allies would agree to place only on List III, such as diesel engines, ball bearings, and certain oil exploration and transmission equipment, were either vital to the fulfillment of trade agreements, in which case the United States supported their exclusion, or judged by West European governments to be of dubious strategic significance.[55]

That the West Europeans consented to participate in CoCom should not be taken to imply that they did so without reservation. Most states expressed considerable concern over both the domestic and international implications of what might be *perceived* as a peacetime agreement to engage in economic warfare, even though West European governments clearly did not consider or intend their selective controls as such. These concerns had an important influence on the rules and procedures that characterized CoCom. The participants agreed that CoCom's workings would be kept completely confidential and informal so that members

53. See the telegram from the deputy special representative in Europe for the ECA (Katz) to the administrator of the ECA (Hoffman), November 25, 1949, reprinted in *FRUS*, 1949, 5:174.

54. The formation of the CG/CoCom and its rules and procedures are reviewed in detail in a telegram from the secretary of state to Certain Diplomatic Offices, April 26, 1950.

55. Since it judged West European economic recovery to be of vital strategic significance, the United States was willing to tolerate West European trade in strategic items if necessary to fulfill the terms of existing trade agreements. This policy is set out in a telegram from the secretary of state to the Embassy in France, August 27, 1948, reprinted in *FRUS*, 1948, 4:564–68. This principle continued to inform U.S. policy after the formation of CoCom. See the telegram from the secretary of state to the Embassy in France, April 13, 1950, reprinted in *FRUS*, 1950, 4:81–82.

could publicly deny their participation, if necessary.[56] If CoCom's existence were to be publicly affirmed, France and the Netherlands in particular made it clear that they might be forced by domestic pressure to withdraw. Furthermore, CG and CoCom decisions would have no binding authority. Rather, they would serve as recommendations to the member states, who would be entirely responsible for implementing and enforcing agreed controls through national laws and procedures. CoCom was granted neither treaty status nor the power to impose sanctions against violators. Compliance was in effect voluntary, yet necessary, if the so-called gentlemen's agreement was to survive.

The export control system was also disassociated from other collective Western institutions, particularly NATO and the OEEC. As the latter included neutral nations such as Sweden and Switzerland, member governments deemed it inappropriate for CoCom to be associated with it. With regard to NATO, the West Europeans argued that the newly formed defensive alliance should not be "tainted" by the embargo, which might suggest an offensive posture.[57] Also, although it was expected that West Germany and Japan would participate in CoCom, neither were members of NATO at the time.

For the United States to accept coordination on such informal terms was clearly a compromise. Although confidentiality was preferable politically to West European executives, it created friction within the U.S. government by making it difficult for the Executive to convince a skeptical Congress that West European cooperation in export controls was indeed satisfactory. Congressional discontent led to the passage of the Battle Act, discussed below. More important, U.S. officials would have preferred to subsume CoCom within NATO, so that issues of economic security could be treated as part of political and military strategy.[58] Linking CoCom to NATO would have made it easier for U.S. officials to ensure that defense considerations received prominence in the control process. As it turned out, the CoCom delegations of member states came to be staffed by officials from Economics rather than Defense Ministries. Moreover, granting CoCom treaty status and building domestic political support for it in member countries, as was done with NATO, would have

56. Telegram from the secretary of state to Certain Diplomatic Offices, February 15, 1950, reprinted in *FRUS*, 1950, 4:76–77.
57. The relationship between CoCom and NATO is discussed in a telegram from the first secretary of the Embassy in the United Kingdom to the secretary of state, March 8, 1951, reprinted in *FRUS*, 1951, 1:1056–58. Interestingly, in the early 1950s, European officials feared that association with CoCom would compromise the defensive character of NATO; in the early 1980s, they feared that association with NATO would compromise the defensive character of CoCom (see Chap. 8, below).
58. The U.S. preference that export controls be handled within the context of NATO is expressed in a telegram from Harriman to Hoffman, November 5, 1949, reprinted in *FRUS*, 1949, 5:169–71.

made it easier for the United States to strengthen CoCom institutionally, for example, by giving it the authority to establish standard procedures and penalties for all members to apply in enforcing controls. National discretion in enforcement tended to encourage a lack of uniformity in member policies, which could be exploited by individuals or firms seeking to transship controlled goods to the East.

Nevertheless, CoCom was valuable to the United States. It created an institutional interest within West European bureaucracies in the export control issue. In addition, once items were placed on the list, it obliged America's trade-sensitive allies to bargain multilaterally (in effect, with the United States), rather than to act unilaterally, in order to have them removed. Through the quantitative and the so-called watch lists (International Lists II and III), it provided a means for the United States to monitor how many and what types of potentially strategic items were being traded from Western to Eastern Europe.

Of most immediate significance was that CoCom provided an institutional mechanism for the U.S. effort to expand the control list and move alliance policy from a strategic embargo to economic warfare. Once the limits of allied cooperation on list 1A had been reached, U.S. officials brought a formal proposal to CoCom to review its 1B list. West European governments initially were strongly opposed. Nonetheless, in March 1950, amid much controversy, CoCom members agreed to review the U.S. 1B list.[59]

By January 1951, a substantial portion of the 1B list had been accepted for control by CoCom. A further round of U.S. initiatives was made between April and August, at which point the allies agreed to even more extensive quantitative controls and also added thirty-four of fifty-three U.S. 1A items to CoCom List I.[60] By early 1952, more progress was made, and the differences between the U.S. 1A and 1B lists and CoCom Lists I and II, respectively, were relatively minor. Alliance strategy had become that preferred by the United States—economic warfare.

ECONOMIC WARFARE AS ALLIANCE STRATEGY: COERCION OR CONGRUENT INTERESTS?

Given their serious reservations, what accounts for the willingness of the West European states to cooperate in economic warfare? The dominant explanation has been offered by Adler-Karlsson, who focuses on the ability of the United States to use its economic resources as weapons

59. Telegram from the secretary of state to the Office of the U.S. High Commissioner for Germany, March 31, 1950, reprinted in *FRUS*, 1950, 4:80–81.
60. Telegram from the chargé in France to the secretary of state, August 2, 1951, reprinted in *FRUS*, 1951, 1:164–65.

of coercion against its allies. Adler-Karlsson writes: "In spite of all the West European reluctance, its governments did cooperate in the embargo policy. Thus we must also ask why the West European nations did cooperate as much as they did. *The answer is clearly to be found in the American threats to cut off aid in cases of non-compliance.*"[61] Adler-Karlsson's interpretation has been overwhelmingly accepted by students of East-West trade. Klaus Knorr, for example, argues that in the late 1940s, "the United States used its economic power to impose the policy on its reluctant allies by threatening to cut off economic and military aid to them at a time when American economic aid to the West European countries was several times larger than the total turnover of their trade with the Communist states of Eastern Europe."[62] The Office of Technology Assessment (OTA), in its authoritative background study on East-West trade prepared for the U.S. Congress, makes a similar point and cites Adler-Karlsson.[63] Stephen Woolcock and Gary Bertsch also emphasize the importance of U.S. aid as a coercive instrument.[64] Other analysts, though qualifying the argument somewhat, nevertheless accept the basic validity of the trade-aid link as the key explanatory factor.[65]

The argument for coercion rests on the fact that in 1950 and 1951, the Congress passed successive pieces of legislation that explicitly linked the continuation of U.S. aid to full West European cooperation with U.S. export controls. Riders to the Supplemental Appropriations Bills of 1950 and 1951, known as the Cannon and Kem Amendments, required the Executive to deny economic aid to uncooperative states during any time in which the armed forces of the United States were actively engaged in hostilities.[66] In October 1951, more permanent legislation, in the form of the Mutual Defense Assistance Control Act of 1951, replaced the Cannon and Kem Amendments.[67] The so-called Battle Act linked the provision of both economic and military aid to compliance with U.S. export controls and was to be applied in peacetime as well as

61. Adler-Karlsson, *Western Economic Warfare*, p. 45, emphasis added.
62. Klaus Knorr, *The Power of Nations* (New York: Basic Books, 1975), p. 142.
63. U.S. Congress, OTA, *Technology and East-West Trade* (Washington, D.C.: GPO, 1979), pp. 114, 154.
64. Stephen Woolcock, *Western Policies on East-West Trade* (London: Royal Institute for International Affairs, 1982), p. 8, and Gary Bertsch, *East-West Strategic Trade, CoCom, and the Atlantic Alliance* (Paris: Atlantic Institute for International Affairs, 1983), p. 10.
65. See Richard J. Ellings, *Embargoes and World Power* (Boulder, Colo.: Westview Press, 1985), pp. 80–84, and Beverly Crawford and Stephanie Lenway, "Decision Modes and International Regime Change: Western Collaboration on East-West Trade," *World Politics* 37 (April 1985), 388. Bruce Jentleson similarly accepts that the threat of aid denial played a significant role in shaping West European behavior, yet he stresses the positive sources of U.S. leverage as well. See *Pipeline Politics: The Complex Political Economy of East-West Energy Trade* (Ithaca: Cornell University Press, 1986), pp. 71–73.
66. The amendments are reprinted in Adler-Karlsson, *Western Economic Warfare*, pp. 26–27, 200.
67. The text of Public Law 213, 82d Congress, is reprinted in Mutual Defense Assistance Control Act of 1951, *Reports to the Congress*, hereafter *Battle Act Reports*, no. 1, October 15, 1952, appendix A.

wartime. It directed the Executive to terminate assistance to any nation that exported primary strategic materials (items on the U.S. 1A list) or "other materials" (in effect, items on the U.S. 1B list) to the Soviet bloc, unless the president determined, in exceptional cases, that it was in the national interest not to end aid.

The Battle Act and the amendments that preceded it provided the Executive with a legislative obligation to coerce allies that refused to cooperate fully with U.S. controls. Adler-Karlsson and others argue that because U.S. economic assistance in the 1950–1953 period was more valuable to Western Europe than was East-West trade (see Table 2), the allies buckled under U.S. pressure and agreed, despite their own preference, to join the United States in pursuit of economic warfare.

Despite the existence of the legislation and their dependence on U.S. assistance, there are strong reasons to doubt that the West Europeans were coerced into participation in economic warfare by the threat of aid denial. First, the legislation itself was a matter of serious dispute between the Congress, which initiated and passed it, and the executive officials who were directed to implement it. Both the State Department, which conducted the export control negotiations, and the Economic Cooperation Administration (ECA), which was responsible for administering the aid program, had profound reservations about the desirability of coercion.[68] Officials in those agencies believed that linking U.S. aid to West European compliance in CoCom would lead to less effective implementation of controls and, more important, might jeopardize the willingness of the allies to cooperate in CoCom altogether. As State Department officials argued confidentially: "There is no intention of using the threat of withholding ECA aid to force the acquiescence of European governments in U.S. policies on export controls, for U.S. policy in the long run will be infinitely more effective if based on the spirit and principle of cooperation and a common recognition of the danger in developing the military potential of the Soviet Union and satellites."[69] The State Department also expressed the fear that West European governments, faced with the choice between U.S. aid and East-West trade, might in the interest of asserting their sovereignty refuse the aid and thereby bring about the destruction of the Mutual Security Program.[70] Although executive officials were willing to expend considerable diplomatic resources in an effort to convince the allies of the desirability of economic warfare, they explicitly rejected cross-issue

68. See the telegram to the administrator of the ECA, May 12, 1949, reprinted in *FRUS*, 1949, 5:113–14, and the telegram from the secretary of state to the embassy in France, April 13, 1950, reprinted in *FRUS*, 1950, 4:81–82.

69. Ibid., pp. 81–82.

70. See the letter from Acheson to the chairman of the Senate Foreign Relations Committee, June 9, 1952, reprinted in *FRUS*, 1952–54, 1:847–49.

Table 2. U.S. assistance to Western Europe, compared to East-West trade

	1949	1950	1951	1952	1953	1954	1955
Western Europe							
U.S. economic aid	6276	3819	2268	1349	1265	637	466
U.S. military aid	—	37	605	1014	2867	2226	1541
Total exports to Eastern Europe	832	653	746	743	791	974	1100
Total imports from Eastern Europe	1012	813	1010	995	909	1039	1358
Total aid	6726	3856	2873	2363	4132	2863	2007
Total East-West trade	1844	1466	1756	1738	1700	2013	2458
England							
U.S. economic aid	1614	955	266	350	410	200	35
U.S. military aid	—	3	22	38	155	171	107
Total exports to Eastern Europe	109	72	45	43	43	70	103
Total imports from Eastern Europe	149	177	266	235	207	207	294
Total aid	1614	958	288	388	565	371	142
Total East-West trade	258	249	311	278	250	277	397
France							
U.S. economic aid	1313	701	435	263	397	86	3
U.S. military aid	—	16	346	486	1108	684	499
Total exports to Eastern Europe	65	35	39	39	51	74	126
Total imports from Eastern Europe	72	34	54	58	41	63	84
Total aid	1313	717	781	749	1505	770	502
Total East-West trade	137	69	93	97	92	137	210

Source: Adler-Karlsson, *Western Economic Warfare, 1947–1967* (Stockholm: Almquist and Wiksell, 1968), p. 46. All figures in millions of U.S. dollars.

linkage. If the allies could not be persuaded, executive officials were prepared to tolerate some differential in the scope of U.S. and CoCom controls.

The majority in Congress took a different view. As reflected in the Battle Act, the prevailing attitude was that no differential was tolerable; those allies unwilling to replicate U.S. controls fully were unworthy of U.S. aid. Congress was more conservative than the Executive on this matter for two important reasons. First, because of the secrecy of CoCom, many in Congress were unaware of the extent to which West European governments had already participated in an export control program by mid-1950. Executive officials were placed in an unenviable position: they had to convince the Congress that the allies were cooperating yet were unable to provide detailed evidence for their assertion publicly.[71] Not surprisingly, executive reassurances frequently met with skepticism in light of routine press reports that "war materials" were making their way to the East from Western Europe.

Congressional sentiment also reflected a larger discontent with the substance and process of U.S. foreign policy. Some of the most vocal proponents of coercive legislation (e.g., Senators James Kem and Kenneth Wherry) were isolationists, still uncomfortable with the multilateral thrust of U.S. policy. Although skeptical in any case of the military and economic aid programs, they believed that at the very least the beneficiaries should display their gratitude by giving full support to U.S. policy.[72] In their eyes, for the United States to expect full cooperation in the denial of strategic goods, particularly when U.S. soldiers were dying in Korea, was clearly not to ask too much of the allies.

The Battle Act was symptomatic of a larger struggle over the control of foreign policy. By 1950, members of Congress, reacting to Truman's unilateral dispatch of U.S. forces to Korea and his decision to commit the United States to the defense of Europe through the North Atlantic Treaty, grew increasingly concerned over what they perceived as the dangerous usurpation of foreign policy power by the Executive. In East-West trade, the Executive was again exercising unilateral control over policy while keeping Congress at bay. Even for those in Congress who did not consider this particular issue all that significant, the Battle bill represented a way for Congress to reestablish its rightful position in the foreign policy process.

Given the divergent perspectives of the Executive and Congress, the key point is that ultimately the Executive had responsibility for implementing the coercive legislation. And, despite the fact that the West Europeans were in continual violation, executive officials *never* withdrew aid in accordance with its provisions. Both the Cannon and Kem Amendments allowed for some discretion in their implementation, of which the Truman administration took full advantage. After the passage of the Cannon rider, the National Security Council (NSC) granted a blanket exception, determining that to cut off aid to any foreign country would be contrary to U.S. security interests. It argued that "national security should be interpreted broadly enough to permit situations where an

71. For a discussion of role of CoCom secrecy in the interbranch dispute, see the letter from Representative Battle to the director of Mutual Security, September 29, 1952, reprinted in *FRUS*, 1952–54, 1:896–901.
72. For the Senate debate over the Battle Act, see *Congressional Record* (Senate), August 27, 1951, pp. 10661–78, 10700–15.

assistance program might properly be continued even though trade with the Soviet bloc in security items continues."[73] Similarly, immediately following passage of the Kem Amendment, the NSC approved a general interim exception for all countries, pending a case-by-case examination of the situation of each. After its review, and despite the strong protests of Senator Kem and others, the NSC granted country exceptions to each NATO member and to Japan.[74]

Under the terms of the Battle Act, aid could only be continued to countries that exported U.S. list 1A items if "unusual circumstances" dictated that an aid cutoff would be detrimental to U.S. security interests.[75] Apparently the circumstances were perpetually unusual, as aid was never withdrawn, despite repeated violations, and exceptions were granted in every year from the inception of the Battle Act in 1952 through the end of the decade. For example, exceptions were granted in 1952 to Denmark for tankers; to Italy for grinding machines; to the Netherlands for oil-drilling equipment; and to Britain and France for various machine tools, ball bearings, and chemical and electrical equipment. In 1953, West Germany, France, Norway, and Britain were permitted to sell ball bearings, aluminum, and locomotive equipment to the East.[76]

Finally, there is evidence to suggest that West European governments were not only aware of the conflict between the Executive and Congress but also actively assisted the Executive in evading the intent of congressional legislation. How to deal with the problems posed by the Battle Act was the subject of extensive discussion in CoCom. By mid-1951, the relatively few items on the U.S. 1A list that had not been placed on CoCom List I (e.g., rail-transport and coal-mining equipment, tin, aluminum, and ball bearings) were economically critical to the allies in that their export was required to fulfill existing trade contracts with the East and thus to obtain vital imports. The British, speaking for the West

73. See "NSC Determinations under Public Law 843, Section 1304: The Cannon Amendment," Statement of Policy by the National Security Council, undated, included in a note by the executive secretary of the NSC (Lay) to the NSC, and reprinted in *FRUS*, 1950, 4:254.

74. A list of the countries granted exceptions by the NSC, and supporting arguments, is found in "Report by the National Security Council regarding a Review of its Determinations under Section 1302 of the Third Supplemental Appropriations Act, 1951," October 23, 1951, reprinted in *FRUS*, 1951, 1:1203–10.

75. The exception clause is found in Mutual Defense Assistance Control Act of 1951 (Public Law 213, 82d Congress), sec. 103. For an account of the exceptions granted, see *Battle Act Reports*, nos. 1–14, October 15, 1952 through December 20, 1960.

76. See *Battle Act Reports*, no. 1, October 15, 1952, pp. 47–57; no. 2, January 16, 1953, pp. 77–86; and no. 3, September 27, 1953, pp. 73–77. The only instance in which Battle Act sanctions were applied did not involve a CoCom member but Ceylon, which exported rubber to China in exchange for rice. Ceylon was not receiving U.S. aid but was declared ineligible to receive it. The restriction was imposed in 1952 and dropped in 1956. See Berman and Garson, "United States Export Control Policy," p. 837.

Europeans in CoCom, made it clear that unconditional restrictions on such items would be unacceptable.[77] The State Department was sympathetic. Yet, the Battle Act required that West European states replicate U.S. controls in full to continue receiving aid.

With the passage of the Battle Act imminent, State Department officials proposed in CoCom a new round of control negotiations, along with the introduction of an "exceptions" procedure, that would allow any member state to export a given item on the CoCom embargo list so long as no other member state objected to the sale in that particular instance. U.S. officials proposed that the necessary list additions be made, with the understanding that the United States would grant exceptions when necessary—even for primary strategic goods—in order to permit the allies to fulfill their existing contracts with the East. State Department cables indicate that U.S. officials presented these proposals as a "package" with regard to the U.S. legislative situation and that West European officials were aware of the Executive's "problem" and were willing to help resolve it.[78]

By early 1952, the deal had been struck. West European officials accepted most of the CoCom list additions proposed by the United States, and executive officials, keeping their side of the bargain, subsequently granted the exceptions noted above to West European exporters.[79] These were publicly reported to the Congress, with executive officials paying careful attention to the timing and wording of these reports in order to minimize their domestic political impact.

Although this arrangement resolved an immediate political problem for CoCom, it also inadvertently worked to the long-run detriment of the multilateral regime. By legitimizing an exceptions procedure, the allies compromised what had been an unconditional embargo on List I items. In future years CoCom members would request exceptions not only for hardship cases but also to develop export markets in the East or, in the case of the United States, to advance particular foreign policy interests. The politicization of the exceptions procedure contributed to the weakening of CoCom during the 1970s and 1980s.

In any event, in light of executive officials' obvious distaste for coercion, their failure to impose sanctions despite blatant violations, and the explicit collaboration of the allies in avoiding sanctions, it is difficult to

77. Britain's attitude is relayed in a telegram from the ambassador in the United Kingdom (Gifford) to the secretary of state, July 17, 1951, reprinted in *FRUS*, 1951, 1:1151–53.

78. Telegram from the U.S. ambassador in Britain to the secretary of state, July 16, 1951, reprinted in *FRUS*, 1951, 1:1148–49, and from the U.S. ambassador in France to the secretary of state, July 20, 1951, reprinted in ibid., pp. 1157–58.

79. Theodore Kent Osgood found that by early 1952, only four items on the U.S. 1A list were not also controlled by the allies. See Osgood, "East-West Trade Controls and Economic Warfare" (Ph.D. diss., Yale University, 1957), p. 52. According to the OTA, the number of items on the CoCom embargo list reached its peak of 285 in January 1952. See *Technology and East-West Trade*, p. 156.

accept that the threat of aid denial was a credible and effective instrument of coercion. Although the Battle Act received great publicity and created considerable political resentment in Western Europe, it ultimately had little to do with the shift in alliance policy to economic warfare.[80]

Moreover, to emphasize coercion as the primary explanatory factor leads one to obscure the fundamental national security interests of the United States, as defined by executive officials. The concept that best captures those interests in the early postwar period is multilateralism.[81] As executive officials repeatedly argued, the export control program was part of a larger multilateral security effort, the cornerstones of which were the creation of a stable transatlantic economy and the maintenance of an adequate NATO defense effort.[82] The threat of aid denial would jeopardize the political solidarity and cohesion of the nascent Atlantic Alliance, while actual denial would harm both West European economic recovery and the development of NATO at critical points in their infancy. For both the Truman and the Eisenhower administrations, this price was simply too high to pay. Forced to choose between their preferred export control strategy and the maintenance of alliance solidarity, they would opt for the latter. As I discuss below, they had the luxury of avoiding that choice in 1950, but not in 1954.

Rather than as a consequence of effective coercion, the adoption of economic warfare as alliance policy is best explained by the combination of U.S. leadership and a shift in the East-West trade preferences of the West European states that took place during 1950. Before the middle of 1950, persistent efforts by U.S. officials to convince the allies of the need for economic warfare, on the grounds that the Soviet Union was an aggressive state whose economy primarily served as a mobilization base for the military sector, were unsuccessful. The crucial turning point, however, was the outbreak of the Korean War.

The North Korean invasion of the South generated a profound, if temporary, security crisis in Western Europe. It led West European officials (and their U.S. counterparts) to a reassessment of the nature of the Soviet threat and how best to meet it. The dominant belief before this invasion was that the Soviet Union would attempt to expand its influence by peaceful means, that is, by subversion. The Korean attack signaled an

---

80. Thomas Schelling is probably correct, however, in suggesting that the Battle Act did force West European governments to focus greater attention on the export control issue than they otherwise might have. See *International Economics* (Boston: Allyn and Bacon, 1958), p. 495. See also Jentleson, *Pipeline Politics*, p. 71.

81. A recent discussion that emphasizes multilateralism as the dominant theme in U.S. national security policy is Robert Pollard, *Economic Security and the Origins of the Cold War*.

82. See, for example, the statement of policy by the National Security Council on Economic Defense (NSC-152/2), July 31, 1953, reprinted in *FRUS*, 1952–54, 1:1010.

alarming shift in Soviet methods to overt aggression.[83] That change, in turn, suggested to West European leaders the possible realization of their worst security fear: a massive conventional Soviet attack, for which they were largely unprepared.[84] Many U.S. and West European officials even viewed the Korean attack as a conscious Soviet ploy, designed to draw U.S. forces to the East and clear the way for a Soviet thrust against the far more attractive target of Western Europe.[85] In the wake of Korea neither the European Recovery Program, which had been the primary defense against Soviet subversion, nor the relatively modest U.S. nuclear deterrent appeared adequate to West European leaders to provide for their security.

The main alliance response to the increased likelihood of direct military conflict with the Soviet Union was rearmament. Before the attack, West European leaders had resisted a defense buildup, on the grounds that economic recovery should be given priority, in the interest of sustaining domestic political stability. The Korean invasion provided the impetus for West European officials to accept that economic considerations had to be subordinated, at least in part, to defense needs.[86] Plans were adopted by most West European governments to increase defense spending and periods of military service. Multilaterally, NATO began its transformation from a "paper" alliance, designed to boost West European morale, to one that would make a serious effort to provide an adequate defense against Soviet attack. The provision of U.S. military aid (and troops) played a major role in that process, as tangible evidence of the U.S. commitment to share the burden of rearmament.[87] Indeed, West European leaders viewed U.S. aid as a precondition of the rearmament effort. In light of this view and the importance U.S. officials placed on getting the allies to rearm, the idea that they would cut off aid in accordance with the Battle Act appears all the more implausible.

83. See Robert Osgood, *NATO: The Entangling Alliance* (Chicago: University of Chicago Press, 1962), p. 69. Osgood provides one of the best discussions of the impact of Korea on Western perceptions. The perceived shift in Soviet methods is also noted in "A Proposal for Strengthening Defense without Increasing Appropriations," April 5, 1950, enclosed in a memo from the undersecretary of the Army to the secretary of state, April 10, 1950, reprinted in *FRUS*, 1950, 3:45.

84. Osgood, *NATO: The Entangling Alliance*, pp. 37, 68–69. The fear of attack may not have been well-founded, but as Osgood notes (p. 68), "the estimate of a potential aggressor's intentions is peculiarly subject to sudden shifts from complacency to alarm."

85. See Bernard Brodie, *War and Politics* (New York: Macmillan, 1973), pp. 63–64; "East-West Trade Controls in Relation to the Upcoming Foreign Ministers' Meeting," Guidance Paper for the Use of the U.S. Delegation to the Four Power Foreign Ministers' Meeting, Geneva, October 1955, dated October 5, 1955, annex 4, p. 2; and Trachtenberg, "A Wasting Asset," pp. 16–18.

86. See the Agreed Minute of the United States, British, and French Foreign Ministers, September 19, 1950, reprinted in *FRUS*, 1950, 4:187–88, and Osgood, *NATO: The Entangling Alliance*, pp. 65–68. U.S. officials had struggled, with little success, to get the allies to accept a reordering of priorities prior to Korea. See "Building Up the Defensive Strength of the West," May 3, 1950, paper presented in the Office of European Regional Affairs, as Background for the May North Atlantic Council Meetings, reprinted in *FRUS*, 1950, 3:86.

87. Osgood, *NATO: The Entangling Alliance*, pp. 40–41.

The combination of an increased security fear and the consequent defense build-up in response enhanced the attractiveness of economic warfare to West European officials. First, the belief that the Soviets were preparing for an imminent conventional war lent credence to the conception of the Soviet economy as a "war economy," one devoted primarily to serving the needs of the defense sector. Trade in industrial goods and raw materials that in peacetime would be expected to serve civilian needs was now seen, under the assumption of imminent military conflict, as making an important contribution to the war mobilization of an adversary. This argument, linking Soviet intentions, the Soviet economy, and industrial (i.e., list 1B) trade, had been made by U.S. officials in CoCom since 1949. Only in the atmosphere of heightened tension created by the Korean War did it gain credibility in the eyes of West European officials.[88]

Second, the rearmament decision helped to pave the way for the eventual adoption of economic warfare by prompting West European leaders to allow security to outweigh economic considerations. As long as economic recovery was the overriding priority, West European leaders could, and did, resist economic warfare on the grounds that East-West trade made an important contribution to meeting recovery goals. Once the general decision was made to grant priority to security concerns, it was a small step to allow the same ordering to apply specifically to East-West trade policy. Economic warfare became a complement, in the economic realm, to a new approach to dealing with the Soviets; or, as Acheson reported to Truman, defense mobilization and the extension of export controls went "hand-in-hand as constructive responses" to the "unmistakable warning to the free world" sounded by the Korean invasion.[89]

The formal decision to adopt economic warfare was taken in September 1950. As of May, U.S. officials had made little progress in gaining CoCom agreement on list 1B items—only 45 of 288 were accepted for CoCom Lists I or II—largely because West European CoCom officials had not been authorized by their governments to abide by the criteria of economic warfare.[90] Agreement on criteria required acquiescence at the

88. The impact of Korea on West European receptivity to economic warfare is discussed in "State Department Position with Respect to Export Controls and Security Policy," report by the executive secretary to the NSC, August 21, 1950, reprinted in *FRUS*, 1950, 4:170.

89. "Report to the President on United States Policies and Programs in the Economic Field which May Affect the War Potential of the Soviet Bloc," enclosed in a letter from Acheson to Truman, February 9, 1951, reprinted in *FRUS*, 1951, 1:1027. Western leaders also determined that in the context of rearmament, "in many cases the goods required by Western Europe for defense needs might be of the same character as those to be controlled on strategic grounds." See the Report on the London Tripartite Conversations on Security Export Control, October 17–November 20, 1950, reprinted in *FRUS*, 1950, 4:236. Thus, rearmament and economic warfare policies complemented each other in a specific fashion, as well as more generally.

90. The results are reported in a telegram from Acheson to the Embassy in France, August 5, 1950, reprinted in *FRUS*, 1950, 4:162–63.

highest levels of government. In September, Foreign Ministers Acheson of the United States, Bevin of Great Britain, and Schuman of France agreed that "*in the present world situation* . . . strategic considerations [as opposed to economic] should be predominant in selecting items for international export control." The Western allies should employ effective export controls "to limit the short-term striking power of the Soviet bloc, and to retard the development of its war potential in the longer term." These controls would entail new restrictions on exports "of selected items which are required in key industrial sectors that contribute substantially to war potential."[91] In addition to items of direct military application, the concept "war potential" covered "industrial sectors that served to support the basic economy of a country and which therefore support either a peacetime or wartime economy."[92]

Agreement at the ministerial level set the stage for reconsideration of the U.S. 1B list, which took place first among the three major powers and then in CoCom. In November, Britain and France agreed to add 175 items to CoCom Lists I and II (102 and 73, respectively) and exchange information on 69 others. In January 1951 these tripartite recommendations were accepted by the remainder of the CoCom members.[93]

U.S. officials thus achieved what they sought in 1950—alliance support for economic warfare. Their victory was not, however, the result of effective coercion in the face of conflicting preferences. Rather, it was the product of U.S. leadership in the context of congruent preferences. U.S. officials brought control proposals to CoCom and engaged in an intense and sustained diplomatic effort to convince other members of their merits. By focusing on Britain and France, they adopted a "divide-and-conquer" tactic that helped to alleviate the constraint on cooperation brought about by the existence of large numbers of participants.[94] They set a clear domestic example by curtailing all but the most innocuous U.S. trade with the East. They helped to ease the economic burdens of economic warfare by allowing exceptions for export to the East of those relatively few items that were especially critical to West European economic recovery. And, as I discuss in Chapter 4, U.S. officials assured that non-CoCom suppliers—in particular, Sweden and Switzerland—informally observed the CoCom restrictions.

In the absence of these U.S. efforts and in light of the economic costs, it is unlikely that West European governments would have taken the

91. See the Agreed Minute of the Foreign Ministers, September 19, 1950, pp. 187–88, emphasis added.
92. The definition is found in the Guidance Paper, October 5, 1955, p. 4.
93. The results are reported in an editorial note, printed in *FRUS*, 1951, 1:1012.
94. Oye, "Explaining Cooperation under Anarchy," p. 21.

initiative to adopt economic warfare. At the same time, however, U.S. efforts only bore fruit in the context of a shift in West European export control preferences, one triggered by the Korean War and the concomitant perception of an immediate military threat. Without that shift, U.S. officials would have been forced to adopt cross-issue linkage to extract West European compliance. As the discussion below suggests, both U.S. leadership and congruent preferences were necessary for the adoption of economic warfare as alliance strategy.

The Retreat from Economic Warfare

West European support for economic warfare was short-lived. By late 1953, America's CoCom partners, led by Great Britain, sought a dramatic liberalization of alliance export controls and called for the expansion of East-West trade. Prime Minister Churchill's public statement of February 25, 1954, was a key catalyst. He stated, in part:

> The more trade there is between Great Britain and Soviet Russia and the satellites, the better still will be the chances of our living together in increasing comfort. . . . the more the two great divisions of the world mingle in the healthy and fertile activities of commerce, the greater is the counterpoise to purely military calculations. . . . I do not suggest that at the present there should be any traffic in military equipment, including certain machine tools, such as those capable only or mainly of making weapons, but a substantial relaxation of the regulations affecting manufactured goods, raw materials and shipping.[95]

In private, British officials proposed to their U.S. counterparts a 50 percent cut in the control list and, more important, a shift in the governing criteria to that of a strategic embargo.[96]

U.S. officials reacted with alarm. They contended that although the Korean War had ended, the Soviet Union and China still posed serious threats to Western security. Moreover, it would be politically imprudent to relax the control system, a valuable weapon for the West in the cold war. U.S. officials were willing to accept some reductions in controls, but not the "wholesale downgrading" envisaged by the British. With regard to criteria, in contrast to the British call for restrictions only on items of "military or near military character," they urged that the West "continue

95. Churchill is cited by Harold Stassen in U.S. Congress, Senate, Committee on Government Operations, Permanent Subcommittee on Investigations, *East-West Trade*, hearings, 84th Cong., 2d sess., February 15–17, 20, and March 6, 1956, p. 450.
96. British proposals are discussed in the Notes from the Ambassador in the United Kingdom to the government of the United Kingdom, December 3, 1953, and to the Department of State, March 1, 1954, both reprinted in *FRUS*, 1952–54, 1:1062–64, 1082–84.

to control categories of items important to key sectors of the industrial base underlying the war potential of the Soviet bloc."[97]

The United States and other CoCom members, most of whom supported the British, attempted to settle their differences between March and August of 1954, in a series of high-level talks involving American, British, and French officials, and subsequently in negotiations involving all members of CoCom. The outcome represented a clear victory for the British position. CoCom members agreed to slash the multilateral control list by nearly 50 percent, from 474 to 252 categories of items to be controlled.[98] Equally important, the reductions reflected a change in criteria to that of a strategic embargo. The allies formally abandoned their emphasis on Soviet bloc war potential—the control of items of major industrial as well as military significance—and instead agreed for the future to focus more narrowly on items used mainly or primarily in military production or whose impact would be felt in the Soviet defense rather than civilian sector.[99]

For the generally accepted explanation of this outcome, I again turn to Adler-Karlsson, whose analysis once more emphasizes the relative utility of U.S. coercive power. Adler-Karlsson depicts West European governments as facing a clear choice between U.S. aid and East-West trade. In 1950 the value of aid outweighed that of trade, so the allies accepted economic warfare; by 1954, however, with aid declining and the potential to expand East-West trade, they were in a position to adopt an alternative approach. As U.S. economic aid became a waning coercive asset, the allies could defy the United States, follow their political and economic instincts, and initiate the restoration of trade with the East. In accounting for the 1954 revision, Adler-Karlsson points out that "the economic pressure that the U.S. could bring to bear on Western Europe lost its force and almost vanished with the disappearance of aid."[100] Table 2 indicates the gradual decline of U.S. economic aid, which reflects the utilization and completion of the European Recovery Program.

Accepting Adler-Karlsson's argument for 1954 presupposes that one accepts his claim that aid was an effective coercive instrument before that date. Yet even leaving the problems with that aside, the argument for 1954 must acknowledge that the decline of *economic* aid did not leave

97. The U.S. position is described in the telegram from the Chief of Operations Mission in the United Kingdom to the Department of State, November 10, 1953, reprinted in *FRUS*, 1952–54, 1:1039–49, and in the note from the U.S. ambassador to the government of the United Kingdom, March 1, 1954, pp. 1063–64.

98. CoCom List I was reduced from 270 to 167 items; List II from 80 to 23 items; and List III, the surveillance list, from 124 to 62 items. The final results are discussed in the Memorandum of Discussion at the 210th Meeting of the National Security Council, August 1, 1954, and in the Report to the NSC by the secretary of state and director of foreign operations, August 30, 1954, both reprinted in *FRUS*, 1954, 1:1235–55.

99. *Battle Act Reports*, no. 5, November 23, 1954, p. 43.

100. Adler-Karlsson, *Western Economic Warfare*, p. 47.

U.S. officials devoid of potential levers. As Table 2 indicates, American *military* assistance replaced economic aid in significance, and its volume outweighed Western Europe's total East-West trade turnover in both 1953 and 1954, when the CoCom negotiations took place. Executive officials were empowered by the Battle Act to cut off military aid, if they so desired. Moreover, they possessed other coercive instruments as well. The Export Control Act allowed executive officials to apply so-called non-frustration controls and deny U.S. exports to states that traded the same or similar goods to the East. Even if one accepts that U.S. aid diminished in utility, it is probable that West Europeans perceived U.S. trade to be more important than East-West trade during this period.

Such potential instruments of coercion were no more useful in 1954 than they had been in 1950. U.S. officials continued to be constrained by their own commitment to maintain the political cohesion, and economic and military strength, of the Western alliance. Military aid, for example, was as critical to the achievement of U.S. and alliance defense objectives in 1954 as it had been during the initial rearmament effort of the Korean War. By 1954, the domestic economic strains and political discontent generated by increased defense spending were placing considerable pressure on West European governments to reduce significantly their commitment to NATO's conventional forces. American aid, like U.S. troops, was a symbol of the willingness of the United States to share the burden of Europe's defense and had served since 1950 as a tacit quid pro quo for the allies to endure the economic sacrifice of defense spending. Without aid, West European governments would be tempted to rely even more fully on the U.S. nuclear deterrent, allowing conventional forces to deteriorate further. To U.S. officials, the outcome would be strategically precarious; despite the shift to nuclear deterrence, they still believed that NATO required an adequate conventional capacity in order to avoid recourse to nuclear weapons in the event of war.[101]

Similarly, employing the aid weapon would create profound resentment in Western Europe and thus strain the political cohesion of the alliance. U.S. officials recognized by 1954 that the cold war was largely evolving into a political struggle; for the United States to create a rift in the alliance by using threats and intimidation would result in a major propaganda victory for the Soviets. Defending before a hostile Congress the administration's decision not to deny aid, one executive official suggested that the result of denial might be "a successful operation, but the patient would be dead."[102] The theme of multilateral solidarity was par-

---

101. See Osgood, *NATO: The Entangling Alliance*, pp. 106–16, 125. The nuclear strategy called not for the replacement of conventional forces but for the integration of tactical nuclear weapons into existing NATO units.

102. Statement of John Barton, Department of Commerce, in *East-West Trade*, 84th Cong., 2d sess., p. 356.

ticularly important to Eisenhower, who continually lectured his more combative cabinet aides about putting the allies first as "the best defense against Communism."[103]

Without military aid (and as discussed below, U.S. exports) as usable coercive instruments, the ability of U.S. officials to achieve their desired East-West trade strategy again depended on the preferences of other CoCom members. By 1954, the strategic, economic, and political conditions that had led West European governments to adopt economic warfare no longer existed. Strategically, the circumstances surrounding the outbreak of the Korean War proved to be truly extraordinary: this period was perhaps the only time in the postwar era that West European governments actually feared an imminent Soviet conventional attack. That attack never came, and with the death of Stalin, the armistice in Korea, and the beginning of a thaw in East-West relations, the fear subsided. NATO, in fact, officially decided in 1953 that "the tensions upon which its war plans were based should be regarded as of infinite duration."[104] Without the threat of an imminent war, the justification for denying the Soviets exports that contributed to their war potential was severely weakened. It permitted West European CoCom members to argue, for example, that it was unnecessary to control the export of machinery and equipment that primarily served peaceful purposes yet would be converted by the Soviets to military production *if* war broke out.[105]

More generally, if war was not inevitable, the plausibility of the conception of the Soviet economy as being devoted overwhelmingly to war preparation was open to challenge. Once it became arguable that the Soviet economy could serve peaceful or civilian needs, it became possible to justify the sale of items otherwise considered strategic, such as copper and petroleum. For these items, the allies argued during the 1954 review that Soviet domestic production was sufficient to meet military needs, and thus exports would serve civilian development. These arguments were assisted by the concurrent increase in Soviet demand for Western consumer goods, which enhanced the image of the Soviet economy as being devoted to the pursuit of civilian needs.[106]

These strategic considerations, which marked a shift from intensified to stable military competition, took on greater significance in light of another major development by 1954: the beginning of West European economic recovery. Recovery signaled the need for export outlets, and the traditional markets of Eastern Europe were a primary target. Churchill stressed the need for Britain to "keep open our trade in every

103. See the Memorandum of Discussions at the 197th Meeting of the National Security Council, May 13, 1954, reprinted in *FRUS*, 1952–54, 1:1163.
104. "East-West Trade Controls," October 5, 1955, annex 4, p. 2.
105. *Battle Act Reports*, no. 5, November 23, 1954, p. 11.
106. On copper and petroleum, see *East-West Trade*, 84th Cong., 2d sess., pp. 240, 420.

possible direction," particularly in view of the economic revival of German and Japanese competition. In the absence of an immediate strategic threat, West European governments saw no justification for allowing security concerns to outweigh economic considerations in the control process. They believed, in the words of a U.S. negotiator, that "the risk of general war is not great enough to warrant foregoing the gains from trade which would follow from a relaxation of controls."[107] As Churchill's February 1954 remarks suggested, West European officials anticipated political gains as well, in that increased trade would reinforce the relaxation of East-West political tensions. Overall, the export control strategy judged essential in 1950 to protect national security appeared to West European officials by 1954 to be strategically unnecessary and politically counterproductive, as well as economically costly.

Despite the 1954 CoCom revisions, the United States continued to practice economic warfare unilaterally. The Department of Commerce maintained its policy of requiring validated licenses for virtually all shipments to the Soviet bloc and of approving license requests for only a modest range of clearly nonstrategic items.[108] Importantly, however, the rationale underlying economic warfare shifted after 1954. With the threat of war having receded, strategic considerations no longer occupied a position of prominence. And in any event, without West European support, comprehensive U.S. controls could not have a meaningful impact on the Soviet economy and thus on the allocation of resources to the military sector. Rather than on strategic factors, U.S. economic warfare came to be based more squarely on political considerations. As long as "tensions between the free world and the bloc were still acute," as the State Department put it, economic warfare was an important symbol of U.S. cold war resolve.[109] It demonstrated, to the Soviets and the rest of the international community, the profound discontent of the United States with Soviet foreign and domestic policies and the willingness of the United States to stand in opposition to them. U.S. officials continued to judge the economic costs of maintaining more comprehensive controls unilaterally, essentially as a political statement, as minimal. And, despite the 1954 review, U.S. firms still did not express much enthusiasm for an expansion of East-West trade.[110]

107. See the National Intelligence Estimate, "Consequences of a Relaxation of Non-Communist Controls on Trade with the Soviet Bloc," March 23, 1954, reprinted in *FRUS*, 1952–54, 1:1131.

108. U.S. Department of Commerce, *Export Control*, 36th Quarterly Report (2d Quarter, 1956), p. 2. The 1954 revision left the CoCom List at 252 items; the U.S. control list contained 474 items (the pre-revision CoCom figure), plus 113 more controlled unilaterally by the United States. See "United States Security Export Controls," Report for the N.S.C. by the N.S.C. Planning Board, June 11, 1954, reprinted in *FRUS*, 1952–54, 1:1193.

109. *Battle Act Reports*, no. 12, April 20, 1959, p. 2.

110. Ibid., no. 5, November 23, 1954, p. 15. See also the testimony of business executives in *East-West Trade*, 84th Cong., 2d sess.

Economic Warfare's Last Stand: The China Differential

The growing importance of political, as opposed to strategic, considerations in U.S. export control policy is further indicated by U.S. efforts to apply controls differentially against members of the Communist bloc. Differential control policies initiated by the United States had important consequences for multilateral coordination in CoCom.

The most controversial initiative concerned China. Because of its direct role as an aggressor during the Korean War, Western export controls on trade with China had been even more restrictive than those on trade with the European Soviet bloc. In 1952, a China Committee (Chincom) was formed as a working group of the CG in order to administer the China controls.[111] In 1954, CoCom allies agreed to continue economic warfare against China, despite the relaxation of controls with regard to the Soviet Union and Eastern Europe. The resulting disparity in the control lists, known as the "China differential," was comprised of approximately two hundred items.

The China differential was accepted by other CoCom members in 1954 only with great reluctance, as a political sop to U.S. conservatives who were discontent with the relaxation of controls. By 1955, however, Great Britain, France, and Japan began to lobby the United States for the elimination of the differential. Their concern reflected the prevailing sentiment in CoCom that economic sacrifices should be justified by clear strategic necessity. The China differential meant forgoing opportunities in a potentially vast market and, equally important, did not make sense strategically. Items covered by the differential were routinely transshipped from the West through the European Soviet bloc to China.[112] Given the close ties between the two Communist giants, unless CoCom members were willing to reimpose economic warfare against the Soviet Union—and clearly they were not—more extensive controls targeted solely at China were doomed to impotence.

U.S officials recognized the transshipment problem and that China could not ultimately be denied items covered by the differential. They urged its retention nevertheless, on political and pyschological grounds. In the eyes of U.S. officials, to lift the differential would be to reward China for its belligerence in Korea and its continued aggressive stance toward Formosa. It would also signal, to non-Communist states in the Far East, a weakening of U.S. resolve to conduct the cold war in Asia.[113]

111. On the history of the China control system, see *Battle Act Reports*, no. 9, June 28, 1957, pp. 31–43.
112. See the telegram from the Office of the Permanent Representative at the North Atlantic Council to the Department of State, October 6, 1955, reprinted in *FRUS*, 1955–57, 10:259–62.
113. Telegram from the Department of State to the Permanent Representative at the North Atlantic Council, October 1, 1955, reprinted in *FRUS*, 1955–57, 10:255–56.

Conflict over the differential attracted attention at the highest levels of government. In January 1956 President Eisenhower promised British prime minister Anthony Eden that the United States would review the differential with the purpose of relaxing, if not completely eliminating it.[114] The administration, however, proved slow to deliver on its commitment. By the middle of 1956, the British expressed their frustration by publicly announcing that "more use will be made of the exceptions procedure to permit reasonable exports in appropriate cases to China of goods which are not on the Soviet list."[115] The United States sought to placate the British and others with offers of limited decontrol but continued to resist any fundamental reconsideration of the differential. Britain responded in May 1957 by announcing that it was unilaterally abolishing the differential and from that point would apply uniform controls to the Soviet Union and China. Other members followed Britain's lead, and in late 1957, Chincom was quietly disbanded.[116]

The case of the China differential provides an early example of the inability of the United States to lead consistently and effectively in CoCom. By insisting, essentially for political reasons, on strategically indefensible controls, U.S. officials compromised the integrity of the embargo and unnecessarily increased its economic burden on CoCom members. The effectiveness of CoCom suffered as a result. Exceptions for export to China increased dramatically, from $3 million in 1954 to $79 million in 1956.[117] More important, the British, French, and other governments made arbitrary use of the exceptions procedure (i.e., exported items without the prior approval of other members), in violation and defiance of CoCom's rules.[118] And, as long as the political conflict over the differential persisted, the United States was unable to convince other members to strengthen the strategic embargo by addressing enforcement problems or accepting list additions. Even more disturbing was that some member governments made public statements questioning the continued existence of CoCom. In July 1956, State Department officials involved in the negotiations reported that U.S. intransigence on the China issue was "being used by the UK and others as an excuse for

114. See memorandum of a conversation, White House, January 31, 1956, reprinted in *FRUS*, 1955–57, 10:308–12.
115. *Battle Act Reports*, no. 9, June 28, 1957, p. 39.
116. Ibid., no. 10, January 24, 1958, pp. 15–18, and letter from Prime Minister Macmillan to President Eisenhower, May 29, 1957, reprinted in *FRUS*, 1955–57, 10:467–68. The NSC discussed, and rejected, the option of imposing sanctions on CoCom allies for their defection from Chincom. See Memorandum of Discussion at the 336th Meeting of the National Security Council, September 12, 1957, reprinted in *FRUS*, 1955–57, 10:491–94.
117. See the memorandum from the deputy assistant secretary of state for economic affairs to the secretary of state, March 5, 1957, reprinted in *FRUS*, 1955–57, 10:420–21.
118. See memorandum from the steering committee of the Council on Foreign Economic Policy to the chairman of the council, July 13, 1956, reprinted in *FRUS*, 1955–57, 10:377–80.

possible abandonment of the entire multilateral control system on the ground that constant US opposition 'to the views of the majority' makes the system increasingly unworkable."[119]

Given the costs and risks to CoCom, why was the Eisenhower administration unwilling to abandon the differential? A key reason involves the fragmentation of the U.S. export control process. During the early 1950s, a consensus existed within the Executive on the need for economic warfare against both the Soviet Union and China. By 1955, fissures emerged. State Department officials proved willing to accommodate the allies by adjusting the China differential as political irritation increased. Defense Department officials, however, feeling that the United States had conceded too much in the list revision of 1954, were determined to "hold the line" on China, a position consistent with the prevailing sentiment in Congress. Defense officials were especially adamant on this issue, in part because U.S. soldiers had only recently been killed in direct combat with the Chinese.[120] The result was a stalemate. Despite the president's order to the secretary of state at the meeting with Eden to "get in everyone, get in the Defense people and the others, and see what we can do to back away from this thing," the administration could not manage to fashion any compromise proposals sufficiently timely or attractive to Britain and France to deter them from acting unilaterally, in defiance of CoCom.[121]

In addition, the United States was constrained by its insensitivity to the economic costs of controls. In explaining to Eisenhower the decision by Britain to abandon the China differential, Prime Minister Macmillan wrote "we live by exports—and by exports alone."[122] That statement reflected a sentiment shared by British government and industry, but by neither in the United States. Unlike their counterparts in other CoCom countries, U.S. firms did not lobby for the removal of the China differential and thus failed to provide a counterweight to those within the United States who argued for its maintenance. For its part, the U.S. government decided to retain the China differential even after all other CoCom members abandoned it. The economic costs, estimated by the administration at $40–75 million annually, were judged to be insufficient to abandon an important symbol of U.S. resolve.[123]

119. Memorandum from the deputy assistant secretary of state for economic affairs to the deputy under secretary of state, July 18, 1956, reprinted in *FRUS*, 1955–57, 10:381.

120. Ibid., p. 383, and memorandum of a conversation between Secretary of State Dulles and the British ambassador, April 13, 1956, reprinted in *FRUS*, 1955–57, 10:339–41. Dulles noted that Defense Department opposition had been pitched in very emotional terms.

121. Memorandum of a conversation, White House, January 31, 1956, p. 311.

122. Letter from Prime Minister Macmillan to President Eisenhower, May 29, 1957, p. 467.

123. The estimate is in a memorandum from the undersecretary of commerce to the chairman of the Council on Foreign Economic Policy, August 13, 1957, reprinted in *FRUS*, 1955–57, 10:484–87.

ENFORCEMENT AND THE INSULATION
OF THE WESTERN TRADING SYSTEM

By 1953, the manner in which controls were enforced in Western
Europe on trade with the Soviet Union and Eastern Europe had become
of major concern to the United States. Given the informal nature of the
CoCom arrangement, the less enthusiasm governments felt for the controls agreed to, the less they could be expected to enforce them with
vigilance. As the control lists became more extensive, West European
enforcement was noticeably less effective.[124] The problem was particularly severe in West Germany because of the strength of its traditional
trade links to the East, its special relationship with East Germany, and
the fact that its formal controls were initially more restrictive than those
applied in the rest of Western Europe. Until May 1950, for example,
there were no controls on interzonal trade, that between West and East
Germany. Strategic goods could be channeled from West to East Germany
and then on to the Soviet bloc from there. In 1951, a Senate subcommittee
which investigated the enforcement problem concluded that West Germany was a "veritable open channel" for the flow of strategic goods to the
East and that "the will to enforce export controls is totally lacking in the
West German government."[125] It estimated that the annual leakage of
strategic goods from West Germany was even greater than the level of
West German exports to the East reported in official statistics.[126]

A related problem was that of transit trade, considered one of the
major legal loopholes in the allied control system. At free ports, such as
Hamburg, Rotterdam, and Antwerp in Europe and Hong Kong and
Macao in the Far East, traders could bring goods, reload them, and re-
ship them to controlled destinations without interference from the government within whose territory the port was located.[127] For the smaller
states of Europe, such as Belgium, Denmark, and the Netherlands, transit trade was a major source of income. Despite the leakage to the East
and urging by the United States, these governments were reluctant to
interfere with the flow of trade through their free ports.[128]

124. Adler-Karlsson, *Western Economic Warfare*, p. 64.
125. See U.S. Congress, Senate, Committee on Interstate and Foreign Commerce, *Export Controls and Policies in East-West Trade*, report, 82d Cong., 1st sess. (Washington, D.C.: GPO, 1951), pp. 13, 22.
126. U.S. officials estimated that illegal strategic goods shipped from West Germany to the East total over $100 million annually; West German exports to the East reported officially were $75 million in 1950 and $64 million in 1951. See *Export Controls and Policies in East-West Trade*, p. 22, and Adler-Karlsson, *Western Economic Warfare*, p.228.
127. *Export Controls and Policies in East-West Trade*, pp. 47, 76.
128. Chuthasmit, "The Experience of the United States in Controlling Trade with the Red Bloc," pp. 139–40; *Battle Act Reports*, no. 5, May 17, 1954, p. 28; and Adler-Karlsson, *Western Economic Warfare*, p. 65.

ECONOMIC CONTAINMENT

The ineffectiveness of allied enforcement created a special problem for the United States. Strategic goods exported from the United States to Western Europe could be transshipped to the Soviet bloc, thus in effect frustrating U.S. control efforts. This raised the danger of what U.S. officials referred to privately as a potentially "explosive" problem: having to apply non-frustration controls against Western Europe, thereby disrupting intra-Western trade, to protect the integrity of national export controls directed against the East.[129] The mere mention of such controls prompted indignation; other CoCom members considered it "paternal and authoritarian" and not in keeping with the multilateral spirit of CoCom.[130] In addition to generating political friction, the systematic application of non-frustration controls would jeopardize the emerging links between U.S. firms, as sources of supply, and their West European counterparts. Those transatlantic links created prosperity for both economies and provided the foundation for Western economic liberalism, which reinforced and helped to cement cooperative political and strategic relationships among the governments of the Western alliance.

The overriding interest of U.S. officials in insulating the Western trading system from the spillover effects of East-West export controls led to a search for responses to the transshipment problem. One involved the application of a reexport control system. Under the Export Control Act, U.S. officials claimed the right to control U.S.-origin products and components even after such items had left U.S. territory. Western recipients of U.S.-origin items were required by U.S. law to obtain permission from American authorities before reexporting them to other destinations. Failure to comply could result in U.S. sanctions, the most important being the placement of the firm in question on a U.S. blacklist and the denial of U.S. exports to that firm.[131]

Although reexport controls may have afforded additional protection to U.S.-origin goods, their appeal was limited by the fact that they involved only Western *firms* (more precisely, contractual agreements be-

129. The concern of executive officials over the possibility of having to apply non-frustration controls is indicated in the memorandum prepared in the Economic Defense Advisory Committee (EDAC) for the NSC Planning Board, March 9, 1954, reprinted in *FRUS*, 1952–54, 1:1107. U.S. officials exercised such authority as early as 1948, when Section 117(d) of the Export Cooperation Act directed the administrator to refuse delivery of strategic commodities to Western Europe that would be reexported to the East. In practice, U.S. officials permitted the export of strategic goods to its allies so long as the receiving government was willing to assure that such goods, the products of such goods, or "similar" goods would not be reexported, or to provide information as to why such reassurance was impossible. See the memorandum by the assistant secretary of state for economic affairs (Linder) to the secretary of state, April 9, 1951, reprinted in *FRUS*, 1951, 1:1067.
130. Memorandum by the secretary of state and the director of mutual security to the executive secretary of the National Security Council, April 23, 1952, reprinted in *FRUS*, 1952–54, 1:837.
131. Reexport control authority and its use in the 1950s and 1960s is discussed in Berman and Garson, "United States Export Control Policy," pp. 817, 851–62.

102

tween U.S. and Western firms) and not Western *governments*. Reexport controls could not be used to enlist the cooperation or improve the enforcement of other Western governments. On the contrary, other CoCom members viewed such controls as an extraterritorial infringement upon their sovereignty, in defiance of international legal norms. Thus, at worst, reliance on reexport controls could be taken as a vote of no confidence in CoCom and risked sparking a diplomatic confrontation in the alliance; at best, it left the U.S. government to police Western Europe unilaterally to enforce its national controls, without the assistance of other CoCom governments.

While maintaining the reexport control system, U.S. officials sought ways to involve other CoCom governments in the protection of U.S.-origin goods and, more generally, to improve their enforcement efforts. First, in 1952 the United States obtained the support of other CoCom governments, and neutrals such as Sweden and Switzerland, for the IC/DV (Import Certificate/Delivery Verification) system.[132] The system required an importing firm to obtain a certificate from its government, to be presented to the exporting firm's government on request, stating that a given item would not be reexported without the explicit authorization of the importer's government. To ascertain that the item actually arrived at its stated destination, the exporter's government was given the right to request a delivery verification from the importer, again to be signed by the importer's government. The IC/DV system thus committed other CoCom governments to being responsible for preventing the unauthorized diversion of U.S.-origin controlled goods to the East. Though it increased the confidence of the United States that its products would be protected, the IC/DV system was not foolproof, and importers sometimes evaded it through forgery or fraud.[133]

Second, as an explicit part of the 1954 agreement to reduce the control list, U.S. officials obtained a commitment from other CoCom governments to improve their enforcement. During the 1954 negotiations, other Western governments contended that they could not improve enforcement unless the control list was shortened; the United States responded by claiming it would not accept a shorter list without improved enforcement. Since CoCom rules stipulated that unanimity was required for items to be removed from the list, the regime context made it possible for U.S. officials to extract something in return for their acquiescence. Other CoCom members pledged to strengthen the IC/DV system, in some cases by requiring that import certificates be transmitted from government to government, rather than through commercial channels, in order to discourage forgeries. Each also agreed to require formal

132. The system is described in *Battle Act Reports*, no. 5, May 17, 1954, p. 26.
133. Adler-Karlsson, *Western Economic Warfare*, p. 65.

authorization from the government of an exporter, in the form of a Transit Authorization Certificate, before allowing controlled goods in transit to pass through their territories en route to the Soviet bloc.[134] Finally, transaction controls were implemented to prevent firms in an allied country from arranging financially for the sale of strategic goods from a third (nonparticipating) country to the East. Where necessary, CoCom members enacted new legislation to provide the legal basis for these enforcement initiatives.[135]

The enforcement agreements served to facilitate intra-West trade because U.S. officials accepted them as evidence of the willingness of West European governments to protect against unauthorized diversion. U.S. officials felt sufficiently confident to reduce the Positive List drastically, from 1468 to 787 items.[136] U.S. firms exporting to Western Europe were no longer required to secure export licenses for the items removed and thus could avoid the associated delays and uncertainty. And, given improvements in the IC/DV system, licenses for those items remaining on the Positive List could be processed more quickly. In 1956, Secretary of Commerce Weeks argued that "although the revision of our controls benefited American business and industry, by removing what was an impediment to American competition with free world countries, it did not in any way weaken our economic defense position."[137] For U.S. firms, the fact that their competitive position in intra-Western trade was strengthened made their inability to compete in Eastern markets all the more tolerable.

Overall, the 1954 revision marked the emergence of a tacit arrangement between the United States and its Western allies, one designed to maximize the effectiveness of the East-West export control system and foster the expansion of intra-Western trade. In exchange for ready access to American products and technology and the ability to export nonstrategic items to the East, West European governments agreed to abide by multilateral export controls and enforce them satisfactorily. In exchange for assurances that U.S. (and other Western) strategic technology would be protected, U.S. officials took responsibility to minimize the administrative burden of the control system (i.e., to shorten the control list) and to facilitate the access of the allies to U.S. technology (i.e., to prevent reexport or non-frustration controls from interfering significantly with intra-Western trade). The deal was somewhat precarious in that it was framed in the context of an informal gentlemen's agreement and within that context depended on subjective U.S. assessments of West

---

134. *Battle Act Reports*, no. 9, June 28, 1957, p. 21.
135. Ibid., no. 12, pp. 18–33.
136. See the statement of Commerce Secretary Weeks in *East-West Trade*, 84th Cong., 2d sess., p. 371.
137. Ibid.

European enforcement performance—assessments that might be influenced by the changing political climate of the United States. Nevertheless, it held together remarkably well until the late 1970s.

CONCLUSION

Patterns established in the first postwar decade had a decisive effect on the future development of U.S. and alliance export control policy. U.S. executive officials obtained extraordinary powers to control private commerce in peacetime. That authority was granted in the midst of a national security crisis, yet quickly became institutionalized. Foreign economic policy was thus not immune from the expansion of power and authority, in the federal government generally and the executive branch in particular, that accompanied the onset of the cold war and the emergence of a national security state in the United States.[138]

As a result, throughout the postwar period there has existed an uneasy tension, and potential conflict, between the liberal and nationalist aspects of U.S. trade policy. This tension is all the more significant because U.S. officials have committed themselves to promoting and preserving liberalism in the world economy. During the 1950s, they managed to reconcile the responsibilities of liberal leadership with the strategic necessities of export control policy. By the 1980s, however, the problem was becoming more intractable, as executive officials drew on their delegated authority to expand the scope of export controls, with regard to purpose and destination.

The development of CoCom was also decisively shaped by the experience of the early cold war. Fear of provoking the Soviets or generating domestic discontent led West European governments to accept CoCom as only an informal gentlemen's agreement. CoCom has survived, in essentially that form, for over four decades. That it survived (a feat some observers consider noteworthy in itself) provided U.S. officials with the opportunity to maintain export control coordination in circumstances less accommodating to export denial than in the early postwar period. At the same time, that CoCom survived in that form complicated U.S. efforts to strengthen it, which had important implications for Western security and intra-Western technology transfer.

In terms of strategy, early conflicts in CoCom demonstrated the limits on West European export control preferences and on the ability of the United States to alter them. West European governments proved willing to engage in economic warfare against the Soviet Union only when East-West tensions were at a peak and when confronted with the possibility of

138. On the overall emergence of the postwar national security state, see Yergin, *Shattered Peace*.

direct Soviet aggression against them. In the absence of such an immediate security threat to Western Europe, U.S. officials demonstrated an inability to determine alliance strategy without jeopardizing the viability of CoCom or the cohesion of the alliance itself. Similarly, in the absence of a clear strategic rationale, the United States could not sustain alliance support for economic warfare against China. Although U.S. officials judged the economic sacrifice worth bearing in order to signal strong disapproval of Chinese behavior, other CoCom members clearly did not.

Finally, CoCom's first decade displayed the inconsistency in U.S. leadership that would plague the multilateral regime over the next three decades. During 1949–1954, effective leadership enabled the United States to establish a multilateral strategic embargo and, as long as West European preferences allowed, to coordinate the more ambitious strategy of economic warfare. During 1955–1957, however, the United States exhibited some of the behavioral tendencies that would later significantly constrain its ability to lead. U.S. officials urged controls on other CoCom members that could be justified politically but not strategically; subsequently, bureaucratic conflicts within the Executive prevented the United States from adjusting those controls. Cooperation in CoCom suffered as a result.

Throughout the 1960s, U.S. officials were less concerned with shaping Western Europe's export control preferences and more concerned with adjusting U.S. policy to the failure of economic warfare and to a changing international political and economic environment. The manner in which the United States adjusted its policy ultimately had profound consequences for CoCom's strategic embargo and for the relationship between export controls and intra-Western trade.

CHAPTER FOUR

The Consolidation of CoCom's Strategic Embargo and the Struggle to Adjust U.S. Policy, 1958–1968

The failure to convince its allies to pursue economic warfare left the United States, by the late 1950s, in a quandary. In response to Eastern demand, West European and Japanese exports of capital goods, machinery, and equipment increased rapidly. This trade undermined the potential effectiveness of the unilateral controls, beyond the CoCom consensus, that the United States continued to employ. Given the relative self-sufficiency of the Communist states, even with the support of other CoCom members it would have been difficult for economic warfare to retard the economic progress of the Soviet bloc significantly. Yet without that support, a unilateral U.S. attempt was doomed to certain failure. Economic warfare brought the United States neither economic nor strategic benefits, leaving U.S. officials to consider whether, and how, to adjust their East-West trade strategy.

One option was for U.S. officials to renew their efforts to obtain the support of other CoCom members for economic warfare. In light of the experience of the 1950s, however, executive officials considered such an approach to be futile and counterproductive. With one significant exception, that of wide-diameter steel pipe in 1962, the United States refrained from making the denial of exports of *economic* importance to the Soviet Union and Eastern Europe a major alliance issue during the 1958–1968 period.

U.S. officials were left with three other possibilities. First, they could simply refuse to adjust U.S. policy, thereby accepting a differential between U.S. and CoCom controls. Such a strategy would be economically irrational: it would deny U.S. firms the benefits of trade but do little to affect the economic position of the Soviet bloc. It might serve symbolic political purposes, however, by continuing to demonstrate American resolve in light of the cold war in general and, by the mid-1960s, the conflict in Vietnam in particular.

Second, the United States could adjust its export controls fully and unconditionally to conform to the level agreed in CoCom. It could maintain a strategic embargo, abandoning economic warfare and allowing U.S. firms to compete on an equal footing with their West European and Japanese counterparts. This strategy would presumably bring the United States the security benefits of a selective embargo and the economic benefits of expanded trade.

Third, the United States could adjust its policy conditionally in order to serve political objectives. It could attempt to use the very act of adjustment, together with the promise of trade expansion, to gain political benefits in its relations with the East. Trade would primarily be an instrument of politics; whether it coincidentally served economic objectives would be of secondary importance.

The third option, that of tactical linkage, ultimately prevailed, although it was not fully implemented until the early 1970s. Both the Kennedy and the Johnson administrations favored this strategy, but their ability to carry it out was constrained by the Congress, which, in the context of the Vietnam War, favored little or no trade with the East as a demonstration of U.S. political and moral resolve. The adjustment of U.S. policy did not come legislatively until a fragile coalition in Congress, seeking the political and economic benefits of détente, passed the Export Administration Act of 1969.

The interbranch conflict emphasized the primacy of political considerations in U.S. East-West trade policy. Unlike the case for Western Europe or Japan, the expansion of trade with the East could not be justified in the United States solely or even chiefly on the grounds that it was commercially profitable. For the Congress, trade was a "favor," of which the East was unworthy; for the Executive, it was a bargaining lever that could be useful in resolving an ongoing stalemate with the Soviet Union and in enhancing U.S. influence in Eastern Europe. Even U.S. firms felt obliged to subordinate economics to politics. Until the late 1960s they generally remained reluctant to trade with the East or even to express publicly their desire to do so on commercial grounds. Those industry officials that did speak out in favor of expanded trade felt it necessary to couch their arguments in terms of political benefits, as if profits were a distant concern that barely entered their calculations.

Although the 1960s was a period of adjustment in American policy, it was one of consolidation in the alliance context. U.S. leadership, buttressed by America's dominant technological position, assured the effectiveness of coordination in CoCom. An effective CoCom, in turn, made it easier for U.S. officials to insulate the expansion of trade and technology transfer among Western states from the East-West export control regime. Developments in the alliance context are discussed below, followed by an examination of the struggle to adjust U.S. national policy.

CoCom and Intra-Western Technology Transfer, 1958–1968

During the 1958–1968 period, CoCom inspired relatively little of the high-level political friction that existed through much of the previous decade. To be sure, there were disputes over the strategic significance of certain items, particularly between the United States and France.[1] Unlike CoCom's first decade, however, the overall purpose of multilateral controls and the question of whether to treat targeted countries differentially were not matters of contention. Member governments accepted the necessity of a narrowly circumscribed, uniformly applied strategic embargo, and its administration became largely a matter of routine.

An important step in the consolidation of the strategic embargo was the list review of 1958. America's CoCom allies believed that the 1954 revisions, although clearly a turning point, were insufficient in that they had permitted the retention on the control list of numerous items of lesser strategic significance. Two prominent examples were steel and ships, both of which gave rise to contention in CoCom after the 1954 review.[2] In the 1955–1958 period, some CoCom members interpreted CoCom restrictions liberally in order to facilitate trade in categories deemed to be overcontrolled.[3]

Prompted by pressure from West European governments, the 1958 CoCom review resulted in a significant liberalization of controls. International List I was reduced from 181 to 118 items, as various machine tools, electrical generating equipment, diesel engines, oil-drilling equipment, industrial diamonds, and numerous other items were decontrolled. The quantitative control list (International List II), which had

1. See, for example, Mutual Defense Assistance Control Act of 1951, PL 213, 82d Cong., *Reports to the Congress* (hereafter *Battle Act Reports*), no. 14, December 20, 1960, and Gunnar Adler-Karlsson, *Western Economic Warfare, 1947–1967* (Stockholm: Alquist and Wiksell, 1968), pp. 98–99. The friction in CoCom between the United States and France is not surprising, given de Gaulle's assertion of nationalism and his distancing of France from NATO during the early 1960s. Indeed, as a demonstration of national resolve, in 1965 the French chairman of the Consultative Group (CG) refused to convene meetings. As a result the CG, which met infrequently in any event, was quietly disbanded. Personal correspondence with William Root, former director of the State Department's Office of East-West Trade, April 15, 1987.

2. Adler-Karlsson, *Western Economic Warfare*, pp. 93–94.

3. On ships, for example, other members refused to abide by strict quotas and instead justified whatever sales were made as a necessary quid pro quo to maintain imports from the East. More generally, the State Department complained in 1956 that quotas in CoCom were "often meaningless," because members set them with commercial as opposed to security considerations in mind. See telegram from the Delegation at the North Atlantic Council Meeting to the Department of State, May 4, 1956, reprinted in *FRUS*, 1955–57, 10:361–63.

Table 3. Value of CoCom exceptions, 1952–1968

Year	Value ($ million)	Year	Value ($ million)
1952	5.8	1960	.8
1953	5.5	1961	.8
1954	4.5	1962	1.4
1955	2.4	1963	1.8
1956	2.6	1964	.7
1957	2.5	1965	2.1
1958	1.1	1966	N.A.
1959	.5	1967	8.0
		1968	11.0

Source: 1952–1965, compiled from the Battle Act Reports, nos. 2–19; 1967–1968, from "Special Report on Multilateral Export Controls," submitted by the president to the Congress, printed in Export Administration Act: Agenda for Reform, U.S. Congress, House, Committee on International Relations, Subcommittee on International Economic Policy and Trade, hearings, 95th Cong. 2d sess., October 4, 1978, p. 52. The figures for 1952–1965 reflect the assumption that all CoCom exceptions were documented, as required by Congress, in the Battle Act Reports.

been utilized in the early 1950s to implement economic warfare, was abolished completely.[4] In light of the abolition of the China differential in 1957, the newly streamlined control list applied to trade with China as well as with the European Soviet bloc. Overall, as Alder-Karlsson noted, the changes adopted in 1958 could "almost be regarded as a final revision in the economic warfare, as the CoCom policy hereafter, with few exceptions, was concentrated on commodities which by all participants were considered to be properly 'strategic'."[5]

CoCom members also agreed in 1958 to conduct list reviews annually and did so from 1958 through 1969.[6] More frequent reviews helped to keep the list updated and to assure that it reflected technological progress of military relevance in the West and the Soviet Union. The most important items added during this period related to computers and integrated circuits, which took on critical significance in American military systems. Other items, for which military applications did not develop in the United States as expected or for which the Soviet Union developed adequate capabilities, were periodically removed from control.[7]

Data on CoCom exceptions, compiled from the Battle Act Reports, are presented in Table 3. The recorded value of exceptions granted between 1958 and 1965 was modest, representing less trade than that which took

4. Battle Act Reports, no. 12, April 20, 1959, and Adler-Karlsson, Western Economic Warfare, pp. 96–97.
5. Adler-Karlsson, Western Economic Warfare, p. 96.
6. List reviews were conducted from October 1959 to January 1960; October 1960 to April 1961; April to July 1962; November 1963 to April 1964; November 1965 to June 1966; November 1966 to February 1967; and October 1968 to May 1969. See Battle Act Reports, nos. 14–20. From 1969 to 1984, CoCom practice was to review the list approximately every three years.
7. William A. Root, "Trade Controls That Work," Foreign Policy, no. 56 (Fall 1984), p. 63.

place through the exceptions mechanism between 1952 and 1957, despite the fact that in the years after 1958, overall Western exports to the East increased significantly (see Table 4, below). The relative paucity of exceptions, at least prior to 1967, suggests that the distinction between what member governments considered strategic and nonstrategic trade was fairly sharp and also, perhaps, that through the list review of 1958 and subsequent reviews, CoCom members were fairly diligent in decontrolling items judged to be of lesser strategic importance.[8] To facilitate further trade in such items, in 1960 CoCom adopted an "administrative exceptions" procedure, which provided member governments with the national discretion to permit the export of CoCom-controlled items of lesser strategic significance without having to obtain the formal approval of the group.[9] Members were required to inform CoCom of such exports, after the fact.

While the available evidence is scanty, the issue of enforcement does not appear to have been one of major contention in CoCom during this period. With regard to agreed controls, American officials expressed little dissatisfaction with the programs and efforts of other CoCom members. A Senate report prepared in 1962 for a subcommittee of the Judiciary Committee, for example, was critical of the allies for neither endorsing economic warfare nor being willing to enforce *unilateral* U.S. controls.[10] With respect to the enforcement of items controlled *multilaterally*, however, the assessment was considerably more benign. The report found little fault in the implementation of CoCom regulations, noting that over time other members had developed national laws and procedures to strengthen the enforcement of controls, were willing to apply and enforce IC/DV procedures and transit controls, and generally cooperated fully with U.S. investigations of suspected diversions.[11] Di-

8. The relatively sharp increase in exceptions value in 1967 and 1968 primarily reflected pressure to sell computers and related equipment to the East, as such items began to have more widespread application in the West. Even these figures are modest in comparison to the value and volume of exceptions recorded in the 1970s and 1980s. (See below, Chaps. 5 and 8.) On the competition to sell computers to the East, see the statement of Hugh Donahue in U.S. Congress, Senate, Committee on Banking and Currency, Subcommittee on International Finance, *East-West Trade*, hearings, 90th Cong., 2d sess., June 4–27 and July 17–25, 1968, pp. 149–62.

9. *Battle Act Reports*, no. 14, December 20, 1960, p. 4. Adler-Karlsson suggests that de Gaulle used the administrative exceptions mechanism to export to the Soviets on a large scale in 1960. See *Western Economic Warfare*, p. 98.

10. U.S. Congress, Senate, Committee on the Judiciary, Subcommittee to Investigate the Administration of the Internal Security Act and Other Internal Security Laws, *Report on Export Controls in the United Kingdom, France, Italy, Federal Republic of Germany, Belgium, and the Netherlands*, report submitted by Senator Thomas J. Dodd and Senator Kenneth B. Keating, April 4, 1962.

11. Ibid., pp. 3, 15–16. This benign assessment reflected the attitude of executive officials, because the report relied heavily on input from State Department officials stationed in Western Europe. Although difficult to discern from the available evidence, the apparent lack of concern in the Executive may have been more a reflection of U.S. technological dominance (see below) than of the high quality of West European enforcement. Because the United States enjoyed unilateral control over many strategic items, it could afford to be relatively less concerned about its allies' enforcement efforts, even if those efforts were not always fully vigilant.

Table 4. Western trade with the Soviet Union, 1958–1969 ($ million)

Country	1958	1959	1960	1961	1962	1963	1964	1965	1966	1967	1968	1969
						Exports						
U.K.	145.5	97.6	148.9	194.4	161.0	178.8	111.3	128.6	141.1	178.8	249.5	233.2
France	75.9	90.1	115.6	110.0	138.1	64.2	64.1	72.0	75.6	155.3	256.5	265.1
F.R.G.	72.2	91.1	185.3	204.0	206.8	153.5	193.6	146.3	135.3	197.9	273.0	406.0
Italy	31.1	43.6	78.6	89.5	102.3	113.6	90.7	98.1	90.1	132.0	179.0	284.5
Japan	18.1	23.0	60.0	65.4	149.4	158.3	181.9	168.3	215.0	157.7	179.0	268.3
U.S.	3.4	7.4	38.4	45.6	19.7	22.6	146.4	45.2	41.7	60.2	57.7	105.5
						Total Trade						
U.K.	312.0	274.1	358.7	432.7	396.5	433.5	382.8	461.6	492.9	515.8	628.9	706.4
France	170.8	191.2	210.2	207.3	248.8	205.3	205.3	218.0	247.2	342.4	439.3	478.3
F.R.G.	164.2	196.5	345.4	401.5	421.8	362.3	430.9	421.6	423.5	472.7	567.1	740.8
Italy	71.5	131.6	204.4	239.6	268.6	289.5	237.9	279.4	280.1	409.7	463.5	531.3
Japan	40.3	62.5	147.0	210.8	296.6	320.3	408.6	408.5	515.4	611.7	642.5	802.5
U.S.	20.8	35.7	61.3	68.4	35.6	43.7	167.1	87.8	91.2	101.2	116.2	157.0

Source: IMF, *Direction of Trade Yearbook,* 1958–1970.

versions did occur, which is not surprising in light of the fact that since 1959, the Soviet Union and its East European allies had engaged in a systematic campaign to acquire strategic technology from the West illegally.[12] The report suggested further, however, that diversions tended to be relatively infrequent and to involve complex schemes requiring the cooperation of illicit traders across several states.[13]

CoCom and U.S. Leadership

The stability enjoyed by CoCom in the 1958–1967 period can be attributed primarily to effective U.S. leadership. For one thing, the United States acted as the "conscience" of CoCom. U.S. officials set a domestic example by minimizing exceptions and devoting sustained attention to enforcement.[14] They worked to preserve the coverage of the control list by bringing virtually all list addition proposals to CoCom and establishing their strategic merit and by utilizing their veto to counter any efforts to relax controls significantly in the aftermath of the major list revisions of 1958.[15]

As it had since the inception of CoCom, the United States took chief responsibility for assuring the compliance of non-CoCom suppliers, through informal agreements with governments and the threat of trade sanctions against private firms. Both methods were extensively employed by U.S. officials to ensure that the main non-CoCom suppliers— Sweden and Switzerland—did not frustrate the intent of CoCom controls by exporting or reexporting strategic items to the East. With the exception of closely monitored "essential trade," both countries were forced by the United States to observe CoCom controls. In the absence of U.S. pressure, the neutral status of Sweden and Switzerland would

12. See Jan Sejna, "Soviet and East European Acquisition Efforts: An Inside View," in Charles M. Perry and Robert L. Pfaltzgraff, eds., *Selling the Rope to Hang Capitalism?* (McLean, Va.: Pergamon-Brassey's International Defense Publishers, 1987), pp. 70–74. Sejna was, during the 1950s and 1960s, chief of staff to the Czechoslovakian minister of defense and secretary of the Military Committee of the Communist party of Czechoslovakia.

13. *Report on Export Controls*, p. 16. Sejna similarly notes that Eastern acquisition agencies were forced to focus their efforts on European neutral and third world states because of the existence of multilateral Western controls. See "Soviet and East European Acquisition Efforts," p. 72.

14. On U.S. enforcement efforts, see, for example, U.S. Department of Commerce, *Export Control*, 63d Quarterly Report (1st Quarter, 1963), pp. 12–14, and 67th Quarterly Report (1st Quarter, 1964), pp. 12–13. On exceptions, the general pattern of U.S. behavior, at least until 1965, was to request them infrequently for Poland and not for other destinations. See "U.S. Differentiation among European Communist Countries," Background Papers prepared for the Miller Committee, March 1965, reprinted in the *Declassified Documents Quarterly* (1977), item 311A, p. 72. Although systematic data is not publicly available, by 1965 the United States probably began to request exceptions for East European destinations other than Poland, particularly in the area of computers. At about that time, IBM began to establish a market presence in Eastern Europe.

15. William A. Root, "Trade Controls That Work," p.63, and "U.S. Policy on East-West Technology Trade: Past, Present, and Future," in John R. McIntyre and Daniel S. Papp, eds., *The Political Economy of International Technology Transfer*, (Westport, Conn.: Quorum Books, 1986), pp. 217–18.

have driven each to cooperate only minimally, if at all, with the Western strategic embargo.[16]

At the same time, U.S. policy helped to minimize the burdens of participation in the control system for other CoCom members. The most important contribution of U.S. officials was to acquiesce in the 1958 reduction of the multilateral control list, despite retaining more comprehensive controls nationally. The United States could have resisted, as it did several years earlier when pressure in CoCom mounted for a relaxation of the China differential. Instead, U.S. officials quickly took up British and French requests for a Consultative Group meeting to refine CoCom's strategic criteria further and revise the list downward in accordance with the new criteria.[17] They did so to avoid the controversy that had plagued CoCom over the China differential and in explicit recognition of the fact that to be coordinated effectively, multilateral export controls needed to take into account both alliance military security and the economic needs of other CoCom members.[18]

Despite the CoCom revisions, U.S. control criteria continued to reflect the concerns of economic warfare. In the words of an authoritative 1963 State Department policy review, they were "based on the broader concept of military-industrial capabilities, rather than the narrowly defined CoCom concept of present military production."[19] In addition to CoCom-controlled items, the U.S. list contained machinery and equipment which could be diverted to military use during wartime or which embodied technology currently of major industrial significance.

The critical point is that U.S. officials did not systematically attempt to impose this broader conception of strategic trade on other CoCom members. For the purposes of CoCom, they accepted that the items to be embargoed were those used primarily in *military* production, those embodying advanced technology of *military* significance, and those which constituted *military* bottlenecks in the bloc.[20] Further, they accepted that

16. Adler-Karlsson, *Western Economic Warfare*, pp. 74–78, and "International Economic Power: The U.S. Strategic Embargo," *Journal of World Trade Law* 6 (September/October 1972), 505–6, 511–17. See also Hendrik Roodbeen, "Trading the Jewel of Great Value: The Participation of the Netherlands, Belgium, Switzerland, and Austria in the Western Strategic Embargo versus the Socialist Countries" (Rijksuniversiteit Leiden, the Netherlands, unpublished manuscript, 1991), chap. 5. Roodbeen notes that while CoCom as a whole expressed some interest in non-CoCom suppliers during the early 1950s, the task of extracting compliance quickly became a matter of unilateral economic pressure exerted by the United States on the countries involved.

17. See memorandum of a conversation between the economic minister of the British Embassy and the assistant secretary of state for economic affairs, December 17, 1957, and memorandum of conversations, December 30–31, 1957, reprinted in *FRUS*, 1955–57, 10:504–8.

18. See NSC-5704/3, Statement of U.S. Economic Defense Policy, September 16, 1957, reprinted in *FRUS*, 1955–57, 10:95–98.

19. U.S. Department of State, Policy Planning Council, "U.S. Policy on Trade with the European Soviet Bloc," memorandum prepared for the president, July 8, 1963, reprinted in the *Declassified Documents Quarterly* (1976), item 289A, p. 25.

20. The criteria were articulated in the context of the 1958 list review. See *Battle Act Reports*, no. 12, April 20, 1959. In certain cases, U.S. officials had been reluctant to accept these principles

to qualify as strategic, the equipment, technology, and materials under consideration needed to be directly related to the *peacetime* military production and capabilities of the bloc.[21] This reflected the underlying strategic conception of the embargo: that total war between East and West was neither imminent nor inevitable and that the West was engaged in peacetime military competition designed to deter war. Items that *would* be used chiefly for military purposes, in the circumstances of war or Soviet bloc war preparation, were not prohibited by the embargo.

By accepting the more narrow criteria and the decontrol of less strategic items, the United States helped to alleviate the economic, administrative, and political burdens of participation in CoCom. The streamlined control list meant that, for other member governments, adherence to CoCom and the significant expansion of East-West trade were not incompatible. Western Europe and Japan could fully participate in a multilateral control regime yet simultaneously accelerate their exports to the East of industrial equipment and turnkey plants. In addition, the task of enforcement was facilitated by enabling governments to focus attention on fewer items, ones that lacked widespread commercial availability. The United States made a particular contribution by proposing the decontrol during the 1960s of many products containing integrated circuits as those products became more widely available.[22] Finally, U.S. acceptance of narrower controls contributed to political harmony in CoCom. Had U.S. officials systematically sought to obtain compliance for more comprehensive controls or had they resisted the decontrol of less strategic items, the resentment of and defiance by other members would have been profound. Their commitment to CoCom, and perhaps CoCom itself, would have been called into question.

That the United States performed adequately in CoCom is not to suggest that its leadership was unqualified or that U.S. East-West trade policy generated no controversy in the alliance context. On the contrary, considerable conflict emerged over the Kennedy administration's wide-diameter pipe initiatives, which involved an isolated effort by the United States to reimpose multilateral economic warfare against the Soviet Union. Conflict was also apparent in the cases of Cuba and China, in which U.S. officials sought to constrain other members' trade by employing its export controls extraterritorially. Both initiatives are discussed below. Although each strained alliance relations significantly, neither

---

prior to the 1958 review. On copper wire, for example, the United States urged strict controls but was rebuffed by Great Britain, which argued in CoCom that the Soviets had adequate supplies for military purposes, even though there were shortages in the civilian sector. See the letter from the British ambassador to Secretary of State Dulles, June 6, 1956, reprinted in *FRUS, 1955–57*, 10:365–66.

21. "U.S. Policy on Trade with the European Soviet Bloc," p. 24.
22. William A. Root, "U.S. Policy on East-West Technology Trade," p. 218.

primarily involved CoCom.[23] Within CoCom, U.S. officials created some discontent by reneging on an offer not to impose reexport control requirements on controlled items that fell within the category of administrative exceptions.[24] U.S. officials also injected foreign policy considerations into the CoCom control process by encouraging and liberally granting to other CoCom members exceptions for the export of controlled items to Poland, in order to reward that country's efforts to distance itself from the Soviet Union.[25]

The overall impact of these initiatives on the functioning of CoCom was modest. They did, however, reflect characteristics of U.S. export control behavior (e.g., the tendency to politicize controls, to push beyond the CoCom consensus, and to act unilaterally) that, in the circumstances of the 1970s and 1980s, rendered problematic U.S. leadership and thus the effectiveness of CoCom.

With due regard to the above qualifications, U.S. leadership was more effective in the 1958–1968 period than during the 1970s, when it neither provided adequate discipline to CoCom nor minimized the burdens of participation, and during most of the 1980s, when it provided more discipline yet simultaneously exacerbated the burdens of the system. In accounting for the relative effectiveness of U.S. performance during CoCom's second decade, two factors deserve emphasis. The first concerns America's overall approach to the Soviet Union, which in the postwar era lurched between confrontation and accommodation. During periods of confrontation, U.S. officials tended to push for export controls beyond the multilateral consensus, thereby increasing administrative and economic burdens. This situation clearly existed during the early 1980s and the mid-1950s. During periods of accommodation, the United States was tempted to relax vigilance and not provide sufficient and consistent discipline to CoCom; this was the experience of the 1970s. In the 1958–1968 period, U.S. policy was in transition from the

23. The items under contention in the Cuba and China cases were only of broad economic significance and thus not subject to multilateral control. A U.S. proposal that Cuba be added to the list of CoCom-controlled target states was rejected by the Consultative Group. See Harold J. Berman and John R. Garson, "United States Export Controls—Past, Present, and Future," *Columbia Law Review* 67 (May 1967), 840. The pipeline conflict, though it involved CoCom briefly, was mainly played out in NATO and in discussions among the senior political officials of the states involved. See discussion below.

24. William A. Root, "CoCom: An Appraisal of Objectives and Needed Reforms," in Gary K. Bertsch, ed., *Controlling East-West Trade and Technology Transfer: Power, Politics, and Policies* (Durham: Duke University Press, 1988), p. 423.

25. In 1959, for example, of the approximately $500,000 of exceptions granted, all but approximately $35,000 were destined for Poland. See *Battle Act Reports*, no. 13, March 15, 1960, pp. 33–36, and no. 14, December 20, 1960. In 1960, all exceptions were for Poland other than a shipment of molybdenum from France to China and a large shipment of communications cable from Italy to the USSR. *Battle Act Reports*, no. 15, March 22, 1962. In general, see the presidential determinations recorded in the *Battle Act Reports*, nos. 14–19, covering 1959–1965. The "Polish differential" on exceptions was not a source of discontent in CoCom, mainly because it involved the expansion, rather than the restriction, of West European trade with the East.

confrontational posture of the early cold war to the more conciliatory stance of détente. Somewhat fortuitously, this transitional policy facilitated U.S. leadership in CoCom. The maintenance of East-West conflict meant that an effective strategic embargo remained a priority, prompting U.S. officials to provide discipline to both U.S. export policy and CoCom. At the same time, movement toward détente and, as discussed below, toward a more liberal U.S. East-West trade policy facilitated accommodation and tolerance by executive officials of relaxed controls on less strategic trade in the multilateral context.

Second, U.S. leadership, and thus the effectiveness of CoCom, was made easier by the clear technological superiority enjoyed by the United States. During the 1950s, when the control criteria were very broad and the list contained items of general industrial significance, other CoCom members could produce and export much of what was considered strategic. Under the narrower criteria of the strategic embargo, however, the control list came to emphasize dual-use items that incorporated very advanced technologies. For most of such items, the United States enjoyed a considerable lead over other CoCom members in development and production. By the mid-1960s, for example, a heavy emphasis of CoCom controls was on computers and electronic items, areas of clear American dominance.[26]

U.S. officials enjoyed similar advantages in the domestic arena. In recognition of the critical importance of technology to national defense, the U.S. government (in particular, the Defense Department) strongly supported research and development in defense-related areas during the 1950s and 1960s. Programs funded by the Department of Defense (DoD) in nuclear technology, electronics, and aeronautics were especially successful. An important consequence of the government's leading role was that U.S. officials could regulate the timing of when and conditions under which many critical technologies would be released to the private sector. Consequently, for such technologies, military applications frequently preceded civilian ones.[27]

The technological dominance of the United States over other Western states, along with the pivotal role of the U.S. government in technology development domestically, facilitated U.S. leadership in CoCom in several ways. It enhanced the ability of U.S. officials to identify critical technologies and to gain support for their control multilaterally. In CoCom, it has always been easier to have items added to the control list *before*, rather than after, commercial applications have developed.[28] To the ex-

26. *Battle Act Reports*, no. 17, December 31, 1964, p. 4. By the end of the 1964 review, 54 of 146 CoCom categories contained electronics items.
27. This point is discussed more fully in Chap. 6, below. See also National Academy of Sciences, *Balancing the National Interest* (Washington, D.C.: National Academy Press, 1987), p. 56.
28. Personal correspondence with William Root. Root notes that during the 1980s CoCom controls on computers and integrated circuits were more comprehensive than those on robotics;

tent they could control the release of militarily relevant technologies to the private sector at home and abroad, U.S. officials could simultaneously provide the strategic justification for controlling such items in CoCom and face minimal resistance on commercial grounds.

Technological dominance also served as a source of leverage in CoCom. Even if the threat was not always explicit, the knowledge that the U.S. government could disrupt the flow of essential technology undoubtedly reinforced the willingness of other alliance members to participate effectively in the multilateral control system.[29] As others have noted, U.S. bargaining leverage in CoCom has depended at least in part on its ability to outpace other members technologically.[30] The same point, of course, applies to non-CoCom suppliers, the number of which remained small—again facilitating the U.S. leadership role—as long as the United States dominated the development of and access to militarily critical technologies.

CoCom and Soviet Capabilities

As noted in Chapter 1, the effectiveness of cooperation in CoCom must be distinguished analytically from the *impact* of CoCom controls on Soviet capabilities. With regard to impact, assessments of CoCom through the 1960s vary considerably. Roger Carrick, a former British CoCom official, has defended and praised the Western embargo, calling it sophisticated and appropriate to its task and claiming that its successes, though perhaps unsung, were nevertheless real.[31] Peter Wiles similarly argued in 1970 that compared to other embargo attempts, CoCom was a "most favorable exemplar" which "shines like a rational, though not very friendly, deed." Its main virtue was its limited and feasible means and objectives; CoCom performed "a minimally hostile act in order to preserve the peace."[32]

Others judged CoCom as having had far less of an impact. Adler-Karlsson wrote of the "crumbling" or "withering away" of the embargo

---

the former were placed on the list prior to widespread commercial application, whereas the latter were not.

29. Adler-Karlsson argues, for example, that in 1964, Britain and West Germany accepted U.S. computer control proposals because of their extreme dependence on U.S. sources of supply and fear that the U.S. government might constrain their access to them. See "International Economic Power," p. 504.

30. Berman and Garson, "United States Export Controls," p. 842, and Henry R. Nau, "Export Controls and Free Trade: Squaring the Circle in CoCom," in Bertsch, ed., *Controlling East-West Trade and Technology Transfer*, pp. 400–402.

31. R. J. Carrick, *East-West Technology Transfer in Perspective* (Berkeley, Calif.: Institute of International Studies, 1978), pp. 29, 44, and, generally, chap. 5.

32. See U.S. Congress, Joint Economic Committee, Subcommittee on Foreign Economic Policy, *A Foreign Economic Policy for the 1970s: Part 6*, hearings, 91st Cong., 2d sess., December 7–9, 1970, p. 1242.

between 1953 and 1967 and viewed the remaining control list as being of minor importance.[33] In his study of Western technology transfer to the Soviet Union, Anthony Sutton took a similar position, condemning the CoCom controls as being neither sufficiently broad in scope nor well enforced by Western officials.[34] Some in Congress echoed these concerns during the 1960s, accusing CoCom of being a glaring example of the inability of the United States to protect its national security.[35]

As others have recognized, this divergence in views is due in part to a failure to clarify precisely the objectives of the CoCom embargo.[36] Adler-Karlsson and Sutton assume, either implicitly or explicitly, that the purpose of CoCom until 1968 was (or should have been) economic warfare. Judged by such a standard, the embargo was a failure; as discussed below, Soviet bloc economic growth was retarded only marginally, if at all, by the controls.[37] Yet by 1954 and clearly by 1958, CoCom was understood by its members as no longer serving the purposes of economic warfare. The control list was not only shortened after 1954 but the objective of the embargo was changed as well.

In considering CoCom a success, Wiles and Carrick are judging it according to the relatively more modest standards of a strategic embargo. Yet even judged as such, it is important to clarify objectives precisely. If the purpose of the controls was to prevent the Soviet Union from becoming a formidable military power, they clearly failed. By the 1960s, the Soviets had fully demonstrated their ability, symbolized most dramatically by the launching of Sputnik in 1957, to produce and maintain a wide array of advanced weapons systems, despite the controls. Soviet success resulted from its willingness to accord the highest priority to the defense sector, at the expense of civilian needs, in the allocation of scarce resources (especially skilled manpower and technology). As the State Department confirmed in a classified report in 1965:

It is significant that during the period 1950–57 when the Western nations achieved a substantial degree of harmony in trade control policy the U.S.S.R. was able . . . to build a military establishment second only to the U.S.; and during the period 1958–1964 when rigid controls remained on military items and on items judged to be important to

33. Adler-Karlsson, *Western Economic Warfare*, chaps. 8 and 9.
34. Anthony Sutton, *Western Technology and Soviet Economic Development, Volume III: 1945–1965* (Stanford: Hoover Institution Press, 1973), p. 54.
35. See, for example, the statement of Congressman Delbert Latta in U.S. Congress, House, Committee on Banking and Currency, *East-West Trade*, hearings, 90th Cong., 2d sess., 1968, p. 24.
36. See Gary Bertsch et al., "East-West Technology Transfer and Export Controls," *Osteuropa-Wirtschaft* 26 (June 1981), p. 126.
37. More precisely, economic warfare failed to achieve strategic objectives as defined in Chap. 2. It may, however, have been effective as a signaling instrument. See David Baldwin, *Economic Statecraft* (Princeton: Princeton University Press, 1985), pp. 235–50.

advanced weapons systems the Soviets made tremendous strides in a variety of missile and space programs.[38]

Yet, if the objective of the controls was to *delay* Soviet acquisition of militarily relevant Western technology in order to assist the United States and the West in maintaining a qualitative advantage in the arms race, then it is reasonable to consider CoCom something of a success. A study carried out by U.S. intelligence agencies in 1965 indicated that Soviet bloc development of a wide range of miniaturized military equipment was hindered significantly by the denial of semiconductor and transistor manufacturing know-how and equipment.[39] The study further noted that "the denial of sophisticated technologies for modern telecommunications has dampened—and in some cases halted—Bloc-wide plans for expanding secure 'land-line' communications networks, without which air defense capabilities, for example, are handicapped."[40] Access to the most advanced Western computer technology would have facilitated Soviet research and development of advanced weapons systems. Similarly, the easy availability of Western metallurgical technology would have allowed the Soviets to reduce the weight and improve the performance characteristics of their military aircraft. Overall, the relaxation of CoCom controls would have given the Soviet military "an improved ability to perform its basic missions."[41]

The above assessment suggests that, although its impact on the overall state of Soviet military capabilities was modest, the strategic embargo did retard to some extent the ability of the Soviet Union to remain competitive with the United States in the application of advanced technologies to military purposes. The maintenance of Western and U.S. lead time came to be seen during the 1960s as the primary purpose of CoCom controls, and the Western allies were willing to make the necessary, albeit modest, economic sacrifices in order to contribute to this aim. The fact that the Soviet Union deemed it necessary to establish a bloc-wide strategy for the illegal acquisition of embargoed items lends further credence to the view that the controls mattered in the overall superpower competition.[42]

38. "Strategic Importance of Western Technology to the Soviet Bloc," Background Papers prepared for the Miller Committee, p. 14.
39. This example and others are developed in "Military Aspects of Export Control of Technology," Background Papers prepared for the Miller Committee, p. 86.
40. Ibid.
41. "Strategic Importance of Western Technology to the Soviet Bloc," Background Papers prepared for the Miller Committee, p. 16.
42. Adler-Karlsson argues to the contrary that the impact of controls on the ability of the United States to maintain lead time "seems not to have been of much importance in the total picture." *Western Economic Warfare*, p. 123. He also suggests, however, that the Soviet system was twice as efficient as that of the United States in operationalizing the military applications of new technology. If that was indeed correct (the evidence provided is scanty), then it is at least plausible to argue that the embargo did matter in the maintenance of U.S. lead time by helping to offset the weakness in the U.S. procurement system.

Export Controls and Intra-Western Trade

The tacit deal struck among the allies in the early 1950s, which helped to insulate CoCom controls from intra-Western trade, remained intact during the 1960s. A report on the obstacles to the free flow of intra-Western trade, prepared for the Commerce Department in 1969, noted that the broadest *potential* area of adverse impact concerned the application of U.S. non-frustration controls.[43] The reason was simple: given the risks of transshipment, the United States maintained licensing requirements on a broad array of U.S. exports to non-Communist destinations.In practice, however, such controls proved to have little detrimental effect on intra-Western trade. The Commerce Department approved the overwhelming majority of license requests for export to Western destinations and processed them rapidly, completing over 90 percent within five working days.[44] To ease the licensing burden further, in 1968 Commerce introduced a distribution or "bulk" license, which covered all transactions between a U.S. firm and eligible Western end users for a given period of time, thereby eliminating the need to secure a license for each individual transaction.[45]

America's intra-Western controls were similarly liberal with regard to technical data. Most unpublished technical data, which required a validated license for export to the East, could be shipped to Western destinations under general license as long as recipients provided assurances that they would not reexport without U.S. permission. Some U.S. officials were uneasy with this arrangement, given the uncertainty over how, and to what extent, West European governments exercised control over technical data exports.[46] Nevertheless, the Commerce Department was unwilling to apply more restrictive controls, declaring the subject to be one that "must be approached with great care, so as not to do unnecessary damage to our freedoms and to the economic growth of our country and the free world."[47] The unrestricted flow of technical data was particularly critical to American multinationals, many of which were establishing and nurturing production and distribution networks with subsidiaries, affiliates, and licensees in other Western countries. In 1957 there were twelve hundred U.S. subsidiaries in the European Community; by 1966 the number had grown to over four thousand.[48]

43. U.S. Department of Commerce, *Factors Affecting the International Transfer of Technology among Developed Countries* (Washington, D.C.: GPO, 1970), p. 17.
44. In 1968, for example, Commerce Department officials testified that 93 percent of intra-Western licenses were processed within five days and 98 percent within ten. *East-West Trade*, 90th Cong., 2d sess., p. 229.
45. Ibid., pp. 403–4, 765, and U.S. Department of Commerce, *Export Controls*, 91st Quarterly Report (First Quarter 1970), p. 18.
46. See, for example, *Report on Export Controls*, p. 11. The problem of technical data controls received far greater attention in the latter half of the 1970s, as I discuss in Chap. 6.
47. *Export Controls*, 61st Quarterly Report (Third Quarter 1962), p. 11.
48. The comparable figures for U.S. subsidiaries in Great Britain are eight hundred in 1957

As an adjunct to the basic control regime, U.S. officials reserved the right to control the reexport of U.S.-origin parts and components and the foreign-produced products of U.S. technical data. The rigorous application of reexport controls had the potential to disrupt intra-Western trade by driving the overseas customers and affiliates of U.S. firms to seek alternative sources of supply. As in the case of non-frustration and technical data controls, however, in practice, reexport controls did not pose a significant constraint. Many of the items for which reexport licenses were required were simultaneously CoCom-controlled, which meant that exceptions were necessary for them to be exported to the East. Because such exceptions were requested and granted only infrequently and already required U.S. review and acquiescence, the additional burden imposed by the reexport licensing requirement was probably modest.[49] For items not controlled by CoCom but by the United States unilaterally, the regulations were potentially more bothersome. Compliance by foreign firms does not appear to have been widespread, however, in part because such firms were not always cognizant of the scope and extent of U.S. controls and also because U.S. officials appeared not to judge nonstrategic (i.e., non-CoCom-controlled) trade sufficiently important to warrant a concerted effort to ensure full compliance.[50] As late as 1967, the Department of Commerce deemed it necessary to issue a public reminder that U.S.-origin components and materials used to produce foreign-made end products were indeed subject to the export control laws of the United States.[51] Overall, reexport controls did not pose a serious obstacle to the West European export of nonstrategic items to the Soviet Union and Eastern Europe. To the extent they did, the technological dominance of the United States rendered it difficult or impractical for West European firms to develop or turn to alternative sources of supply.

---

and twenty-three hundred in 1966. See Alfred Grosser, *The Western Alliance* (New York: Vintage, 1982), pp. 221–22. In general see Robert G. Gilpin, *U.S. Power and the Multinational Corporation* (New York: Basic Books, 1975).

49. Reexport requirements on items falling into the administrative exceptions category did pose an additional burden because the United States did not have the right to review such cases in CoCom until after the fact. It is unclear from the available evidence whether and to what extent such controls posed a meaningful constraint on Western Europe's trade with the East.

50. Even in the early 1980s, when the United States was relatively zealous in its application, the National Academy of Sciences found that independent foreign firms were "either ignorant or casual" in their compliance with reexport controls. See *Balancing the National Interest*, p. 107.

51. See John Ellicott, "Extraterritorial Trade Controls—Law, Policy, and Business," in Martha L. Landwehr, ed., *Private Investors Abroad—Problems and Solutions in International Business in 1983* (New York: Matthew Bender, 1983), pp. 9–14. To say that U.S. reexport controls were not well publicized and that officials did not enforce them comprehensively is not to suggest that they neglected them entirely. U.S. officials reserved the right to prosecute and punish violators with the denial of export privileges, even when foreign firms were unaware of the existence of licensing requirements. See Ellicott's discussion of the Raytheon case, pp. 12–13.

U.S. officials, who possessed the wherewithal to extend controls and significantly disrupt intra-Western trade, refrained from doing so for two reasons. First, during the 1960s, free trade and the expansion of U.S. direct investment continued to constitute major foreign policy objectives for the United States. Noninterference with private transnational links contributed to the realization of the postwar U.S. vision: the formation of a prosperous, interdependent economy in the West, which could provide the economic foundation for the military and political cohesion of the anti-Soviet alliance. The insulation of East-West controls from intra-Western trade added to American national security, broadly defined.

Second, with regard to national security as more narrowly defined, executive officials had reason to be confident that they could adequately protect products and technologies of direct military significance. As noted above, the United States enjoyed a considerable lead over its allies in the development and application of such technologies, and the U.S. government controlled their release to the private sector at home and abroad. When such items were released, U.S. officials could rely on CoCom to provide an additional layer of protection against transshipment to the East. CoCom's importance should not be underestimated; executive officials explicitly recognized that in the absence of common, effective controls, harsher restrictions on West-West trade would be necessary.[52]

For West European governments, dependence on and access to U.S. technology as well as the opportunity to engage in nonstrategic trade with the East reinforced their willingness to participate in CoCom and even their tolerance for the extraterritorial application of U.S. reexport controls. State Department officials clearly understated the discontent of their West European counterparts by noting before Congress that because reexport requirements were part of voluntary contractual agreements between U.S.-based and foreign firms, allied governments did not "often" complain of them as an extraterritorial control.[53] West European resentment was profound, and government officials generally refused to acknowledge or cooperate with U.S. efforts to enforce reexport controls.[54] At the same time, however, those officials did not impede U.S. efforts, despite the fact that they frequently involved the blatant violation of West European sovereignty. Certain U.S. Foreign Service personnel based in Western Europe, for example, were designated "economic defense officers," with responsibility for alerting foreign firms to U.S.

---

52. Department of Commerce, *Factors Influencing the International Transfer of Technology*, p. 19.
53. *East-West Trade*, 90th Cong., 2d sess., p. 224.
54. *Report on Export Controls*, pp. 16–18.

export control regulations and tracking down violators of those regulations to face penalties administered by the U.S. government.[55] As long as reexport controls did not significantly interfere with the ability of West European firms to service Eastern markets or with their access to U.S. sources of supply, however, West European governments were willing to object more in principle than in practice.

When U.S. extraterritorial controls did greatly interfere with allied trade, other Western governments were willing to challenge the United States. During the late 1950s and 1960s, such challenges did not involve trade with the Soviet Union and Eastern Europe but with China and Cuba. At issue were the Treasury Department's Foreign Asset Control Regulations rather than Commerce's reexport authority. Through the exercise of the former, derived from the Trading with the Enemy Act, U.S. officials attempted to maintain the integrity of their virtually total embargoes on trade with China, Cuba, North Korea, and North Vietnam by regulating exports to those destinations by U.S. subsidiaries based in other Western countries.[56] In 1957, Canada resisted U.S. efforts to prevent the Canadian subsidiary of Ford from exporting trucks to China, and in a similar case in 1965, the French government forced the U.S. to back down with the assistance of the French courts.[57] In confrontations over trade with Cuba, the United States was again forced to retreat, and Treasury was left to rely on a program of voluntary compliance, in which U.S. firms were asked to discourage their subsidiaries from pursuing the trade.[58]

Although not involving CoCom per se, these conflicts generated considerable political friction and foreshadowed the more dramatic confrontation over the Siberian gas pipeline that took place in the early 1980s. They demonstrated the potential problems caused in the alliance context by U.S. efforts to assert and enforce unilateral controls beyond the CoCom consensus.

THE STRUGGLE TO ADJUST U.S. POLICY

As noted at the outset of this chapter, a central concern of the United States from 1958 through 1969 was the adjustment of its national East-West trade policy. Two related factors prompted the need for U.S. adjustment: the growing East-West trade of the allies and recognition by

55. Ibid., pp. 15, 19–20.
56. *East-West Trade*, 90th Cong., 2d sess., pp. 410–15.
57. For discussion of the cases, see Samuel Pisar, *Coexistence and Commerce* (New York: McGraw Hill, 1970), pp. 134–37; Berman and Garson, "U.S. Export Control Policy," pp. 867–76; and Steven Kobrin, "Hegemony, American Multinationals, and the Extraterritorial Enforcement of Export Embargoes," unpublished paper, 1988.
58. Pisar, *Coexistence and Commerce*, p. 137.

U.S. executive officials of the strategic futility of unilateral economic warfare.

Eastern markets became increasingly attractive during the 1960s. While international trade increased at a respectable 8 percent per year, East-West trade grew at an estimated 12 percent.[59] Yet as Table 4 indicates, the great majority of Western trade with the Soviet Union was captured by Western Europe and Japan. With the exception of 1964— trade figures for that year reflect an extraordinary U.S.-Soviet wheat deal—the United States lagged far behind, its exports and total trade representing a mere fraction of that of its allies.

The expansion of trade was due to a conscious decision on the part of the Soviets to import technology and capital goods from the West. For Khrushchev, expanded trade was an important component of the Soviet effort to "catch up and overtake" the capitalist nations economically. The United States was invited to participate from the outset. In 1958, Khrushchev sent President Eisenhower a letter offering normalization and the extensive development of U.S.-Soviet trade. Included within it was a long "shopping list" of machinery and equipment, across a wide range of sectors, including mining, food processing, chemicals, textiles, and transportation.[60] The Soviets in return offered to export manganese, platinum, timber and furs. The United States, however, proved unwilling to take such a step unless and until it became politically expedient to do so.

Western Europe and Japan had no such political reservations about meeting Soviet demand. Their exchange of machinery and equipment for Soviet raw materials grew under the auspices of a series of multiple-year, government-to-government trade agreements. Western exports frequently took the form of "turnkey plants"—complete factories installed and equipped on-site in the Soviet Union with the help of Western engineers and technicians. Thus, the Soviets obtained not only equipment but also Western know-how as a means to accelerate the development of their industrial base.

Because the Soviet market was an attractive outlet for surplus export capacity, competition among Western suppliers was intense. In 1960, the president of the British Council for the Promotion of International Trade stated that "the importance of East-West trade for our capital goods industries is only beginning to be fully appreciated," because "there are few areas of the world to which Britain can export her engineering products on such a scale."[61] By 1964, Britain accounted for 50 percent of all Western sales of chemical plants and equipment to the

59. Statement of Philip Tresize in *East-West Trade*, 90th Cong., 2d sess., p. 45.
60. *East-West Commerce* (London), June 16, 1958.
61. Comments of Lord Boyd-Orr, reprinted in *East-West Commerce* (London), November 30, 1960.

East.[62] For Japan, a $100 million Soviet order in 1963 for tankers, motor vessels, and floating cranes was said by the Tokyo press to "breathe new life into Japan's ship-building industry, which has been suffering from stagnation in exports."[63] As I discuss below, exports of wide-diameter pipe to the Soviet Union were valuable to the West German steel industry, prompting widespread opposition in West Germany to the NATO embargo of 1962.

By 1964, Western export competition became government-guaranteed long-term credit competition. The Soviet Union made no secret of its desire to finance its capital goods imports with official credits and of its willingness to favor those firms that could offer the most attractive terms. To the United States, long-term credits were a form of financial aid that allowed the Soviets to increase the resources devoted to their civilian sector without penalizing the military.[64] U.S. officials worked hard to maintain an informal agreement among the allies to limit export credits in East-West trade to a maximum of five years. In 1964, under the weight of economic competition, the consensus collapsed. The British offered the Soviets a fifteen-year $300 million credit for the construction of chemical plants, and the French followed with a ten year credit in excess of that amount. Other states quickly followed suit; Fiat, for example, struck an $800 million deal with the Soviets, which included a $350 million credit from the Italian government extending for up to fourteen years.[65]

Among the beneficiaries of this credit competition, and of the expansion of East-West trade in general, were the West European subsidiaries of U.S. multinationals. The State Department estimated that affiliates of U.S. firms located in Western Europe exported manufactures to Eastern Europe in the amount of $150 million in 1965 and $225 million in 1967.[66] These figures compared with only $23 million in 1965 and $63 million in 1967 of manufactures shipped directly from the United States. That the U.S. government did little to curb these exports by U.S. subsidiaries in Western Europe made it all the more difficult to justify on economic grounds the continuation of unilateral economic warfare.

Unilateral economic warfare was equally difficult to justify on strategic grounds. By the early 1960s, U.S. policy-makers were forced to recog-

62. CIA, "Acquisition of Chemical Equipment and Technology by the Soviet Bloc from the Free World, 1964," memorandum, undated, reprinted in *Declassified Documents Quarterly* (1978), item 20A, p. 1.

63. *East-West Commerce* (London), February 25, 1963.

64. See the statement of Secretary of State Dean Rusk in U.S. Congress, Senate, Committee on Foreign Relations, *East-West Trade*, hearings, 88th Cong., 2d sess., March 13, 16, 23, April 8–9, 1964, pp. 15–16.

65. A good discussion of the Western export credit regime is Beverly Crawford, "Western Control of East-West Trade Finance," in Bertsch, ed., *Controlling East-West Trade and Technology Transfer*, pp. 280–312.

66. *East-West Trade*, 90th Cong., 2d sess., pp. 87–88.

nize that the strategy had not achieved what had been hoped for it at the beginning of the cold war. Soviet economic growth had clearly not been crippled by the embargo; on the contrary, as Secretary of State Rusk publicly conceded in 1964, the Soviet Union had become "a largely self-sufficient economy with a broad industrial base and a well-developed technology."[67] The Soviet economy had grown at a very respectable rate of approximately 10 percent annually. Indicative of this success was the fact that the Soviets were now exporting some of the items the West attempted to deny it during the early 1950s, such as petroleum and certain types of machine tools.

Not surprisingly, economic warfare was also a failure with regard to its ultimate strategic objectives. Given the priority accorded by the Soviets to defense preparation, the negative impact of trade denial on the aggregate level of Soviet resources was most likely felt by the civilian sector rather than by the military. Moreover, economic warfare failed to drive a wedge between the Soviets and their East European satellites. If anything, it enhanced the ability of the Soviets to draw the bloc closer together and to consolidate its control over the economies of the smaller East European states. According to one analyst, the embargo did more damage to the latter, forcing them to become economically more dependent on the Soviet Union.[68]

In 1961, the CIA produced an estimate of the probable impact of a reimposition of multilateral economic warfare on the Sino-Soviet bloc. It concluded that "because of the highly diversified resources of the Sino-Soviet bloc economies, and their generally advanced state of industrialization, the economic effects of a trade and transport embargo would be *minimal in the long run*."[69] An effectively enforced NATO and Japanese embargo would impose an economic loss of approximately $4 billion on the bloc, most of which would be felt in the first six months. This was not an insignificant loss. Given the strategic rationale underlying economic warfare during the 1950s, however, the more important CIA finding was that even an effective Western embargo would have no impact on Soviet military programs or preparedness.[70] The brunt of any economic dislocation would be borne by the Soviet consumer, as it had been during the 1950s. In terms of the analysis developed in Chapter 2, the CIA believed that should economic warfare reduce the economic resources available to Soviet leaders, Soviet preferences would be adjusted to maintain the pre-embargo level of military output.

67. Statement of Secretary of State Dean Rusk in *East-West Trade*, 88th Cong., 2d sess., p. 5.
68. Nicholas Spulber, "East-West Trade and the Paradoxes of the Strategic Embargo," in Alan Brown and Egon Neuberger, eds., *International Trade and Central Planning* (Berkeley: University of California Press, 1968), p. 120.
69. CIA, "Estimated Impact of Western Economic Sanctions against the Sino-Soviet Bloc," memorandum, July 16, 1961, reprinted in *Declassified Documents Quarterly* (1975), item 55E, p. 2.
70. Ibid., p. 6.

Since alliance-wide cooperation was unlikely, the CIA also estimated the impact of continued *unilateral* U.S. controls. It found such a strategy to be an exercise in futility. A U.S. embargo would have marginal economic impact because alternative suppliers in Western Europe and Japan could provide virtually anything denied to the Soviets by the United States. Acting by itself, the United States could do very little either to weaken the Soviet economy or to reduce the aggregate resources available to the Soviet military. Actually, it could not even prevent the allies from *strengthening* the Soviet industrial base through expanded trade in machinery and equipment and financed by export credits on generous terms.

Wide-Diameter Pipe

The CIA did determine in 1961 that wide-diameter (i.e., forty-inch) steel pipe was one of a very few commodities for which an effective *multilateral* embargo would impose significant economic costs on the Soviets.[71] In 1962–1963, the Kennedy administration triggered a major crisis in the alliance with its attempt to deny the Soviets that item.[72]

The U.S. initiative can be understood as one last attempt to gain alliance cooperation for economic warfare, but only with regard to one critical commodity in a sector crucial to Soviet economic development and foreign policy, energy. By the Soviets' own admission, wide-diameter steel pipe was an important bottleneck item. Soviet output of forty-inch pipe from 1963 to 1965 was projected to be only 850,000 tons; in order to fulfill the requirements of the Seven-Year Plan (1959–1965), they would need to import an additional 700,000 tons.[73] For the Kennedy administration, the increase in East-West tensions generated by the Berlin crisis, along with the renewal of pressure from Congress for more extensive trade restrictions (see below), helped to justify applying pressure on the allies to control the export of a critical commodity, albeit one of more economic than direct military significance.

The import of steel pipe would enhance the ability of the Soviets to export oil to Western Europe. Wide-diameter pipe was required for the construction of the so-called Friendship Pipeline, which was to carry oil from the Baku fields, through Eastern Europe, to the West. In the eyes of U.S. officials, the rapid increase in Soviet oil exports to the West—from 116,000 barrels per day in 1955 to 486,000 in 1960—had geo-

71. Ibid., pp. 4–5.

72. For detailed discussions, see Angela Stent, *From Embargo to Ostpolitik: The Political Economy of West German-Soviet Relations, 1955–1980* (Cambridge: Cambridge University Press, 1981), chap. 5, and Bruce W. Jentleson, *Pipeline Politics: The Complex Political Economy of East-West Energy Trade* (Ithaca: Cornell University Press, 1986), chaps. 3 and 4.

73. Stent, *From Embargo to Ostpolitik.*, p. 101.

political implications.[74] Opening congressional hearings on the subject in 1962, Senator Keating claimed the Soviets were planning "to drown (the West) in a sea of oil . . . not for economic but for political and military reasons."[75] Pipe exports would not only improve Soviet economic capabilities but might also provide the Soviets with a future source of political leverage over the allies, particularly energy consumers such as Italy and West Germany. By 1961, the Soviets already possessed 22 percent of the Italian oil market, a sufficiently large share for U.S. officials to react with alarm.[76]

The United States also had economic stakes in an embargo. Whether correctly or not, U.S. oil firms considered Soviet oil, which was allegedly being dumped, as a threat to their market position in Western Europe. The major oil companies favored restrictions on Soviet-West European energy trade, and the Kennedy administration was well aware of their sentiments.[77]

Wide-diameter steel pipe had been removed from the CoCom list in 1958, as part of the consolidation of the strategic embargo. A U.S. proposal in 1961 to return it to the list was rejected by the allies, despite U.S. efforts to depict the item as having direct military significance. U.S. negotiators argued that in the event of war, the pipeline would be used to supply oil to Red Army divisions in Eastern Europe and would also thereby release Soviet transport vehicles for other military activities.[78] For most CoCom members, neither the increase in East-West tensions nor Khrushchev's economic and diplomatic offensives justified a return to what they perceived to be selective economic warfare. Only France, which shared American economic concerns as an oil exporter, offered strong support.[79]

Economically, West Germany stood to suffer greatly from an embargo. The export of pipe was clearly more significant than the import of oil; two-thirds of Soviet imports of pipe were from West Germany. West German firms expected Soviet orders to help overcome sluggish demand in the steel sector, and in October 1962, three large concerns signed a contract to provide the Soviets with 163,000 tons of pipe, valued at $28 million.[80]

---

74. Spulber, "East-West Trade and the Paradoxes of the Strategic Embargo," p. 121.

75. Keating is quoted in Stent, *From Embargo to Ostpolitik*, p. 99.

76. Jentleson, *Pipeline Politics*, p. 90. In addition to seeking restrictions on pipe exports, the United States sought to convince its allies to reduce Soviet oil imports. On U.S. success with Italy, see ibid., pp. 107–12.

77. Adler-Karlsson, *Western Economic Warfare*, p. 130, and Jentleson, *Pipeline Politics*, pp. 100–103.

78. Adler-Karlsson, *Western Economic Warfare*, p. 130.

79. Within the U.S. government, Secretary of State Rusk opposed the embargo, arguing that the United States should abide by the narrow criteria of the strategic embargo in the alliance context. See Jentleson, *Pipeline Politics*, p. 104.

80. Stent, *From Embargo to Ostpolitik*, p. 104.

Lacking support in CoCom, the U.S. moved its initiative to NATO, a forum which did not require unanimity and which U.S. officials hoped would be more sympathetic to its broad geopolitical concerns. In November 1962, NATO passed a resolution calling for an alliance embargo on all exports to the Soviets of pipe over nineteen inches in diameter. Italy, faced with U.S. threats to apply extraterritorial controls, agreed to reduce, though not eliminate, its pipe deliveries. Japan, pressured diplomatically by the United States, observed the NATO embargo. Sweden, a member of neither NATO nor CoCom, flatly refused to cooperate. Most damaging politically was that the British government defied the United States and refused to endorse the resolution, despite U.S. requests at the highest levels of government and the fact that Britain was only a minor supplier of wide-diameter pipe.[81]

Whether to comply with the embargo resolution caused major political problems in West Germany. Because it was to be applied retroactively, West German firms were to be forced to break the contracts they had signed in October. In the face of intense U.S. pressure and despite widespread domestic opposition, the Adenauer government accepted this fate, choosing loyalty to the United States over national economic interests. To avoid a Bundestag vote that would have overturned German compliance, the Christian Democrats, in a constitutionally dubious maneuver, removed themselves from the legislative chamber. In the midst of much controversy, the decision to comply was upheld.[82]

The steel pipe episode demonstrated the ability of the United States to shape alliance (more precisely, Italian and West German) policy on a specific East-West trade issue, albeit at a high political cost. Given the overall failure of the United States to gain alliance support for economic warfare after 1953, the steel pipe encounter should be viewed more as an isolated instance of the effective exercise of U.S. leverage than as indicative of an enduring U.S. capability. Moreover, the resulting political costs to the United States in its relations with Britain and West Germany likely outweighed the economic harm inflicted on the Soviet Union. While Soviet efforts were hampered by the denial of wide-diameter pipe, the Friendship Pipeline was eventually completed and the NATO resolution was formally lifted in 1966.

The pipe dispute also sharply raised, for the first time, West European suspicions of U.S. economic motives in the utilization of East-West trade controls. The allies were well aware of the interests of U.S. oil firms in keeping West European markets clear of Soviet oil. More important, shortly after the passage of the NATO pipe resolution, the United States

81. Adler-Karlsson, *Western Economic Warfare*, pp. 131–32, and Jentleson, *Pipeline Politics*, pp. 118–23. See also Alan P. Dobson, "The Kennedy Administration and Economic Warfare against Communism," *International Affairs* (London) 64 (Autumn 1988), pp. 599–616.
82. Stent, *From Embargo to Ostpolitik*, p. 107.

consummated a major sale of surplus wheat to the Soviet Union.[83] The historical parallel between this sequence of events and that of 1982 is striking, to the point of U.S. officials contending in both instances that wheat exports absorbed Soviet economic resources, whereas pipe exports and oil imports created them.

The Emergence of a Linkage Strategy

In the aftermath of the steel pipe conflict, it was all the more apparent to the Kennedy administration that U.S. East-West trade policy required adjustment. The president asked the State Department to provide a comprehensive review of U.S. policy and to devise a strategy for the long run. The task, performed by the Policy Planning Council under Walt Rostow, was completed in July 1963.[84] Because this lengthy review captured the logic and substance of the strategy preferred by the Kennedy and Johnson administrations and was eventually implemented by the Nixon administration, it is worth considering in some detail.[85]

As its initial premise, the Policy Planning Council accepted that the Soviet bloc was largely self-sufficient and that bloc trade with the Western allies was "beyond the power of the U.S. to affect" (p. 5). More important, it rejected the "resource release" argument, contending that increased Western trade had virtually no impact on the strategic posture of the USSR.

> In theory, trade with the West could ease the resource allocation problem of the U.S.S.R. and in this sense facilitate its effort to maintain or increase military expenditures. In view of the insignificance of the trade to the U.S.S.R. in aggregate terms, and the overriding priority the U.S.S.R. accords to military expenditures, *the strategic importance of this factor must be heavily discounted, if not completely disregarded.* Soviet military expenditures are most importantly influenced by the attempt to keep up with the U.S. Both recently, and repeatedly in the past, Soviet leadership has demonstrated its willingness to accelerate mili-

83. See Dobson, "The Kennedy Administration and Economic Warfare," pp. 613–14.

84. U.S. Department of State, Policy Planning Council, "U.S. Policy on Trade with the European Soviet Bloc," memorandum prepared for the president, July 8, 1963, reprinted in the *Declassified Documents Quarterly* (1976), item 289A. Page numbers are cited in the text.

85. The review is also analyzed in Dobson, "The Kennedy Administration and Economic Warfare," pp. 610–12; Jentleson, *Pipeline Politics*, pp. 126–28; and Philip J. Fungiello, *American-Soviet Trade in the Cold War* (Chapel Hill: University of North Carolina Press, 1988), pp. 144–47. As Fungiello demonstrates (pp. 125–31), the lengthy paper produced by Rostow's staff was not the first in the Kennedy administration to call for the liberalization of export controls and other East-West trade measures. A paper prepared by Kennedy's transition team had made many similar arguments; it was never made public because of the administration's concern over the likely reaction in Congress. Undersecretary George Ball and Rostow produced a paper along similar lines in mid-1961, but their recommendations were overtaken by the Berlin crisis and Vienna Summit.

tary expenditures irrespective of requirements in other fields. (p. 35, emphasis mine)

This conclusion is critical: in contrast to official U.S. thinking in the 1950s, yet in line with the implications of the 1961 CIA study, the Policy Planning Council in effect conceded that increasing the volume of Western exports to the Soviet Union—presumably, even of capital goods and technology—did *not* constitute a threat to the military security of the United States or the Western alliance.

Economically and strategically, the Soviet Union was impervious either to unilateral U.S. economic warfare or to trade liberalization. The Policy Planning Council asserted that "neither full access to, nor complete denial of, trade with the U.S. can affect the Soviet capability to wage war . . . or in any meaningful sense, the performance or potential of the Soviet economy" (p. 4). The United States was a potential alternative supplier, but not a unique one. Yet if bilateral trade was economically inconsequential to the Soviets, the same held true for the United States. U.S. trade controls did not bring great harm to the U.S. economy, and even if the United States expanded its trade to match the level of its allies, "the whole affair would continue to be of minor economic significance" (p. 7).

At the same time, the Policy Planning Council was convinced that trade liberalization could bring the United States significant *political* benefits. If carried out properly, the adjustment of trade controls would give U.S. officials effective leverage in political bargaining with the Soviets. Going even further, it suggested that trade adjustment could be the most important bargaining instrument the United States possessed (p. 86).

The analysis thus far poses something of a puzzle. If U.S. trade was economically insignificant to the Soviets and could easily be replaced by that of the allies, how could trade liberalization be an effective bargaining lever? Why would the Soviets make political concessions to obtain trade that they could procure elsewhere?

The Policy Planning Council found the resolution of this paradox in the *symbolic*, rather than economic, significance of U.S. trade controls and liberalization. For the Soviets, trade restrictions symbolized the overall unwillingness of the United States to treat them with the political respect to which their status as a great power entitled them. To be the target of trade discrimination was a political insult, a suggestion that the Soviets were second-class citizens in the international community. In the history of postwar U.S.-Soviet relations, "no Soviet leader, in either private or public discussions of problems related to improvement of international relations, has failed to point to 'U.S. discriminatory trade practices' as proof of U.S. hostility and as a bellwether of U.S. unwillingness to seek 'a genuine relaxation of tensions'" (p. 57). Khrushchev, seeking normalization during his visit to the United States in 1957, rejected a U.S. offer to make selective trade concessions. The issue, he contended, was discrimination per se; for the United States merely to offer the Soviets equipment to make "shoes and sausages" was an insult to the Soviet people (p. 61).

The Soviet Union attached an importance to normal trade relations out of all proportion to the economic stakes involved. It would, the Policy Planning Council believed, be willing to pay a significant political price for the elimination of U.S. trade restrictions. Moreover, if the United States desired overall political normalization, it had no choice but to inject trade issues into the bargaining process. The Soviets equated "peaceful relations" with "trade relations" and "would consider the issue the decisive test of U.S. intentions in a negotiating process" (p. 77).

In addition to the relaxation of export controls, the United States could take two further steps. First, it could offer the use of official export credits, which the Soviets considered "a natural and necessary corollary of better trade relations" (p. 96). Second, it could grant the Soviets most-favored-nation (MFN) status, the sine qua non of normal trade in the postwar era. The Policy Planning Council warned, however, that the United States should not attempt to extract an explicit political quid pro quo for the removal of its trade restrictions. The Soviets would consider this an affront. Rather, U.S. officials should make clear that their willingness to liberalize trade would be part of a series of negotiations aimed at the resolution of basic political differences (pp. 66–67).

Three crucial points in the analysis merit emphasis. First, the policy review stressed that U.S. East-West trade policy would remain an instrument of statecraft. The primary objective would shift, however, from an effort to weaken Soviet economic capabilities to an attempt to influence Soviet political behavior. Second, given the Soviet attitude that economic discrimination was political discrimination, U.S. officials, in effect, were forced to use trade as an instrument of statecraft. If it desired political détente, the United States would have to initiate economic détente as well. Third, because the source of U.S. trade leverage was symbolic rather than economic, the effectiveness of a tactical linkage strategy, at least initially, was not dependent on the cooperation of U.S. allies. Even though the allies could serve as a satisfactory alternative economically, the Soviet Union was interested in *American* trade. Only the United States, a great power in its own right, could, through more normal economic relations, confer great-power status on the Soviet Union.

The Policy Planning Council recommended that the United States enter a government-to-government trade agreement with the Soviets. Trade was politically too important to be left exclusively to private U.S. firms on the one hand and to Soviet state agencies on the other (p. 12).

By keeping trade relations at a high-priority level, the U.S. government would maximize its bargaining power. U.S. firms would, in essence, be used as instruments of U.S. foreign policy. To ensure that trade provided an ongoing source of leverage, the trade agreement would be negotiated for only one year at a time. Each year, then, U.S. officials could use the negotiations for the renewal of the agreement "to pressure the Soviets to cease undesirable behavior and to encourage desirable developments" (p. 98). Rather than as a "one-shot deal," the link between trade policy and foreign policy would be maintained as a long-run strategy.

Because trade restrictions were symbolically significant to the United States as well as to the Soviets, the Policy Planning Council believed that U.S. officials should not offer a trade agreement until the Soviets had taken a concrete step toward a viable détente. Trade denial was a weapon of the cold war, a symbol of U.S. political and moral resolve; simply to liberalize trade without commensurate Soviet political concessions could not be justified to Congress or the public. Initiating trade merely for the sake of economic benefits would appear to be a "sellout." Thus trade concessions could only be made once the process of "cold war movement" had begun (pp. 82–83).

The Policy Planning Council worried about the implications of adopting a tactical linkage strategy for CoCom. Reducing U.S. controls to the same level as those of the allies, and maintaining CoCom intact, might not be sufficient to satisfy the Soviets. Discrimination was the key issue, and to the Soviets, "discrimination and CoCom are indissolubly linked" (p. 94). CoCom was a clear symbol of U.S. discrimination in civilian trade. The Soviets were willing to trade with the allies, despite their participation in CoCom, because they considered CoCom to be a creation of the United States.

The Policy Planning Council regarded the maintenance of a multilateral strategic embargo to be essential. If the allies would agree, the United States could consider moving alliance export control policy to NATO. If they could not, "the U.S. would have no choice but to insist upon the continuation of CoCom" (p. 96). The position of the Policy Planning Council was clear: the strategic embargo could not be sacrificed in the interest of tactical linkage, and the administration must find a way to realize the objectives of both, simultaneously.

CoCom, of course, was not the only or even the primary constraint operating on both the Kennedy and Johnson administrations. Tactical linkage required the cooperation of U.S. firms, which were to carry out the trade, and of the Congress, which was needed to remove the legislative shackles on the use of trade liberalization as a political instrument.

The Predicament of U.S. Business

U.S. business, one would expect, would be the natural ally of the administration. Its economic interests suggested that it would support trade liberalization in order to compete with other Western firms for its share of the expanding Eastern market. Yet despite the growth of the Eastern market and the trade performance of the allies, until the late 1960s, U.S. business did not place sustained pressure on the government to relax restrictions. Before that time, even in cases where export licenses were easily obtainable, firms were often reluctant to exploit trading opportunities with the East. One analyst noted in 1966 that although the U.S. government was seeking to "unleash" American business on the Communist world, business was "hardly tugging at the leash."[86]

Business inhibitions were due in part to the fact that, despite the attitude of the Johnson administration, a considerable segment of the American public still considered trade with the East to be politically and morally repugnant. These feelings reflected the general tendency among Americans to view U.S.-Soviet relations as an irreconcilable struggle between good and evil; they were also fueled during the mid-1960s by the perception that the Soviet Union and Eastern Europe were providing the weapons of war to the North Vietnamese. Trade with the East continued to be widely regarded as trade with the enemy.

Prominent U.S. firms that overtly engaged in trade with the East risked becoming the victim of organized public disapproval, which could include a boycott of the domestic sales of their products. Conservative political groups, such as the Young Americans for Freedom (YAF), were effective at mobilizing anti-trade sentiment. When it discovered that IBM had sold computers to Eastern Europe and intended to continue, the YAF organized a national "STOP-IBM" project, which included the picketing and harassment of IBM offices locally and of the worldwide meeting of corporate representatives who purchased from IBM.[87] Similarly, it organized a campaign to harass the American Motors Corporation (AMC), simply because a high company official informed a reporter that AMC might be interested in selling autos to the East, even though it had no present plans to do so.[88] Not surprisingly, many U.S. firms that contemplated significant involvement in East-West trade decided that "potential economic gains are outweighed by the possibility of adverse effects on their public relations."[89]

86. Nathaniel McKitterick, *East-West Trade: The Background of U.S. Policy* (New York: Twentieth Century Fund, 1966), p. 29.
87. See the statement of Randall C. Teague, YAF, in *East-West Trade*, 90th Cong., 2d sess., pp. 256–78.
88. Ibid., pp. 994–98.
89. Pisar, *Coexistence and Commerce*, p. 82.

A well-known case of public pressure involved the Firestone Rubber Company, a case particularly important because it concerned an attempt by the Johnson administration to employ tactical linkage. In 1964, the administration proposed to reward Rumania's independent foreign policy initiatives by allowing a U.S. firm to construct a synthetic rubber plant in that country. Firestone expressed an interest in the $50 million deal and late in 1964 signed a preliminary contract with the Rumanians. The deal incited the YAF, which picketed company headquarters in Akron, Ohio, with signs reading "Firestone is building the future war machine of Communism" and which convinced a number of dealers who sold Firestone products to cancel their contracts. In April 1965, Firestone suddenly announced it had terminated negotiations, frustrating both its economic interest and the linkage attempt of the Johnson administration.[90]

The predicament of business—trapped between the pro-trade sentiment of the administration, a judgment it shared, and the anti-trade sentiment of the Congress and public—shaped its posture in the ongoing domestic debate. Corporate officials who were willing to speak publicly in favor of expanded East-West trade generally stressed its potential political benefits. Rather than defending trade on the grounds of economic opportunity, they followed the State Department in emphasizing the utility of trade as an instrument to extract political concessions from, and to bring political change to, the East.[91]

This tendency of business was nowhere more apparent than in the highly publicized Miller Report of 1965.[92] Earlier that year, President Johnson had appointed a special committee to explore the issue of expanding economic ties with the Soviet Union and Eastern Europe. Two-thirds of the committee's twelve representatives, including its chairman, J. Irwin Miller, were high officials in U.S. industry. The remaining members represented labor and academics. Not surprisingly, the committee advocated an expansion of peaceful U.S. trade with the East. It dismissed commercial considerations, however—the potential profits to be made in Eastern trade or the fact that U.S. firms were at a competitive disadvantage vis-à-vis Europe and Japan—as inconsequential. Instead, its argument focused on the use of trade as a political weapon: "Commercial considerations have not been the determining factor in framing

90. Ibid., pp. 84–85. See also McKitterick, *East-West Trade*, pp. 44–47. There is some suspicion that the boycott was encouraged by Firestone's competitor, Goodyear. Goodyear officials publicly proclaimed their disinterest in the deal by stating that "you can't put a price tag on freedom." McKitterick, p. 46.

91. See, for example, the report of Donald Hornig, December 9, 1964, reprinted in *Declassified Documents Quarterly* (1981), item 89A. Hornig led a small delegation of U.S. businessmen to the Soviet Union, November 5–19, 1964.

92. See J. Irwin Miller et al., "Report to the President of the Special Committee on U.S. Trade Relations with East European Countries and the Soviet Union," hereafter cited as *Miller Report* (Washington, D.C.: White House, April 29, 1965).

U.S. policy on this subject and should not be now. . . . trade is a tactical tool to be used with other policy instruments for pursuing our national objectives. . . . we can use trade to influence the internal evolution and external behavior of Communist countries."[93]

In order to advocate increased trade, U.S. firms felt compelled to mimic publicly the position of the Kennedy and Johnson administrations. To the extent that both advocated trade liberalization, business and government interests were complementary; in the long run, however, their interests would prove to be in conflict. Unlike in Western Europe and Japan, where increased trade had the official blessing of governments for *economic* (as well as political) reasons, in the United States, trade was a hostage to politics. Politics could lead to trade liberalization but could just as easily justify a return to trade denial. In the sense that politics led economics, tactical linkage would not be the abandonment of economic warfare but a continuation of it by other means.

The Intransigence of Congress

Congress proved to be the major constraint on the implementation of tactical linkage during the 1960s. While the Executive enjoyed considerable discretion over export control policy, congressional authority was required to utilize the other relevant instruments of trade policy—MFN status and export credits. Until 1969, Congress was unwilling to grant that authority, leaving the Kennedy and Johnson administrations with inadequate room to maneuver.

The sentiment prevailing in Congress until 1968 was that the United States should not adjust its trade policy, regardless of the position of the allies. Congressional attitudes were rooted in a vision of the cold war as enduring and intractable. While the Executive viewed trade as part of a movement toward détente, many members of Congress saw it as an important symbol of U.S. determination to confront Communist ideology and aggression. What the Executive regarded as "bridge building," members of Congress considered "aiding the enemy," particularly while the conflict in Vietnam escalated. Although the Executive believed that polycentrism in the Communist world justified overtures to Eastern Europe, members of Congress conceived of Communism as monolithic and considered trade with any Communist state to be immoral.[94]

These attitudes shaped both congressional initiatives in East-West trade and responses to those of the Executive. In 1962, Congress revised

93. *Miller Report*, pp. 3, 5, 8. According to Edward Fried, who served as the executive secretary of the committee, the State Department was actively involved in its deliberations and helped to shape the final report. Conversation with Fried at the Brookings Institution, March 1, 1983.
94. See, for example, the statement of Senator Mundt in *East-West Trade*, 88th Cong., 2d sess., p. 31.

the Export Control Act, calling for restrictions on exports that would make a significant contribution to the military or economic potential of the Communist states. The word "economic" was explicitly added; while the Kennedy administration contemplated moving away from economic warfare, Congress reaffirmed its commitment to it. The more restrictive wording, which in fact reflected long-standing U.S. policy, was a response to the escalation of East-West tensions, the rapid economic growth of the Soviets, and Khrushchev's claim that the socialist world would overtake the capitalist world economically.

Although the directive may have inhibited it to some extent, the Executive retained the discretion to determine which exports would make a significant contribution, militarily or economically. In its public response to the directive, the Commerce Department stressed the importance of dealing with exports of economic significance on a case-by-case basis and suggested that in some cases it might be advantageous to the national security and welfare of the United States to approve such exports. Commerce went even further and interpreted an additional congressional declaration—that the United States should "use its economic resources and advantages in trade with Communist-dominated nations to further its national security and foreign policy objectives"—as authorizing the Executive to vary the scope and severity of export controls to particular countries.[95] Rather than treating the 1962 amendments as a call to renew economic warfare, the Executive took the opportunity to defend the legitimacy of a tactical linkage strategy. Executive officials continued the practice of differentiating in its export controls to Communist countries, and in 1966, the Johnson administration used its discretion to remove four hundred items from the U.S. control list, a signal of its desire to encourage peaceful trade with Eastern Europe, Congress notwithstanding.[96]

Unfortunately for them, executive officials enjoyed less discretion over MFN status and export credits, both of which were required for tactical linkage to be employed effectively. In 1962, Congress removed executive discretion in selectively granting MFN status to Communist countries. The Trade Agreements Extension Act of 1951 had given the president the authority to grant MFN to those states determined not to be "dominated by the world Communist movement." The Truman administration made such a determination for Yugoslavia in 1951 and the Eisenhower administration for Poland in 1960. The Trade Expansion Act of 1962, however, required denial of MFN to the products of any country "dominated or controlled by Communism."[97] A rider attached to the bill

95. *Export Controls*, 61st Quarterly Report (Third Quarter 1962), pp. 4–5.
96. *Export Controls*, 79th Quarterly Report (First Quarter 1967), p. 5.
97. Thomas A. Wolf, *U.S. East-West Trade Policy: Economic Warfare vs. Economic Welfare* (Lexington, Mass.: Lexington Books, 1973), p. 70.

permitted the continuation of MFN to Poland and Yugoslavia, but the president was denied the right to extend it to other East European states, despite President Kennedy's request that discretion be maintained. The initiative removed one of the key weapons in the Executive's arsenal. In terms of immediate impact, the Johnson administration had planned to extend MFN to Rumania, both as a positive signal and to enable the Rumanians to export to the United States in order to pay for the U.S. exports that were to be allowed as part of the overall strategy.[98]

With regard to export credits, the Executive had been allowed to offer official guarantees (though not actual credits) to East European countries, provided that it reported each instance to Congress. This authority was utilized to grant $145 million of medium-term guarantees, mostly to Yugoslavia.[99] In 1966, however, the Johnson administration's attempt to provide $50 million in guarantees for U.S. firms participating in the Fiat-Soviet project met with strong opposition. Congress amended the Export-Import Bank Act in order to prohibit support of the Fiat deal and effectively to remove presidential discretion entirely. The legislation unconditionally prohibited export-import guarantees (and actual credits) to any nation that provided direct government-to-government aid to North Vietnam. All East European nations, with the exception of Yugoslavia, provided such aid.

It was evident to the Johnson administration that removing the shackles imposed by Congress would require a major legislative initiative. The formation of the Miller Committee in 1965 was intended to prepare Congress and the public for such an initiative by giving outside respectability and credibility to the administration's position. The Miller Committee strongly recommended that the Executive be given the power to grant MFN, extend credits, and liberalize exports that did not enhance Soviet military capability.[100]

As a follow-up to the Miller Report, the administration prepared in the summer of 1965 what it hoped would become the East-West Trade Relations Act of 1966. Given the strong opposition in Congress and the fact that the decision to escalate the Vietnam War was made at the same time, it is difficult to imagine the passage of such legislation. Clearly, if it was to have any chance at all, it would require the active personal support of the president. Johnson's interest in trade liberalization was unambiguous. By the end of 1965, however, he apparently decided that to push such a controversial initiative would endanger support for his "Great Society" program, which was making its way through Congress at

98. See Edward Sklott, "The Decision to Send East-West Trade Legislation to Congress, 1965–66," in *Commission on the Organization of the Government for the Conduct of Foreign Policy*, appendix: vol. 3 (Washington, D.C.: GPO, 1976), pp. 76–77.
99. Wolf, "U.S. East-West Trade Policy," p. 97.
100. *Miller Report*, pp. 18–19.

that time.[101] Lacking the full political force of the president, the East-West Trade Relations Act died a quick and predictable death.

Breaking the Stalemate: The Export Administration Act of 1969

The first step toward the resolution of this impasse came with the passage of the Export Administration Act (EAA) of 1969. Unlike its predecessor, the Export Control Act of 1949, the 1969 act explicitly advocated the expansion of peaceful trade with the Soviet Union and Eastern Europe. Support for the bill in Congress was far from overwhelming: a majority in the House and a vocal minority in the Senate initially opposed its passage.[102]

The arguments put forth in the debate were familiar.[103] Opponents of trade liberalization contended it would demonstrate the weakness of U.S. resolve and would have a detrimental impact on the morale of both U.S. forces fighting Communism in Southeast Asia and nationalist reformers fighting it in Eastern Europe. They questioned the morality of trading with Communist countries which were supplying a Communist government that was "shooting at Americans."[104] Proponents of trade countered by arguing that "we must, in our own interest, stop limiting our trade merely for the momentary moral glow we may derive from an act of self-denial."[105] Opponents argued that trade would release resources, allowing the Soviets to enjoy both guns and butter; supporters claimed unilateral U.S. controls had no impact on the Soviet economy, in light of allied trade.

Ultimately, the proponents of increased trade prevailed. The key to their victory was that both political and economic circumstances had come to favor trade expansion. The inception of U.S.-Soviet détente, suggested by the signing of the nonproliferation treaty, the agreement to begin SALT, and the U.S. commitment to wind down the Vietnam War, increased the bargaining power and decreased the political vulnerability of trade proponents. At the same time, many in the Senate had become sympathetic to the plight of business and, more important, had begun to conceive the problem as a national rather than a private one. As the U.S. position on balance of payments and balance of trade eroded, the strong

101. Sklott, "The Decision to Send East-West Trade Legislation to Congress," p. 73.
102. Office of Technology Assessment, *Technology and East-West Trade* (Washington, D.C.: GPO, 1979), pp. 115–17.
103. Contrast, for example, the statements of Strom Thurmond and George Ball in *East-West Trade*, 90th Cong., 2d sess., pp. 4–19 and 29–44. For a sense of the argument of opponents of liberalization, see the brief statements made by House members in *East-West Trade*, House hearings, 90th Cong., 2d sess., pp. 1–72.
104. Statement of Congressman Ben Blackburn in *East-West Trade*, House hearings, 90th Cong., 2d sess., p. 27.
105. Statement of George Ball in *East-West Trade*, 90th Cong., 2d sess., p. 32.

conviction that the United States must become more competitive in international trade spread.[106] Given its past growth and future potential, the United States could no longer afford to ignore Eastern markets.

With support in Congress, U.S. industry could afford to focus explicitly on the economic dimensions of the issue. In public testimony before Congress, industry officials focused their criticisms on the lengthy U.S. control lists, on the delays and uncertainty of the licensing process, and on the general lack of support they received from government, as compared to their counterparts in Western Europe and Japan.[107] Although they still spoke of the political benefits, businessmen were less reluctant to stress the potential profits of expanded trade. It is somewhat ironic that it was the economic arguments, earlier dismissed by the Executive as inconsequential, which ultimately helped to motivate the Congress to act in accordance with the political position of the Executive.

The 1969 act clearly shifted the legislative thrust of U.S. policy in the direction of trade promotion. It eliminated the requirement of the Executive to restrict exports that would make a significant contribution to the economic potential of the Communist states, and it added the following findings: "The unwarranted restriction of exports from the U.S. has a serious adverse effect on our balance of payments. The uncertainty of policy toward certain categories of exports has curtailed the efforts of American business in these categories to the detriment of the overall attempt to improve the trade balance of the United States."[108] The act directed the secretary of commerce to review the Commodity Control List and to revise it so that it might conform more closely to the lists observed by the allies. In making export licensing decisions, the existence of foreign availability was to be given greater weight; U.S. firms would no longer be denied the sale of items available elsewhere in the West. The Executive was directed to consult the business community on the determination of the control list and to keep firms informed as to the status of their individual license requests. In short, the EAA called for the creation of a business-government alliance, one that would allow the United States to compete more effectively with Western Europe and Japan in the drive for Eastern markets.

The EAA of 1969 marked the formal termination of the U.S. attempt to use economic warfare as a weapon of the cold war. Unable to define alliance strategy in the 1950s, the United States was forced to adjust its own policy in the 1960s. The adjustment mandated by the EAA of 1969,

106. Statement of Senator Walter Mondale, who sponsored the more liberal legislation, in ibid., p. 26.
107. See the statements of Hugh Donahue, Control Data Corporation; A. R. Fredriksen, Minnesota Mining and Manufacturing Co.; and the letters and statements submitted for the record by other representatives of industry in *East-West Trade*, 90th Cong., 2d sess., pp. 149–62, 296–306, and 993–1133.
108. Export Administration Act of 1969 (PL 91–184), sec. 2.

however, did not occur immediately. As we shall see, the Nixon administration did not exercise its authority to liberalize trade until it became politically prescient to do so. Just as economic warfare was initiated as part of the broader U.S. political strategy of the cold war, trade liberalization became part of the broader strategy of détente. The transition from economic warfare was not to "normal" trade but to trade in the service of politics, that is, tactical linkage. For the United States, although the content of East-West trade policy may have changed, political objectives continued to be paramount. The nature of the U.S. adjustment and its consequences for the competitive position of American firms and for the effectiveness of CoCom are the subject of Chapter 5.

CHAPTER FIVE

Tactical Linkage, Export Competition, and the Decline of CoCom's Strategic Embargo

The 1970s witnessed a transformation in America's approach to the Soviet Union. In this chapter I explore that transformation's political and economic dimensions and its implications for CoCom. First, I trace the shift in U.S. export control policy from economic warfare to tactical linkage and examine the efforts of executive officials, in both the Nixon and the Carter administrations, to use the expansion of U.S. trade and technology transfer as an instrument for influencing Soviet behavior. I show that the sources of potential leverage were different in the two instances, but that in both the effectiveness of tactical linkage was eventually undermined, in the former by domestic fragmentation and in the latter by the existence of alternative suppliers.

Second, I argue that because U.S. trade expansion was primarily driven by political, as opposed to economic, considerations, U.S. firms entered the Western competition for Eastern markets at a significant long-term disadvantage. U.S. firms did well in the early years of détente, yet by the middle of the decade, they watched the Soviet Union turn increasingly to more reliable suppliers in Western Europe and Japan. This situation led to a business-government (and intragovernment) conflict in the United States and highlighted the persistence of important differences in East-West trade policy between the United States and its Western allies, despite the fact that each was ostensibly committed to a policy of trade expansion.

In the third section, I examine the implications for CoCom of these developments in U.S. policy. I argue that CoCom became less effective and more controversial, mainly because the United States failed to perform its leadership responsibilities adequately. The ability of the United States to do so was constrained, in part, by the erosion of the overwhelming technological advantages it had enjoyed previously relative to other

Western states. Of greater importance, however, was the lack of a clear U.S. policy commitment to CoCom. The United States was inconsistent. On the one hand, it remained the CoCom member relatively most concerned with the maintenance of strategic export controls. On the other, U.S. officials allowed the pursuit of foreign policy (and, at times, economic) objectives in export control policy to take precedence over, and become entangled with, their efforts to uphold a viable strategic embargo. Consequently, the United States proved incapable either of acting credibly as the conscience of CoCom or of minimizing the administrative and economic burdens of participation in the control system.

THE RISE AND DEMISE OF TACTICAL LINKAGE

Although the Export Administration Act of 1969 represented a clear victory for proponents of trade liberalization, it was not until the latter part of 1972 that the U.S. government began in earnest the process of fully normalizing trade relations with the Soviet Union. In crafting the EAA, Congress granted the Executive considerable discretion with regard to the pace and extent of liberalization. The Nixon administration exercised its discretion to the fullest and refused to allow trade progress to precede an improvement in overall political relations. In effect, Nixon and Kissinger followed the recommendation made by the State Department in 1963 that trade concessions should be part of a larger movement of political normalization. The administration considered economic objectives to be of secondary importance. Trade was a political instrument; or, as Kissinger put it, "expanding trade without a political quid pro quo was a gift; there was little the Soviets could do for us economically."[1]

There was, however, much the Soviets could do for the United States politically. Trade concessions could be used selectively to modify Soviet behavior, and the full normalization of economic relations desired by the Soviets could be retained as a reward, to be granted only after the administration was satisfied that its major political objectives had been fulfilled. For Nixon and Kissinger, the granting of export credits and MFN status, and the relaxation of export controls, provided leverage to assure Soviet cooperation across a broad range of political issues.

The primary objectives sought by the administration concerned Vietnam and arms control. By 1969, the United States was prepared to extricate itself from Vietnam, preferably in a manner that minimized the loss of international prestige. The Soviets could assist by pressuring their North Vietnamese allies to be accommodating at the negotiating table in Paris. An arms control agreement would serve the administration's inter-

1. Henry A. Kissinger, *White House Years* (Boston: Little, Brown, 1979), p. 152.

ests by helping to stabilize the nuclear relationship and to placate the demands of the American public and Congress for reductions in defense spending. Additionally, the Nixon administration sought Soviet cooperation in maintaining stability in the Middle East and Europe, particularly Berlin. In accordance with Kissinger's concept of linkage, negotiations in each of these areas proceeded simultaneously, with progress in one being used to reinforce progress in others.

Initially, Nixon and Kissinger resisted private and governmental pressure for what they considered to be the premature opening of economic relations. In May 1969, the president went on record as being opposed to the passage of the EAA until the Soviets agreed to enter negotiations on a range of bilateral problems. Similarly, Kissinger claims to have fought off pressure from the Department of Commerce, which sought to remove items expeditiously from the Commodity Control List (CCL) that were unilaterally controlled by the United States, and to grant U.S. firms export licenses to sell computers to the Soviet Union.[2] The administration also initially declined to request MFN status for the Soviets. Its reluctance to encourage trade was perhaps best demonstrated in April 1970, when the Ford Motor Company announced it was considering an offer from the Soviets to play a major role in the construction of a truck factory on the Kama River. The administration sharply criticized the idea, and shortly thereafter Ford announced it had rejected the Soviet offer, citing public opposition from the secretary of defense as an important factor in the company's decision.[3]

Economic progress was made gradually, with American concessions granted selectively as U.S.-Soviet political cooperation improved. Despite pressure from Congress, the Commerce Department, and the firms involved, the administration refused for two years to allow gear-cutting machinery, worth $85 million, to be shipped to the Soviet Union. After denying three separate license requests, the administration acquiesced in the deal only after a major compromise was reached in the SALT negotiations.[4] Similarly, sales of equipment for Soviet auto and truck factories were licensed in the context of a breakthrough in the Berlin negotiations. In August 1971, $162 million of exports for the manufacture of

2. Ibid., p. 154. The administration also sought to prevent Britain's International Computers Limited (ICL) from exporting computers to the Soviets, by exercising its veto in CoCom over ICL's exception requests. This attempt aroused the indignation of the British government, and the United States eventually backed down. See Gary Bertsch and Steven Elliott, "Controlling East-West Trade in Britain," in Bertsch, ed., *Controlling East-West Trade and Technology Transfer: Power, Politics, and Policies* (Durham: Duke University Press, 1988), p. 210.

3. Henry R. Nau, *Technology Transfer and U.S. Foreign Policy* (New York: Praeger, 1976), pp. 94–95.

4. Kissinger, *White House Years*, p. 840, and "Chronological Summary of Soviet-American Commercial Relations, 1969–74," hereafter cited as *Chronology*, in U.S. Congress, House, Committee on Foreign Affairs, *Detente: Prospects for Increased Trade with Warsaw Pact Countries*, report, 93d Cong., 2d sess., October 24, 1974, p. 31.

light trucks was approved; in September, the Four Power Berlin Agreement was signed, in which the Soviets guaranteed the unimpeded flow of traffic between West Germany and West Berlin.[5] In general, Nixon and Kissinger favored large projects that would take time to complete and would thus offer the United States a continued source of leverage.[6]

In discussions immediately before the May 1972 Nixon-Brezhnev Summit, Brezhnev promised the United States that he would pressure the North Vietnamese, who had been stalling, to make a serious effort to negotiate a settlement.[7] At the summit, the SALT I treaty was signed, and Nixon and Brezhnev also agreed to begin consultations on a European security conference. In this atmosphere of substantial political achievement, the administration proved willing to move from making selective trade concessions to initiating the fundamental transformation of the bilateral economic relationship. The two leaders established a joint U.S.-Soviet Commercial Commission to conduct the negotiations that would lead to a comprehensive bilateral agreement.[8]

The Nixon administration refused to finalize the trade agreement until it saw tangible results from Brezhnev's promise to cooperate over Vietnam. When the Viet Cong indicated in September that they would accept a cease-fire, the administration was satisfied.[9] The United States also insisted that the Soviets repay their long-standing lend-lease debts; agreement was finally reached for a token repayment of $722 million.[10]

The trade agreement, signed on October 18, 1972, called for the mutual granting of MFN status and reciprocal access to the export credit facilities of each side.[11] It made provisions for the establishment of a U.S. government commercial office in Moscow and a Soviet trade representative in Washington. The Soviets also agreed to permit U.S. firms to establish offices in Moscow. Both sides committed themselves to the tripling of bilateral trade over a three-year period, and the Soviets explicitly stated their expectation that "during the period of effectiveness of this Agreement, foreign trade organizations of the U.S.S.R. will place substantial orders in the U.S.A. for machinery, plant and equipment, agri-

5. *Chronology*, p. 31.
6. Kissinger, *White House Years*, p. 840.
7. Ibid., pp. 1146–47, 1153.
8. *U.S.-Soviet Commercial Agreements, 1972: Texts, Summaries, and Supporting Papers* (Washington, D.C.: U.S. Department of Commerce, January, 1973), pp. 1–3.
9. Paula Stern, *Water's Edge: Domestic Politics and the Making of Foreign Economic Policy* (Westport, Conn.: Greenwood Press, 1979), pp. 28–30, 46.
10. The United States initially demanded $2.6 billion. See *U.S.-Soviet Commercial Agreements*, p. 103, and John P. Hardt and George D. Holliday, *U.S.-Soviet Commercial Relations: The Interplay of Economics, Technology Transfer, and Diplomacy*, print, prepared for U.S. Congress, House, Committee on Foreign Affairs, Subcommittee on National Security Policy and Scientific Developments (Washington, D.C.: June 10, 1973), p. 55.
11. *U.S.-Soviet Commercial Agreements*, pp. 75–77, 88–95.

cultural products, industrial products and consumer goods produced in the U.S.A."[12]

After signing the agreement, the Nixon administration finally relaxed its unilateral export controls, as mandated by the EAA of 1969. In October 1972, the CCL contained 550 unilaterally controlled categories of items; by May 1973, only 73 categories remained.[13]

The Nixon administration's preference for tactical linkage was consistent with and driven by America's political, economic, and military relationship with the Soviet Union as it evolved in the late 1960s and early 1970s. Politically, the shift from the confrontational posture of the cold war to a more competitive orientation (i.e., one that recognized the Soviets as a worthy negotiating partner, although the relationship remained a largely adversarial one) provided a context within which the administration could deploy national economic assets in pursuit of political objectives. As Kissinger recognized, that the bilateral economic relationship was deemed important by the Soviets translated into potential leverage for a U.S. administration willing to politicize it. The use of trade as a political instrument was all the more attractive because the United States remained minimally dependent on the Soviets economically and thus relatively insensitive to the potential economic costs of tactical linkage. Finally, the stabilization of the military relationship, symbolized by the SALT and ABM agreements, helped to mitigate the strategic risks of whatever trade expansion did occur in the process of implementing tactical linkage. As I discussed in Chapter 2, the strategic risks of trade expansion tend to be greater in an atmosphere of intensified military competition than when such competition among potential adversaries is stable or declining.

With regard to the sources of U.S. leverage, the Soviets hoped to gain both economic and political (i.e., symbolic) benefits from the bilateral trade relationship. They sought access to American grain, technology and manufacturing equipment, and export credits. With the possible exception of grain, however, Western Europe and Japan could fulfill most Soviet requirements. As was demonstrated throughout the 1970s, other CoCom members were willing and able to provide the industrial plants and equipment, on generous credit terms, desired by the Soviets. U.S. firms, it was often argued at the time, were better prepared than their West European counterparts to engage in industrial projects on the massive scale preferred by the Soviets.[14] Although this reasoning may

12. The quote is from Article 2, #4 of the Treaty. Ibid., p. 89.
13. Statement of Arthur Hausman, President, Ampex Corporation, in U.S. Congress, Senate, Committee on Banking, Housing, and Urban Affairs, Subcommittee on International Finance, *The Role of the Export-Import Bank and Export Controls in U.S. International Economic Policy*, hereafter *Ex-Im Hearings*, hearings, 93d Cong., 2d sess., April 2, 5, 10, 23, 25, and May 2, 1974, p. 352.
14. Hardt and Holliday, *U.S.-Soviet Commercial Relations*, pp. 30–31.

have been true and the Soviets may have preferred in some instances to contract large projects with U.S. firms, it is also apparent that groups of West European or Japanese firms, working collectively, could perform such tasks adequately.

Western Europe and Japan could not, however, use their trade relationship to confer superpower status on the Soviet Union. Only the United States, itself a superpower, enjoyed this capability. As I argued in Chapter 4, a normal trade relationship with the United States was of profound symbolic importance to Soviet leaders. They viewed it as an acknowledgment of their legitimate position as a great power, as an *equal* to the United States. The importance of equality to the Soviets was emphasized by Kissinger; in comparing his experiences with Chinese and Soviet leaders, he noted the powerful inferiority complex of the latter.

> Equality seemed to mean a great deal to Brezhnev. It would be inconceivable that Chinese leaders would ask for it—if only because in the Middle Kingdom tradition it was a great concession granted *to* the foreigner. To Brezhnev it was central. . . . While he boasted of Soviet strength, one had the sense that he was not all that sure of it. Having grown up in a backward society, nearly overrun by Nazi invasion, he might know the statistics of relative power but seemed to feel in his bones the vulnerability of his system.[15]

Soviet dependence on the United States for the symbolic benefits of a trade relationship had two important implications. First, to utilize that dependence for political purposes, the United States did not require the cooperation of its allies. Given that leverage was derived primarily from the political, rather than economic, needs of the Soviets, the United States was, in this sense, a unique supplier. U.S. leverage could not be dissipated by the economic relationship cultivated by Western Europe and Japan with the Soviet Union. Not surprisingly, U.S. economic initiatives in the early 1970s were taken unilaterally, with no attempt to coordinate with other CoCom members the use of trade promotion for political purposes.

Second, the symbolic importance of the trade relationship was the foundation for its substantive development. Once they were convinced that the United States was no longer discriminating against them economically, the Soviets would be willing to expand their trade links with U.S. firms significantly. If, as was anticipated, U.S. firms could indeed outcompete other Western suppliers, then the United States would develop over time a source of leverage based not on symbolism but on concrete economic dependence. Moreover, without the symbolic foun-

15. Kissinger, *White House Years*, p. 1141.

dation provided by the trade agreement, the Soviets were likely to turn away from U.S. firms, even if they were more competitive. Thus, the symbolic dimension of the economic relationship not only had intrinsic utility but also was necessarily the source of any future leverage the United States might have over the Soviet Union after the euphoria associated with the opening of economic relations had faded away.

Congress and Tactical Linkage: The Jackson-Vanik Amendment

Herein lies the significance of the Jackson-Vanik Amendment to the Trade Reform Act of 1974. In 1971, Congress delegated to the Executive the authority to grant export credits to a Communist country if the president determined that to do so would be in the national interest.[16] Nixon made such a determination for the Soviet Union when the trade agreement was completed in October 1972. The granting of MFN status, however, still required explicit congressional approval. The Nixon administration requested it for the Soviet Union early in 1973, as part of its proposals for comprehensive trade reform.[17] In Congress, however, opponents of détente in general and of Soviet human rights policy in particular sought to withhold MFN, and export credits, until the Soviets made clearly stated public concessions on emigration. What the Nixon administration had promised the Soviets as a reward for their cooperation in 1971–1972 and had expected to use to establish an American presence in the Soviet market, Congress threatened to take away by employing tactical linkage for its own purposes.

The Jackson Amendment, cosponsored by seventy-two senators, was introduced in October 1972.[18] It prohibited the extension of credits or MFN to nonmarket economies that restricted or taxed emigration by their citizens. Similar legislation, sponsored by Charles Vanik, was proposed in the House. In an apparent effort to accommodate Congress, in April 1973 the Soviet Union suspended its exit fee for Soviet Jews leaving for Israel. Brezhnev also made informal pledges to Nixon and Kissinger that large, unspecified numbers would be permitted to emigrate.[19] At the same time, Soviet leaders publicly stated that explicit attempts to link trade relations to Jewish emigration would be looked on with disfavor and could jeopardize détente. Congress was neither impressed by the Soviet concessions nor deterred by its threats. In December 1973, the House passed its version of the trade act with the amendment attached.

16. Hardt and Holliday, *U.S.-Soviet Commercial Relations*, pp. 55–56.
17. Stern, *Water's Edge*, pp. 60–61.
18. For a comprehensive discussion of the politics of the amendment, see ibid., chaps. 2 and 3.
19. *Chronology*, p. 37.

The Nixon administration fought hard to prevent passage in the Senate. It argued that the amendment embarrassed the United States, which had already promised MFN to the Soviets without strings attached; that linkage should be reserved to influence Soviet foreign, rather than domestic, policy; and that "quiet diplomacy" was more likely to be effective in achieving the objectives of Congress. To demand explicit concessions on such an "internal" issue would be considered a grave insult by the Soviets and viewed as a violation of their sovereignty. More to the point, it ran directly counter to Nixon and Kissinger's attempt to maximize the symbolic value of the trade relationship. Throughout 1973–1974, as the patience of Soviet leaders wore thin, the administration sought to assure them that it intended to follow through on its commitment of October 1972.

Following the failure of a last-minute compromise attempt by Kissinger, the Senate passed the Trade Reform Act, along with the Jackson-Vanik Amendment, in December 1974.[20] When President Ford signed it early in 1975, the Soviets announced they would not abide by its conditions, thereby making it impossible for Congress to grant them MFN or access to export credits. More important, the Soviets abrogated the entire trade agreement of 1972, contending that the emigration provision was in violation of it.[21] Without the symbolic benefits of the trade agreement, the incentive of the Soviets to cooperate with the United States politically, or favor American firms economically, was significantly reduced. For all practical purposes, the success of the Jackson-Vanik Amendment signaled the defeat of the Nixon-Kissinger tactical linkage strategy.

Rather than representing the isolated opposition of human rights activists, the Jackson-Vanik Amendment reflected a more general sense of uneasiness, in the U.S. Congress and public, regarding the Nixon administration's East-West economic strategy. By 1973 it appeared to the American public that détente meant the Soviets were given concrete and immediate economic benefits, and in return, the United States received vague promises of "cooperative" Soviet political behavior. Moreover, administration officials frequently emphasized that significant U.S. political benefits would be seen primarily in the long run.[22] They claimed that

20. Kissinger asked the Soviets to pledge privately to liberalize emigration and for Congress to review Soviet performance after eighteen months. Neither side accepted the arrangement. See Stern, *Water's Edge*, pp. 161–63, and the statement of Kissinger in U.S. Congress, Senate, Committee on Finance, *Emigration Amendment to the Trade Act of 1974*, hereafter *Emigration Hearings*, 93d Cong., 2d sess., December 3, 1974, pp. 78, 89, 101–2.

21. Hertha W. Heiss, Allan J. Lenz, and Jack Brougher, "United States-Soviet Commercial Relations since 1972," in *Issues in East-West Commercial Relations*, A Compendium of Papers submitted to U.S. Congress, Joint Economic Committee, 95th Cong., 2d sess., January 12, 1979, p. 195.

22. See the statements of Ex-Im Bank director William Casey and counselor to the Department of State Helmut Sonnenfeldt, *Ex-Im Hearings*, pp. 75 and 498–500. As discussed below, while U.S. officials practiced tactical linkage, they also embraced the ideas and rhetoric of structural linkage, namely, that the intensification of a positive trade relationship leads in the long run to structural changes in the target's domestic and foreign policy. See Chap. 2, above.

in the short run, clashes of interest between the superpowers might still take place, and the United States should be prepared to tolerate it without necessarily resorting to trade denial. To employ trade denial early might destroy the fragile foundation of tactical linkage before it was firmly established. The American people, who traditionally viewed the U.S.-Soviet relationship as a simple struggle between good and evil and trade denial as a critical weapon in that struggle, were now being asked to accept a complex and rather subtle strategy that treated the Soviet Union as both collaborator and competitor, without a clear dividing line between the two.[23] The transition was difficult, and with so much having been promised by the rhetoric of détente, the risk of unfulfilled expectations was great.

Tactical linkage was particularly difficult to justify in that it appeared to offer the Soviets extraordinary economic benefits, as suggested most blatantly by the so-called Great Grain Robbery of 1972.[24] Between July and September of that year, the Soviets purchased one-quarter of the U.S. wheat crop, valued at $750 million. Upon entering the market, Soviet buyers managed to conceal their intention to make such a massive total purchase. Because American grain dealers were collectively unaware of Soviet intent, the price offered the Soviets was only $1.63 per bushel. By September, the market price had been driven up to $2.49 per bushel. The difference between selling and market price was provided to U.S. exporters in the form of a $300 million government subsidy. American consumers, however, found themselves paying higher food prices, as total demand eventually outran supply. The United States, virtually a monopoly supplier, appeared to relieve an important Soviet economic bottleneck, an action in the interest of U.S. grain dealers but at the expense of the Treasury and the U.S. consumer.

Many in Congress also viewed with suspicion proposals for long-term U.S.-Soviet cooperation in the energy sector. U.S. firms were to invest approximately $9 billion to develop, as one of the cornerstones of détente, the natural gas resources of Siberia. Two pipelines would be built: one would carry gas from Siberia to the Soviet port of Murmansk, from where it would be shipped to the East coast of the United States (the North Star project); the other would carry gas to the port of Nakhodka, for shipment to Japan and the U.S. West coast (the Yakutsk project).[25]

23. This line of argument is developed by Alexander George, "Domestic Constraints on Regime Change in U.S. Foreign Policy: The Need for Policy Legitimacy," in Ole Holsti et al., *Change in the International System* (Boulder, Colo.: Westview Press, 1980), pp. 233–62.

24. Hardt and Holliday, *U.S.-Soviet Commercial Relations*, pp. 67–69, and I. M. Destler, *Making Foreign Economic Policy* (Washington, D.C.: Brookings Institution, 1980), chap. 2.

25. Statement of Jack Ray, executive vice-president, Tennessee Gas Transmission Company, and Donald R. Mueller, Yakutsk project director, El Paso Natural Gas Company, in U.S. Congress, Senate, Committee on Foreign Relations, Subcommittee on Multinational Corporations, *Multinational Corporations and United States Foreign Policy*, pt. 10, hearings, 93d Cong., 2d sess.,

U.S. firms would supply tankers and pipeline equipment, the sale of which would be financed by official and private export credits. The Soviets made it clear that long-term financing was essential to the viability of the projects. Repayment would begin only when the pipelines were completed and would take the form of gas deliveries to the United States. The agreement was for twenty-five years, beginning in 1980.

Given the massive scale of the project, critics in Congress wondered whether American energy, capital, and expertise could be better invested elsewhere. They also voiced concern about excessive U.S. dependence on Soviet energy and were skeptical that Soviet gas deliveries would continue, or that Soviet debts would be repaid, if political relations took a turn for the worse. By supplying equipment and credits up front, the United States appeared to bear all the risks associated with the projects.[26] Disapproval was sufficiently strong that Congress, as part of the 1974 renewal of Export-Import Bank lending authority, placed severe restrictions on the use of government financing for Soviet natural gas deals.[27]

Opposition to the granting of official export credits was not confined to Soviet energy deals. Rather than see them as a necessary instrument for U.S. firms to compete with their West European counterparts, congressional critics viewed credits as a form of economic aid that permitted the Soviets to increase their imports without having to make domestic sacrifices. In addition to the energy restrictions, the renewed Export-Import Bank legislation placed an overall ceiling of $300 million on additional credits to the Soviet Union. Furthermore, by making MFN status a prerequisite for access to U.S. government credit facilities, the legislation effectively prevented the Soviets from borrowing at all.[28]

In exchange for what was seen as economic beneficence, congressional critics demanded to know exactly what the United States had gained, or would gain, politically. Although the administration may have been able to point to some diplomatic success in 1971–1972, by 1973–1974 it had become difficult to respond to such questions. The war in Vietnam dragged on inconclusively, Soviet and U.S. interests clashed in the Middle East, and conservatives in Congress believed the SALT treaty to be of dubious value. In these circumstances, the arguments of Kissinger and other administration officials that trade was essential to protect the "process" of détente were not very persuasive.[29]

June 17, 19, and July 17–19 and 22, 1974, pp. 3–90. See also Bruce W. Jentleson, *Pipeline Politics* (Ithaca: Cornell University Press, 1986), chap. five.

26. See, for example, the comments of Senator Church, *Multinational Corporations and United States Foreign Policy*, pp. 1–2 and passim.

27. Heiss, Lenz, and Brougher, "U.S.-Soviet Commercial Relations since 1972," p. 195.

28. Stern, *Water's Edge*, pp. 156–57.

29. See the comments of Kissinger in *Emigration Hearings*, p. 106, and those of Casey and Sonnenfeldt in *Ex-Im Hearings*, pp. 75 and 497–500.

Given the general sentiment in Congress, the broad appeal of the Jackson-Vanik Amendment is easily understandable. For the Soviets to make explicit and tangible political concessions, on emigration or any other political issue, was the least that should be done to justify the economics of détente. Yet what Congress regarded as necessary assurances, the administration viewed as a diplomatic sledgehammer that destroyed the fragile foundation of the tactical linkage strategy.

During 1975–1976, the Ford administration made a number of attempts to revive the normalization of economic relations; these included a presidential letter to four key congressional committees urging a reconsideration of East-West trade legislation.[30] Any possibility of congressional acquiescence was destroyed by election year opposition to détente by presidential aspirant Ronald Reagan and by Soviet and Cuban intervention in the Angolan civil war. In the face of Soviet misbehavior, Kissinger lamented that Congress had left the Executive with neither the carrot nor the stick as a means of influence.[31] Without MFN or credits, the Soviets had no economic incentive to moderate their foreign policy, and the administration could remove neither as punishment.

Tactical Linkage Revisited: The Carter Administration

In part to distinguish his approach from that of his predecessor, President Carter came to office claiming to have abandoned the use of linkage as a foreign policy tool. By early 1978, however, in response to public discontent over Soviet behavior, the administration reversed itself. The impetus for tactical linkage came from the White House and, specifically, the office of National Security Adviser Zbigniew Brzezinski.[32]

The clearest public statement of the administration's "economic diplomacy" was made by Samuel Huntington, a member of the NSC staff in 1977–1978.[33] Huntington criticized the Nixon administration for its laissez-faire approach to U.S.-Soviet economic relations, which, he claimed, permitted the flow of economic benefits to continue even when Soviet behavior did not warrant it. He argued that the United States must adopt a strategy of "conditioned flexibility" rather than one of laissez-faire or pure denial. This strategy required U.S. officials to "review and revise existing lists of embargoed goods and technologies so as to require, on foreign policy grounds, validated licenses for those items of machinery and technology for which the Soviets have a critical need, and for which they are largely dependent on U.S. supply."[34] Such items

---

30. Heiss, Lenz, and Brougher, "U.S.-Soviet Commercial Relations since 1972," pp. 196–97.
31. Henry Kissinger, *Years of Upheaval* (Boston: Little, Brown, 1982), pp. 225, 998.
32. John A. Hamilton, "To Link or Not to Link," *Foreign Policy*, no. 44 (Fall 1981), p. 135.
33. Samuel Huntington, "Trade, Technology, and Leverage: Economic Diplomacy," *Foreign Policy*, no. 32 (Fall 1978), pp. 63–80.
34. Ibid., p. 76.

could be sold to the Soviets, but only conditionally; the United States needed to be prepared to deny their export as a direct and immediate response should the Soviets misbehave.

A crucial difference between the Nixon-Kissinger and Carter administration concepts of tactical linkage lies in the presumed source of U.S. leverage. The former sought, at least in part, to exploit the symbolic dependence of the Soviets on U.S. trade, whereas the latter relied on their substantive dependence. Huntington suggested that trade could be an effective weapon, given the economic needs of the Soviets and the technological superiority of the United States.[35] If the Carter administration was to pursue tactical linkage, however, in reality it had very little choice. Because Congress could not be convinced to reconsider MFN status and export credits, the symbolic dependence of the Soviets simply could not be exploited.

Reliance on the substantive economic dependence of the Soviets meant that along with possible domestic opposition, the pursuit of tactical linkage might also confront international constraints. Although the United States may have been a unique supplier of symbolic benefits, it was only one of many potential suppliers of economic benefits. Thus, to be effective, tactical linkage required multilateral coordination. Given their economic stake in the Soviet market, however, other Western suppliers were not likely to cooperate. At the time of the Jackson-Vanik Amendment, America's major allies made it clear that they did not condone the use of trade denial for political purposes.[36] Huntington sought to evade the international constraint by urging that U.S. foreign policy controls be focused on those export items for which the United States enjoyed a virtual monopoly, such as advanced computers and oil and gas exploration equipment. As U.S. business leaders argued, however, even if U.S. products had quality advantages in certain areas, there was little the United States possessed that could not also be supplied from Western Europe or Japan.[37]

The Carter administration unveiled its tactical linkage strategy in the summer of 1978, in response to Soviet intervention in the Ethiopian-Somalian dispute and to Soviet human rights abuses. It placed energy exploration and production equipment on the CCL, which meant a validated license was required prior to its export. Such items were considered particularly critical because a CIA assessment, released in 1977, claimed that Soviet energy would significantly decrease unless advanced

35. Ibid., pp. 64–65.
36. For the German reaction, for example, see OTA, *Technology and East-West Trade* (Washington, D.C.: GPO, 1979), p. 175.
37. See the statements of business officials in U.S. Congress, Senate, Committee on Banking, Housing, and Urban Affairs, *Use of Export Controls and Export Credits for Foreign Policy Purposes*, hearings, 95th Cong., 2d sess., October 10–11, 1978.

methods of extraction were adopted.[38] The administration also retracted a validated license already granted to the Sperry Corporation to export a computer system to the Soviets. The system was intended to enhance the data processing capabilities of TASS, the Soviet press agency, during the 1980 Moscow Olympics.

In the Sperry case, the Carter administration requested the cooperation of its major allies. All were reluctant and expressed disapproval both of the request and of the attempt to inject short-term political considerations into ongoing trade relationships. As the London daily *The Guardian* commented, "President Carter has angered the French and Germans, and embarrassed the British, by what his allies regard as the clumsy handling of the Soviet computer deal."[39] In order to avoid a conflict with the United States, the British government attempted to sidestep the issue by claiming that no British firm was willing to supply the computer. The French and Germans were more forthright and denied the U.S. request.[40] A French Foreign Ministry official stated that "it is not the French practice to subordinate the sale of civilian industrial material to political considerations."[41] The German Foreign Ministry publicly stated that it would never link exports to domestic developments in the USSR.[42] By early in 1979, TASS had signed a contract with Honeywell-Bull and Thomson of France for equipment to replace the Sperry order.

The West Europeans were similarly against imposing controls on the export of energy equipment and were more than willing to meet Soviet demand.[43] That the West Europeans were willing to step in increased the opposition of U.S. business to sanctions, because oil and gas technology was one of a very few areas in which U.S. firms had a clear advantage over their Western competitors in supplying the Soviet market. As a result of business pressure, supported by the Commerce Department, not one of over one hundred licenses requested by U.S. exporters to sell such equipment was denied after the new controls went into effect.[44]

---

38. CIA, *Prospects for Soviet Oil Production* (Washington, D.C., 1977). The CIA findings are discussed in Richard F. Kaufman, "Western Perceptions of Soviet Economic Trends," Staff study prepared for the U.S. Congress, Joint Economic Committee, March 6, 1978, pp. 3–4.
39. *Guardian* (London), August 17, 1978.
40. Ibid., and *Frankfurter Allgemeine Zeitung*, July 21, 1978.
41. *International Herald Tribune*, July 17, 1978.
42. OTA, *Technology and East-West Trade*, p. 183, and Angela Stent Yergin, *East-West Technology Transfer: European Perspectives*, (Beverly Hills, Calif.: Sage, 1980), p. 33. The West German firm Siemens asked to bid on the computer deal but was refused by the West German government, which sought to avoid a diplomatic dispute with the United States.
43. The United States made no attempt to coordinate its controls. See the statement of Commerce Secretary Kreps in U.S. Congress, Senate, Committee on Banking, Housing, and Urban Affairs, Subcommittee on International Finance, *U.S. Export Control Policy and the Extension of the Export Administration Act*, pt. I, 96th Cong., 1st sess., March 5, 6, 1979, pp. 56–57.
44. Ibid., p. 66.

Many of the licenses were subject to long delays, however, and in at least one case, a Soviet order, valued at over $200 million, was switched from a U.S. to a West European firm because of the inability of the former to secure an export license promptly.[45]

Because the Soviets experienced little difficulty in acquiring Western computing systems or energy technology, the Carter administration's 1978 foreign policy controls did not provide leverage. As the Afghanistan sanctions of 1980 would reaffirm, the Carter tactical linkage strategy proved to be ineffective in influencing Soviet behavior.[46] Unlike the Nixon-Kissinger strategy, however, it was undermined primarily by international, as opposed to domestic, factors.

Export Competition
and Government-Business Relations

The energy equipment and Sperry computer cases demonstrated the growing conflict between U.S. business and the government over East-West trade policy. U.S. firms with an interest in Eastern markets approached détente with high expectations for profits. Yet they found their objectives frustrated by both the government's failure to streamline the export control process and its manipulation of trade for political purposes.

The conflict was not immediately apparent: by 1971 the Nixon administration's interest in laying the foundation for détente was compatible with the interest of business in entering the Soviet market. The administration was a forceful advocate of an active business role; it considered the relationship between private firms and Soviet officials to be significant in initiating and sustaining the momentum of détente. Private firms were, in effect, used as instruments of statecraft. The administration gave its full support to large-scale industrial projects, including those in the energy sector which alarmed Congress. During the Nixon-Brezhnev Summit of 1973, the president went as far as to urge publicly that firms work out specific proposals for the energy projects, and he indicated that the administration would give serious and sympathetic consideration to the financing of them.[47]

45. Statement of Marshall Goldman, professor of economics, in U.S. Congress, House, Committee on International Relations, Subcommittee on International Economic Policy and Trade, *Extension and Revision of the Export Administration Act of 1969*, hereafter *Extension and Revision*, 96th Cong., 1st sess., February 15, 1979, p. 47.

46. It has been suggested that the U.S. Afghanistan sanctions were probably effective in signaling American intentions and helping to deter future aggressive behavior by the Soviets. See David Baldwin, *Economic Statecraft* (Princeton: Princeton University Press, 1985), pp. 261–78.

47. Statement of Robert Ingersoll, Department of State, in *Multinational Corporations and United States Foreign Policy*, p. 109.

The Nixon administration also encouraged firms to sign science and technology agreements with the Soviet Union, so as to provide a framework for on-going industrial cooperation.[48] Many of the largest, most technologically advanced American companies complied, including Bendix, Boeing, Control Data, General Electric, Hewlett-Packard, and Lockheed. Official export credits were also made available, and between January 1973 and June 1974, sixteen loans were approved by the Ex-Im Bank, totaling $469 million.[49] The White House kept close watch over the activities of the bank, and in at least one instance President Nixon intervened personally to override congressional opposition and ensure that a $180 million loan to the Soviets was approved.[50]

West European firms viewed American entry into the competition for the Soviet market with apprehension. Their fear of being outbid was justified by the size and resources of U.S. firms, the apparent Soviet preference for American technology and management, and the political commitment of the U.S. government. The potential for the United States to dominate trade was demonstrated between 1971 and 1973, when it jumped from last place to first among the major Western states in total exports to the Soviet Union. Although the great majority of U.S. exports were agricultural, sales of industrial machinery and equipment also increased significantly. Between 1971 and 1974, U.S. firms landed major contracts worth $400 million for chemical fertilizer manufacturing equipment and distribution facilities; $350 million to equip the Kama River truck plant; $200 million for four ammonia plants; $88 million for iron ore pelletizing plants; $80 million for the Moscow trade center; and $47 million for a friction ball bearing plant.[51]

The success of U.S. firms was short-lived, however. Although the United States continued to dominate in the sale of agricultural products, by 1975, Soviet imports of machinery and equipment had turned decisively to Western Europe and Japan.[52] The decline in the relative position of the United States was a result of the decline of U.S.-Soviet détente and Soviet retaliation for the Jackson-Vanik Amendment.

Although no U.S. manufacturing firms were heavily dependent on the Soviet market, those firms that had invested time, effort, and money were determined to protect their market share. They may have entered the market as the beneficiaries of politics; once involved, they opposed having their position held hostage to politics. By 1975, business officials were increasing pressure on the government to undertake the trade

48. See "Basic Principles of Relations," appendix 1 in Hardt and Holliday, *U.S.-Soviet Commercial Relations*, pp. 81–83.
49. A complete list is provided in *Emigration Hearings*, pp. 60–62.
50. *Chronology*, p. 42.
51. *Emigration Hearings*, pp. 60–62, and Heiss, Lenz, and Brougher, "U.S.-Soviet Commercial Relations since 1972," p. 191.
52. Michael Kaser, "Soviet Trade Turns to Europe," *Foreign Policy*, no. 19 (Winter 1975), 123–35.

Table 5. Western trade with the Soviet Union, 1970–1979 ($ million)

Country	1970	1971	1972	1973	1974	1975	1976	1977	1978	1979
Exports										
U.K.	248	222	225	237	264	510	540	508	978	1236
F.R.G.	356	484	689	1037	1838	2700	2627	2366	2880	3483
France	319	313	423	610	718	1110	1224	1226	1426	1826
Japan	345	396	523	506	1027	1738	1820	1959	2317	2522
Italy	313	291	413	714	1093	933	1072	1256	1316	
U.S.[a]	115	144	557	1390	747	2027	2657	1711	2340	3793
Total Trade										
U.K.	713	672	673	971	1176	1329	1634	1809	2232	2905
F.R.G.	604	776	1010	1691	3002	3984	4176	4026	5072	6868
France	459	529	657	980	1244	1797	2251	2339	2655	4003
Japan	724	815	983	1351	2224	2665	2813	3118	3385	3968
Italy	524	550	559	843	1504	1977	2352	2553	2884	3288
U.S.	179	204	649	1577	981	2217	2921	2080	2714	4328

Source: IMF, *Direction of Trade Yearbook,* 1970–1980.

[a]The fraction of U.S. exports that consisted of agricultural products was 79% in 1972; 77% in 1973; 49% in 1974; 62% in 1975; 64% in 1976 and 1977; and 75% in 1978. See Heiss, Lenz, and Brougher, "United States-Soviet Commercial Relations Since 1972," p. 191.

promotion measures necessary for U.S. firms to compete effectively. Exporters had the strong support of the Commerce Department, but the government as a whole was unwilling to take decisive action. Thus, the latter half of the 1970s witnessed a persistent conflict between business and government and, within the government, over East-West trade. On some issues business leaders allied with the Executive, or parts of it, against the Congress and on others, with parts of the Congress against the Executive.

The conflict with Congress focused on export credits. Given the limited ability of the Soviets to generate hard currency, firms that could rely on official export financing, at subsidized interest rates, could put together a more attractive overall package and thus had the best chance to land Soviet contracts for large-scale industrial projects. Each of the major projects referred to above was financed in part by official export credits.[53] In 1974, Soviet trade minister Patolichev warned U.S. officials that if credits were cut off, the Soviets would favor West European and Japanese producers.[54] The threat apparently was carried out: in 1976, U.S. treasury secretary William Simon testified that approximately $700 million in industrial contracts which would have gone to U.S. firms were instead channeled to Western Europe.[55] A Soviet trade official went even further, claiming that $1.6 billion in contracts, for the 1976–1980 Five-Year Plan, was diverted from U.S. suppliers. A Commerce Department survey taken in 1975 of two hundred U.S. firms engaged in East-West trade found that 80 percent claimed to have lost significant business as a direct result of credit restrictions.[56] Nevertheless, Congress maintained its restrictions on credits, its link between credits and MFN status, and its refusal to grant MFN to the Soviets without public concessions on the emigration issue.

The lack of credits was most damaging in the competition for large-scale projects. For discrete sales of machinery and equipment, however, the ability to secure an export license was a more important factor. Export licensing was the focus of business's conflict with the Executive. As one industry representative put it in 1979, "As currently structured and administered, our export licensing procedures are a bureaucratic horror . . . : self-defeating, unfathomable to many exporters, often capricious and unclear, and characterized by delay, uncertainty, and lack

53. In the usual arrangement, the Ex-Im Bank financed 45 percent of a project, private banks 45 percent, and the importer 10 percent. When official credits were made available, private banks were usually more forthcoming with regard to their share.

54. *Chronology*, p. 40.

55. U.S. Congress, Senate, Committee on Commerce, *America's Role in East-West Trade: Problems and Prospects, 1976–1980*, hearings, 94th Cong., 1st and 2d sess., December 11–12, 1975 and January 30 and February 4, 1976, p. 8.

56. Ibid., pp. 23–24. See also the statement of Donald Kendall, *Ex-Im Hearings*, pp. 275–76, 291–92, and William Verity, chairman of Armco, "Taking Politics Out of Trade with the Soviet Union," *New York Times*, January 2, 1979.

of accountability."[57] Despite the significant reduction of 1973, the U.S. CCL remained longer than the CoCom list.[58] Also, the United States proved more likely than other CoCom members to deny the export of an item on the list.

Far more important, however, was the problem of licensing delays. The U.S. government took significantly longer than other Western governments to determine whether or not an export license would be granted in any particular case.[59] Delays, and the resulting uncertainty that pervaded the licensing system, were considered by numerous U.S. firms to be the most important constraint on their ability to compete effectively.[60] Because U.S. firms could not guarantee the completion of a contract or the delivery of equipment in a timely, predictable fashion, Eastern importers found it more convenient to deal with Western Europe or Japan. Many smaller U.S. firms that lacked the resources necessary to absorb the costs of obtaining a license were altogether deterred from attempting to compete.[61]

It should be kept in mind that the number of cases subjected to significant delays was relatively small, representing only about 3 percent of total license requests.[62] This low amount is due to the fact that most requests were for exports to the non-Communist world. In 1978, for example, the Department of Commerce processed over 65,000 cases, and only 2,400 were for exports to the East. Requests for exports to non-Communist destinations were typically processed within one week, in accordance with the commitment of the United States to prevent East-West export controls from becoming a serious obstacle to intra-Western trade. Requests for exports to the East, although only a small percentage of total requests, made up the great majority of delayed cases. And, the number of cases delayed increased over time. In 1976, the number pending for over ninety days was 689; by 1978, it had grown to 1,988.[63]

There were two major reasons for the delays and uncertainty that characterized the licensing process. The first concerned the nature and

57. Statement of J. Karth, president, American League for Exports, in *Extension and Revision*, p. 205.

58. Statement of Elmer Staats, GAO, in *U.S. Export Control Policy and Extension of the Export Administration Act*, pt. I, p. 9. As of March 1979, the CoCom list contained 105 categories of items, whereas the U.S. list, over 32 percent longer, contained 143.

59. Statement of Thomas Christiansen, Hewlett-Packard, in *Ex-Im Hearings*, pp. 360, 411, and OTA, *Technology and East-West Trade*, pp. 138–41.

60. For a sampling of business complaints, see the statements of business officials in *Use of Export Controls and Export Credits for Foreign Policy Purposes*, and in U.S. Congress, House, Committee on International Relations, Subcommittee on International Economic Policy and Trade, *Export Licensing: CoCom List Review Proposals of the United States*, 95th Cong., 2d sess., June 14 and 26, 1978.

61. Statement of Edward Loeffler, National Machine Tool Builders Association, in *Export Licensing*, pp. 13–14, and statement of Donald Kendall, *Ex-Im Hearings*, pp. 388–89.

62. OTA, *Technology and East-West Trade*, p. 139.

63. Ibid.

Table 6. Volume of export license applications

Fiscal year	Applications
1969	145,369
1970	132,498
1971	107,615
1972	78,561
1973	64,070
1974	65,883
1975	52,600
1976	54,359
1877	58,967
1978	63,476
1979	71,067
1980	75,960

Source: 1969–1978, *Technology and East-West Trade*, p. 140; 1979–1980, *Proposed Legislation to Establish an Office of Strategic Trade*, pp. 109–11. From 1969 to 1974, the Department of Commerce counted applications received; after 1975, they counted applications processed.

capabilities of the administrative system. The Commerce Department found it difficult to process quickly a rapidly growing number of license requests during the latter half of the 1970s. As Table 6 indicates, total requests decreased dramatically in the 1969–1975 period as a result of the sharp reductions in the CCL. From 1975 to 1979, however, the number increased yearly and at a faster rate than the rate of increase in the number of Department of Commerce personnel processing those requests.[64] The problem was compounded by the fact that as more trade came to involve advanced technology, the cases themselves became more complex and thus took longer to process.

Equally important, the Commerce Department was not the only agency involved in the licensing process. State, Defense, the CIA, and, during the Carter administration, the NSC staff, the Department of Energy, and the Office of Science and Technology Policy—all played a role in licensing decisions. The more actors involved, the longer it took to reach decisions in complicated cases, and the fact that different agencies had their own bureaucratic interests increased the uncertainty and inconsistency associated with the process. For example, although the Commerce Department was largely sympathetic to business, the Defense Department during the early 1970s continued to evaluate requests according to the criteria of the cold war.[65] A study by the General Accounting Office

64. Ibid., pp. 139–40.
65. See Robert Klitgaard, "Limiting Exports on National Security Grounds," in *Commission on the Organization of the Government for the Conduct of Foreign Policy*, vol. IV, pt. VII (Washington, D.C.: GPO, 1975), p. 445.

(GAO) in 1978 found that too many agencies were involved, licensing responsibilities were overly diffused, and the result was a serious inability of the system to respond to the needs of exporters.[66] The GAO recommended centralizing and streamlining the process, but the executive agencies involved were unwilling to contemplate fundamental organizational changes.

Second, delays and uncertainty were a direct result of the government's attempt to use trade for political purposes. The ideal requirements of tactical linkage ran directly counter to those of export promotion. To maximize export promotion required the shortest possible control list, along with predictability and efficiency in the licensing process. Firms needed a clear idea of whether a license was likely to be granted before they spent the time, money, and effort to cultivate a market and obtain contracts. Once they made the investment, they were best served by prompt licensing decisions so that customers were not kept waiting. In contrast, tactical linkage required a lengthy control list; in this way the government could maximize its ability to interfere with trade if political circumstances warranted. From the perspective of the exporter, licensing decisions could not be predictable or consistent, because they were conditional on the political behavior of the target. The same license request might be denied in one set of political circumstances, yet approved in another. Finally, tactical linkage required the ability to delay license requests so that trade might be "held hostage" pending changes in the target's political behavior. Potential leverage would be constrained if an administration were forced, in order to satisfy the needs of exporters, to make its decision within an established time frame.

Business officials generally expressed harsh criticism of the use of export controls for foreign policy purposes. The sharp reaction of the business community to the 1978 extension of the CCL, as well as to the injection of the NSC staff into the licensing process, provides a clear example.[67] The frustration of corporate officials was compounded by the knowledge that when the United States employed foreign policy controls, it did so unilaterally.

Attempts by exporters, and their supporters in Congress, to make the licensing system more accommodating found their way into legislation throughout the 1970s.[68] The 1972 amendments to the Export Administration Act called for the establishment of technical advisory commit-

66. GAO, *Administration of U.S. Export Licensing Should Be Consolidated to Be More Responsive to Industry* (Washington, D.C.: GPO, October 31, 1978).
67. See *Use of Export Controls and Export Credits for Foreign Policy Purposes.*
68. A summary of the relevant legislation is found in OTA, *Technology and East-West Trade*, pp. 112–26.

tees (TACs), which were to be comprised of industry and government specialists in high-technology sectors.[69] The TACs were intended to facilitate closer consultation between industry and government on the construction and revision of the control lists. Congress also ordered the secretary of commerce to remove from unilateral U.S. control all commodities freely available outside the United States in sufficient quantity and quality unless the absence of controls would damage national security. This action was important to exporters in that it legitimized the use of "foreign availability" as an appropriate justification for the granting of a validated license.

The 1974 amendments addressed the problem of delays by determining that licensing decisions must be reached within ninety days of the receipt of an application. If they were not, the applicant was to be provided in writing with an explanation for the delay and the estimated time before a decision would be reached. In 1977, the amendments heightened the importance of foreign availability by directing the president to negotiate with non-U.S. suppliers prior to adopting unilateral controls. Where foreign availability existed, the burden was to be placed on the government wishing to control rather than on the firm wishing to export. The 1977 amendments also strengthened the TACs by allowing them to participate in the multilateral as well as the national list review process. Congress, however, refused to require the government to accept the recommendations of the TACs, despite business complaints that their proposals were being ignored.[70]

The EAA was revised again in 1979, in the wake of the Carter administration's extension of foreign policy controls. The concerns of business were reflected in the act's declaration that a policy objective of the United States was to minimize uncertainty in export controls. Before employing foreign policy controls, the president was directed to consider the likely effect of the controls on the U.S. economy; the reaction of other countries; the probability that such controls would be effective, given such factors as foreign availability; and the ability of the U.S. government to enforce the controls. No such conditions were attached to the use of national security export controls, a clear indication of the attempt by Congress to distinguish the two and to make it more difficult for the president to use trade for foreign policy purposes.

The triumph for proponents of export promotion as represented by the EAA of 1979 was short-lived. Only a few months after its passage, foreign policy controls were utilized and national security controls were

69. Ibid., p. 119. Eight TACs, in the following sectors, were established: semiconductors, semiconductor manufacturing and test equipment, numerically controlled machine tools, telecommunications equipment, computers, computer peripherals, and electronic instrumentation.
70. Statement of Thomas Christiansen, *Export Licensing*, pp. 61–63.

tightened in response to the Soviet invasion of Afghanistan, which effectively eliminated any possibility of U.S. exporters regaining the position they had enjoyed in the early 1970s.

Western Europe and Japan: Business-State Collaboration

As during the 1960s, West European and Japanese firms continued to compete for Eastern markets without the constraints faced by their U.S. counterparts. Rather than placing obstacles, their governments generally promoted trade. Export credits at subsidized rates were made available, and the licensing process functioned predictably and efficiently, typically without major delays.[71] The primary objectives served by East-West trade were economic rather than political; government and industry worked closely to increase the national share of Eastern markets. Throughout the decade a steady flow of trade, finance, and energy ministry officials from West European countries journeyed to Moscow in search of increased trade, and the status of bilateral economic relations was usually on the agenda in high-level discussions between Soviet and West European political leaders.

That economic objectives were considered primary should not be taken to imply that West European governments considered trade with the Soviet Union to be politically insignificant. On the contrary, in accordance with the logic of structural linkage, it was generally accepted, particularly in West Germany, that positive economic relations contributed to the development and maintenance of positive political relations with the Soviets.[72] Economics was an important part of the larger détente process. At the same time, West European governments rejected the notion that their trade relationships should be dependent on a particular pattern of Soviet behavior. To use their trade position for explicit political purposes was considered "naive and unrealistic," as well as economically costly.[73] Aside from mutually agreed strategic items, West European governments considered trade with the Soviets to be "normal" trade—essentially the same as trade with any other sovereign state.

Competition for Eastern market shares was intense and took on greater significance because of the 1973 oil crisis and the deep recession in the West during 1974–1975. In the face of sluggish growth and high import bills, sales to the East could help to maintain employment and macroeconomic stability. The aggressiveness of the competition for market shares is suggested by the military metaphors used by West European press and government officials to characterize it. The West Germans were depicted as "forging ahead on the Eastern trade front," whereas

71. OTA, *Technology and East-West Trade*, chap. 9.
72. Ibid., and Stent Yergin, *East-West Technology Transfer*, pp. 15–16.
73. Kaufman, "Western Perceptions of Soviet Economic Trends," p. 11.

the French government reportedly ordered its firms to "attack to the East" in order to reestablish France's balance of payments in the aftermath of the OPEC crisis.[74]

The importance attributed to Eastern markets by Western Europe and Japan is not reflected by aggregate trade figures. Even for the most successful exporters, West Germany and Japan, exports to the Soviet Union comprised only slightly more than 2 percent of total exports. The Eastern bloc as a whole accounted for only 6 percent of West German exports and 4 percent of those of France. For certain sectors or industries, however, the East took on greater significance. In 1975, trade with Communist countries accounted for 25 percent of German machine tool orders, 12 percent of industrial plant orders, and 36 percent of steel pipe orders.[75] Some smaller machine tool firms depended on the East for over half their total sales and thus, in effect, their ability to stay in business.[76] The Office of Technology Assessment argued in 1979 that the survival of the medium-sized machine tool industry in France during the 1970s was dependent on access to Eastern markets.[77]

Given that the Soviets sought Western technology in order to modernize their leading industrial sectors, Western export competition focused on the sale of turnkey factories. Many sales went to the chemical industry, with the largest orders secured by France's Rhône-Poulenc ($1.2 billion in 1976 for plant and equipment over a ten-year period), Italy's ENI ($1 billion over five years for plant and equipment) and Montedison ($800 million in 1973 for eleven plants), and France's Technip ($500 million in 1976 for two plants). British and West German firms were also major exporters of polyethylene, ammonia, and methanol plants.[78]

The aluminum and steel sectors also benefited from Western technology. In 1976, the French concern of Péchiney-Ugine-Kuhlmann won contracts, worth close to $1 billion, to construct on the shores of the Black Sea what the Soviets claimed would be "the largest aluminum complex in the world." In 1972, a West German consortium reached an agreement, worth between $1 billion and $3 billion, to construct at Kursk the world's largest steel complex.[79] In the transportation sector,

74. *Times* (London), April 1, 1975, and *Observer Foreign News Service*, November 15, 1975.
75. *Financial Times*, September 22, 1976.
76. See the special report on German trade with the East in *International Herald Tribune*, April 16, 1975. It noted that "although East-West trade continues to make up a fraction of total German trade, Eastern Europe has become the one area where hard-pressed and recession-conscious German industry knows it can expand."
77. OTA, *Technology and East-West Trade*, p. 185.
78. A comprehensive listing for the 1970s of large-scale Western exports to the Soviet Union, including the supplier firm, the type of import, and its value, is provided by Eugene Zaleski and Helgard Wienert, *Technology Transfer between East and West* (Paris: OECD, 1980). See Table A-33, "Identified Soviet Compensation Agreements with the West," pp. 364–71.
79. On the Péchiney deal, see *Financial Times*, May 4, 1976; on Kursk, see Angela Stent, *From Embargo to Ostpolitik: The Political Economy of West German-Soviet Relations, 1955–1980* (Cambridge: Cambridge University Press, 1980), p. 223.

the Kama River Plant, also the largest of its kind in the world and the single most important project in the 1971–1975 Soviet Five-Year Plan, was expected to increase Soviet heavy truck production by 25 percent when operating at full capacity.[80]

West European firms also supplied advanced communications and information processing technology. For example, France's Thomson won a major contract in 1979 to set up a turnkey factory to manufacture computer-controlled exchanges in the USSR.[81] The deal, worth over $100 million, was the first important overseas sale for Thomson of its advanced digital switching equipment. As discussed below, the technological sophistication of the export caused it to become a major source of controversy in CoCom between the United States and France.

Finally, Western firms played a major role in the development of Soviet natural resources, particularly the exploration, production, and transmission of oil and gas.[82] Mannesmann of West Germany exported over $1.5 billion of steel pipe to the Soviets between 1970 and 1974. In order to improve the Soviets' own pipe-making capabilities, in 1979 Mannesmann and Creusot-Loire agreed to construct a seamless pipe plant, complete with welding and other special treatment equipment and based on Mannesmann's technology.[83] Britain's most important Soviet deal of the decade, valued at over $200 million, was won in 1976 by the Cobberrow consortium, which fought off stiff competition from U.S. General Electric to supply the compressor units for the Chelysbinsk pipeline project.[84]

To finance much of the above trade, France, Japan, and West Germany each extended over $3 billion in official credits to the Soviet Union; Britain, whose exports were relatively less in demand, provided slightly under $1 billion.[85] In an attempt to prevent official rates from dropping too far below market rates, in 1976, countries belonging to the Organization for Economic Cooperation and Development (OECD) agreed to increase to 7.75 percent the official rate for large-scale exports to the Soviet Union.[86] This "gentlemen's agreement" was short-lived, however, largely because of pressure exerted by the Soviets on Western firms. By the end of 1977, France and Italy conceded to the Soviets and agreed to extend credits at rates below that of the consensus. Great

80. Nau, *Technology Transfer and U.S. Foreign Policy*, p. 87, and Herbert E. Meyer, "A Plant that Could Change the Shape of Soviet Industry," *Fortune* (November 1974).
81. *Le Monde*, January 16, 1979, and *Financial Times*, March 5, 1979.
82. For a comprehensive list of energy equipment sales, see OTA, *Technology and Soviet Energy Availability* (Washington, D.C.: GPO, 1981), pp. 204–18.
83. *Financial Times*, April 11, 1979, and *Le Monde*, April 7, 1979.
84. *Financial Times*, October 7, 1976, and December 10, 1976.
85. OTA, *Technology and East-West Trade*, p. 40. The West German government provided guarantees, but generally not interest subsidies.
86. See Joan Pearce, *Subsidized Export Credit*, Chatham House Papers (London: Royal Institute of International Affairs, 1980), chap. 5.

Britain attempted to maintain the consensus, but the government buckled under the pressure of complaints brought by British firms. By mid-1978, Britain had cut its rates so as to move into line with Italy and France, despite strong domestic opposition to lower rates from Conservatives in Parliament. They contended that the British government, because it raised money on the open market, actually took a loss by subsidizing exports to the Soviets.[87] The counterargument, as demonstrated by the U.S. experience, was that the availability and rate of credit was frequently the determining factor in whether a firm successfully negotiated a large contract with the Soviets. In 1976, the Japanese reportedly lost four major deals worth $220 million because they chose to abide, while others did not, by the OECD consensus rate.[88]

Western Europe and Japan developed considerable stakes during the 1970s in the large-scale export of industrial plant and equipment to the Soviet Union, which was financed by long-term credits. Once that commitment is appreciated, it becomes easy to understand the alliance conflict that emerged after the Soviet invasion of Afghanistan, when U.S. officials made the denial of "process know-how" (i.e., turnkey plants) and export credits the centerpiece of their renewed economic warfare strategy.

The National Security Dimensions of Trade Liberalization

During the 1970s the U.S. government departed considerably from the principles that had provided the conceptual foundation for its economic warfare strategy. U.S. officials were willing to permit, even promote, trade that could contribute significantly to the Soviet economy. As we have seen, U.S. officials justified trade promotion in terms of its actual or expected political benefits. In American political discourse, however, East-West trade was still widely viewed as trade with a potential military adversary. Thus, it was incumbent upon U.S. officials to demonstrate that export promotion contributed to, or at least did not detract from, U.S. national security, which was traditionally defined in terms of military competition.

The established American worldview, which was not fully discredited by the onset of détente, had been based on three related premises: that U.S.-Soviet trade benefited the Soviet economy but was of marginal economic importance to the United States, that contributions to the Soviet economy were translated into improvements in Soviet military power,

87. *Financial Times*, May 27, 1978.
88. *Japan Times*, October 4, 1976.

and that increases in Soviet military power were detrimental to U.S. security. Given the embedded appeal of these arguments, failure to address them would have left the architects of détente vulnerable to the charge of having compromised U.S. long-term security in the interests of short-term profits and political gains. The Nixon administration, assisted by other advocates of détente, responded by challenging the traditional assumptions and providing an alternative conception of the relationship between East-West trade and U.S. security.

First, they argued that while trade clearly benefited the Soviet economy, it was no longer of only marginal importance to the United States. Trade had become a two-way street and brought mutual economic benefits. The United States had entered a new era, one characterized by the growth of interdependence, in which no state could afford not to participate fully in the international division of labor. The U.S. trade deficit, Secretary of Commerce Peterson noted, was a "melancholy reminder" that the United States no longer enjoyed a monopoly in the production of a wide range of goods and services.[89] The Nixon economic shocks of 1971 and the OPEC oil revolution of 1973 reinforced this conception of the United States as an "ordinary country" that needed to compete for export markets.

Second, administration officials challenged the traditional logic of the relationship between the Soviet economy and military. Kissinger made the argument abstractly, contending that in the contemporary era power was no longer "homogenous." He argued that "economic giants can be militarily weak, and military strength may not be able to obscure economic weakness."[90] The implication, for those willing to draw it, was that the United States did not need to be alarmed if trade contributed to Soviet economic capabilities. More frequently, the argument focused concretely on the pattern of resource allocation in the Soviet economy. Trade would release resources, proponents of détente claimed, but the benefits would be enjoyed by the Soviet consumer rather than by the military. Soviet priorities had shifted; at the Twenty-Fourth Party Congress in 1971, party leadership committed itself to investing heavily in new technology in order to modernize the civilian economy and enhance the welfare of the average citizen.[91]

Interdependence with the West would encourage these trends. Western technology, particularly if accompanied by intimate and ongoing personal contact, would transfer managerial skills, incentives to innovate, and the general entrepreneurial style of the West to the Soviet

89. Peter G. Peterson, *U.S.-Soviet Commercial Relationships in a New Era* (Washington, D.C.: Department of Commerce, August 1972), p. 13.
90. Henry A. Kissinger, "Moral Purpose and Policy Choices," *Department of State Bulletin*, October 29, 1973, p. 529.
91. Hardt and Holliday, *U.S.-Soviet Commercial Relations*, chap. 3.

Union. Over time, this transfer of technology would help to strengthen the domestic political position of the Soviet "managerial and technical elite," which preferred economic diversity and decentralization to centralized control and an overwhelming emphasis on heavy industry. Trade would lead to reforms, and reforms to increased trade, in a self-reinforcing cycle. Soviet citizens would adopt Western consumption habits and place additional pressure on the regime to emphasize civilian investment and consumption at the expense of the military. Proponents of détente stressed the importance of tourism, which would maximize direct personal contact between the American and Soviet publics.[92]

Once the Soviets committed themselves, through trade, to modernizing the civilian economy, they might even be forced to divert existing resources away from the military sector. A significant increase in the production of automobiles, for example, would require additional outlays for highways, gas stations, and repair facilities. Some analysts went as far as to suggest that given Soviet labor shortages, some degree of military demobilization might be necessary to sustain industrial projects.[93] And because the completion of such projects typically required long lead time, the ability of the Soviets to shift resources easily from civilian back to military projects would be reduced. In short, although it might enhance Soviet economic capabilities, trade would simultaneously induce a shift in the allocation of Soviet resources. The Soviets would ultimately devote less, rather than more, to the military; or, as the fervent French advocate of détente Samuel Pisar put it, "the benefits of cooperation would go to their midriff, rather than to their biceps."[94] Such Soviet complacency would be encouraged by the overall atmosphere of détente; the minimization of the external threat would make it possible to relax internal vigilance.

Third, even if it did contribute indirectly to an increase in their military power, trade would also make the Soviets less inclined to exercise it in a manner detrimental to U.S. security. Over time the Soviets would acquire an interest in détente's economic benefits, which they would be reluctant to jeopardize by an aggressive foreign policy. As Secretary Peterson argued, "a nation's security is affected not only by its adversary's military capabilities, but by the price which attends the use of those capabilities."[95] In other words, even if the benefits of trade were channeled to its biceps rather than its midriff, the need to maintain a positive

92. See the statements of Richard Allen and Alan Brown in U.S. Congress, Joint Economic Committee, Subcommittee on Foreign Economic Policy, *A Foreign Economic Policy for the 1970s*, pt. 6, hearings, 91st Cong., 2d sess., December 7–9, 1970, pp. 1111 and 1217–20.
93. Hardt and Holliday, *U.S.-Soviet Commercial Relations*, pp. 28–29.
94. Samuel Pisar, "Let's Put Detente Back on the Rails," in Fred Warner Neal, ed., *Detente or Debacle: Common Sense in U.S.-Soviet Relations* (New York: Norton, 1979), p. 7. The same point is made in less prosaic language by William Casey in *Ex-Im Bank Hearings*, p. 43.
95. Peterson, *U.S.-Soviet Commercial Relationships in a New Era*, pp. 3–4.

bilateral relationship would assure that the Soviet Union remained a muscle-bound power.

The above discussion suggests that while the Nixon administration practiced tactical linkage, administration officials also employed the arguments and rhetoric of structural linkage, depicting a long-term connection between unconditional trade expansion and positive developments in Soviet behavior that benefited U.S. security. The political appeal of such arguments was readily apparent: they articulated a grand strategy through which the Soviets could be induced over time to become more like and less of a threat to the United States.

Whether the Nixon administration intended for the United States actually to pursue structural linkage as a long-term strategy, or simply articulated its arguments for public consumption, is not altogether clear. It is clear, however, that any effort to develop and sustain the strategy faced formidable domestic constraints. As the earlier discussion of Jackson-Vanik demonstrated, significant segments of the U.S. Congress and the public viewed unconditional trade expansion with the Soviets as neither economically nor politically desirable, and they were skeptical of any strategy that appeared to exchange short-term concessions for possible long-term benefits, particularly when the realization of the latter depended on far-reaching structural changes in Soviet domestic and foreign policy.

Even if intended mainly for public consumption, the arguments of structural linkage could enjoy credibility in the United States only as long as the Soviets were not perceived to be building their military strength or projecting it internationally. By the mid-1970s, proponents of structural linkage were vulnerable on both counts. Not surprisingly, the Carter administration did not cast its linkage variant in terms of grand strategy but instead promised pragmatic short-term results.

*U.S. Leadership and the Decline of CoCom's
Strategic Embargo*

Despite détente and the expansion of East-West trade, during the 1970s the Western allies remained formally committed to a multilateral strategic embargo, one aimed at prohibiting exports that could make a direct and significant contribution to Soviet military capability. CoCom was not abandoned; it declined significantly, however, in effectiveness. First, member governments frequently failed to interpret CoCom's controls uniformly, which led some members to export items that others deemed to be prohibited. Second, the enforcement of controls was lax, enabling the Soviet Union to acquire illegally, without great difficulty, items considered strategic. Third, the control list failed to keep pace with technological change, and it maintained coverage over strategically outdated items, which created friction among member governments and their firms and reduced their incentives to comply with controls. Fourth, in part as a response to the overly comprehensive list, member governments requested and approved embargo exceptions at a faster rate and in greater quantities than they had previously. The manner in which these exceptions were processed and the pattern of approvals led members to question one another's motives and commitments and to accuse one other of hypocrisy. Overall, CoCom was a source of controversy among its members, and by the end of the decade, there was some question as to whether it could continue to serve effectively as a strategic embargo instrument.

The decline in CoCom's effectiveness can be traced, in part, to the general relaxation of East-West tensions and to the acceleration of Western competition to service Eastern export markets. These factors exerted pressure on the allies while they sought to maintain effective coordination of national security export controls. A far more critical factor, however, concerned the changing role and behavior of the United States. During the 1970s, U.S. leadership in CoCom was highly problematic. The United States failed to provide consistent discipline to CoCom, to maintain the support of non-CoCom suppliers, or to alleviate the administrative or economic burdens of the multilateral control system. Moreover, to other CoCom members, U.S. behavior seemed inconstant and unpredictable.

America's failure to lead was mainly a consequence of the shift in its national export control policy, in particular, the effort to use trade expansion for foreign policy purposes. The United States remained, in relative terms, the CoCom member most concerned with the restriction of trade for national security reasons. At the same time, however, the U.S. strategy of trade expansion became entangled with, and often took precedence over, the pursuit of a strategic embargo, nationally and multilaterally. Within the U.S. government, the potential foreign policy (and to a lesser extent, economic) benefits of expanded trade took on primary significance; narrowly defined national security considerations, traditionally dominant, were still important though no longer an overriding concern.

This shift in priorities is suggested by the fact that during détente, U.S. officials permitted not only exports that would enhance Soviet economic capabilities but also those likely to be put to military use and/or with potential military significance. The two most infamous cases involved Kama trucks and Bryant Grinder ball bearings. From 1971 to 1974, the Nixon administration granted numerous licenses and approved the full participation of U.S. firms in the Kama project, despite clear evidence that some of the trucks produced would be used by the Soviet military.[96] Political considerations—the opportunity to link trade to the evolution of détente—overrode the strategic reservations of the Defense Department.[97] Similarly, U.S. firms were permitted to export computers and other equipment to the Zil plant, even though it produced 100 missile launchers annually and 25 percent of its annual truck output of 200,000 went to the military.[98] The State and Commerce Departments approved the exports, despite their full awareness of the plant's military role and objections by the Defense Department that the exports' military contribution was unacceptable.

The Kama and Zil cases demonstrated U.S. tolerance for the diversion of the products of American technology to military *use*. In the Bryant Grinder case, U.S. officials approved exports of probable direct military *significance* to the Soviets. The precision antifriction bearings produced by the Bryant machines had a wide range of civilian uses but were also essential to improving the guidance mechanisms of MIRVed ballistic missiles. Some analysts have claimed that the Soviets could not even have gone into production of MIRVed missiles without the 164 Bryant machines. This assertion is unlikely because prior to the introduction of the machines, the Soviets were able to produce small amounts of precision bearings manually and actually began to MIRV their missiles.[99] Nonetheless, U.S. intelligence officials determined that the machines enabled the Soviets to mass-produce greater quantities of precision bearings, which had been in short supply for both civilian and military uses.[100] By allowing the export of the grinding machines, U.S. officials may have helped to alleviate a critical bottleneck in an important Soviet military program.

The formal justification given by the Nixon administration for approving the sale was the existence of foreign availability. Similar grinding

96. See the statement of Assistant Commerce Secretary Stanley Marcuss in U.S. Congress, Senate, Committee on Banking, Housing and Urban Affairs, Subcommittee on International Finance, *Trade and Technology, Part II: East-West Trade and Technology Transfer*, hearings, 96th Cong., 1st sess., November 28, 1979, pp. 59–70, and the statement of Undersecretary of Defense William Perry in U.S. Congress, Senate, Committee on Governmental Affairs, Permanent Subcommittee on Investigations, *Transfer of Technology to the Soviet Bloc*, hearings, 96th Cong., 2d sess., February 20, 1980, pp. 52–53.

97. Statement of Eric Hirchhorn, deputy assistant secretary of commerce, in U.S. Congress, Senate, Committee on Governmental Affairs, Permanent Subcommittee on Investigations, *Proposed Legislation to Establish an Office of Strategic Trade*, hearings, 96th Cong., 2d sess., September 24–25, 1980, pp. 112–113.

98. See the interagency memo cited by Senator Garn and reprinted in ibid., p. 24.

99. See Dan Morgan, "U.S. Reportedly Sold Soviets Means to Make MIRV Part," *Washington Post*, January 26, 1976, and Thane Gufstafson, *Selling the Russians the Rope? Soviet Technology Policy and U.S. Export Controls* (Santa Monica, Calif.: RAND Corporation, 1981), pp. 10–14.

100. See the statement of Edwin Speaker, Defense Intelligence Agency, in U.S. Congress, House, Committee on International Relations, Subcommittee on International Trade and Commerce, *Export Licensing of Advanced Technology: A Review*, pt. II, hearings, 94th Cong., 2d sess., April 12, 1976, p. 18.

machines were produced by Swiss companies. Yet, in a previous policy guidance statement, the administration had explicitly excluded foreign availability as a factor when the export in question had significant military applications.[101] Rather than foreign availability, the timing of the decision suggests that foreign policy was the major consideration. Bryant Grinder's license had been denied early in 1972; the administration reconsidered and approved it shortly after the completion of the May 1972 summit meeting between Nixon and Brezhnev.

Other cases similarly suggest a less vigilant attitude with respect to trade hitherto deemed strategic. Between 1970 and 1978, U.S. officials permitted the export of array transform processors to the Soviet Union for underground energy exploration, yet they apparently overlooked their potential application to Soviet antisubmarine warfare efforts.[102] In 1978 a U.S. firm was granted licenses to export technical data and equipment for the manufacture and application of advanced metal-coating compounds. The Defense Department initially recommended denial because the company's technique and products were used by the U.S. Air Force and might significantly improve the performance and reliability of Soviet military jet engines. Furthermore, there was no way to safeguard the techniques against diversion to military use. Nevertheless, DoD officials eventually gave their grudging approval in the interest of cooperation with the Commerce Department, which had advised the firm before the review that its application would receive favorable consideration. When the license was granted, Commerce inexplicably neglected to proscribe the ultimate end user; thus, there were not even provisions on paper prohibiting the Soviets from diverting the technology to military use.[103]

The very fact that U.S. officials were willing to license exports of potential military significance to the Soviet Union if it could be demonstrated that these items would be used for civilian or "peaceful" purposes serves as an additional indicator of the U.S. government's shift in attitude and practices. Traditionally skeptical of the utility of end-use assurances in the Soviet context, U.S. officials relied on them throughout the 1970s to allow expanded trade in controlled items. Such assurances could not provide adequate protection against diversion, particularly with regard to technology. An influential DoD task force argued in 1976 that once manufacturing know-how is transferred and absorbed, its initial end use is irrelevant and further applications cannot be con-

---

101. GAO, *Details of Certain Controversial Export Licensing Decisions Involving Soviet Bloc Countries* (Washington, D.C.: GPO, May 5, 1983), p. 1.

102. Ibid., p. 2. See also the statement of Miles Costick in *Proposed Legislation*, pp. 170–71. Costick claimed that American-made array transform processors were carried aboard Soviet surface ships with antisubmarine warfare (ASW) missions.

103. GAO, *Details of Certain Controversial Export Licensing Decisions*, pp. 4–7. The case is also discussed by Senator Garn in *Proposed Legislation*, pp. 10–11.

trolled.[104] As a senior Defense Intelligence Agency official testified in 1982: "Our ability to determine whether or not dual-use technologies are being diverted for military purposes in the Soviet Union or in any other closed society is woefully inadequate. Furthermore, I am not at all optimistic that even with additional resources, we could significantly redress that situation."[105] Another U.S. official frankly conceded that end-use controls were of limited utility but claimed it "made us feel good" to receive the written assurances of the end user.[106]

For the purposes of this discussion, of relative unimportance is whether licensed U.S. exports were actually diverted to military use and, if so, whether they proved to be militarily significant.[107] What does matter is that the United States, in the interest of broad foreign policy considerations, was willing to accept what was traditionally considered strategic risk. Given the customary role of the United States in CoCom, this shift in orientation could not help but have a major impact on the calculations and attitudes of other Western governments, which were pursuing trade liberalization for their own reasons.

Exceptions

The issue of CoCom general exceptions demonstrates a clear link between the shift in U.S. policy and the weakness of CoCom. During the 1950s and 1960s, the United States requested exceptions very infrequently and, until the end of the latter decade, never for export to the Soviet Union. As Table 7 indicates, however, during the 1970s, the United States became the undisputed leader among CoCom members in making requests. It asked for more exceptions than did any other member and, in some years, for more than the total from all other members

104. The argument is made most forcefully in the Bucy Report. See Defense Science Board Task Force, *An Analysis of Export Control of United States Technology: A DoD Perspective* (Washington, D.C.: Office of the Director of Defense Research and Engineering, 1976). The discussion of the report is in Chap. 6.

105. Statement of Jack Vorona, Defense Intelligence Agency, in United States Congress, Senate, Committee on Governmental Affairs, Permanent Subcommittee on Investigations, *Transfer of United States High Technology to the Soviet Union and Soviet Bloc Nations*, hearings, 97th Cong., 2d sess., May 4–6, 11–12, 1982, p. 118.

106. OTA, *Technology and East-West Trade*, p. 167. Written end-use assurances were clearly valuable in the enforcement process, helping to document the illegal diversion of controlled items through various transshipment points.

107. The issue remains one of contention. For the argument that U.S. licensed exports did contribute meaningfully to Soviet military capabilities, see Carl Gerschman, "Selling the Rope to Hang Capitalism," *Commentary* 67 (April 1979), 35–45. At the same time, the State Department claimed in 1979 that of the thousands of CoCom-approved exceptions, only five were likely to have been diverted to military use. See Gary Bertsch et al., "East-West Technology Transfer and Export Controls," *Osteuropa Wirtschaft* 26 (June 1981), 127–28. It is possible, of course, that more diversions took place undetected and, as I discuss in Chap. 6, the entire issue of physical diversion becomes clouded when the export in question involves technical know-how. Whatever the contribution of legally sanctioned trade, the illegal or unapproved diversion of Western technology to Soviet military purposes was clearly the more significant problem (see below).

Table 7. CoCom exception requests

Year	Total	U.S.	U.S. as % of total
1962	124	2	1.6
1966	228	29	12.7
1970	544	153	25.6
1971	635	186	29.3
1972	1085	415	38.2
1973	1268	519	41.0
1974	1369	567	41.4
1975	1798	798	44.4
1976	1039	593	57.1
1977	1044	539	51.6
1978	1680	1050	62.5

Source: Statement of Lawrence Brady, in U.S. Congress, Senate, Committee on Government Affairs, Permanent Subcommittee on Investigations, *Transfer of Technology to the Soviet Bloc,* hearings, 96th Cong., 2nd sess., February 20, 1980, p. 62.

combined. As a percentage of total requests, those of the United States climbed steadily and reached a peak in 1978, when U.S. officials made five of every eight requests.

Not surprisingly, as U.S. requests increased, so too did those of other CoCom members. Given the traditional role of the United States as the conscience of CoCom, its change in behavior legitimized the routine use of the exceptions mechanism and constituted tacit approval of the extension of Western export competition into the realm of CoCom-controlled items. Furthermore, because CoCom decisions were sensitive to precedent, requests by one member to export a particular item not previously allowed prompted others to make similar requests and also to press for approvals at new and higher technical levels.[108] Exception requests mounted rapidly, from an annual level of a few hundred during the 1960s to over one thousand in the 1970s.

The great majority of CoCom exception requests were approved. Between 1974 and 1977, for example, out of over 5,000 total requests, only 101 were denied, roughly 2 percent.[109] Few of the denied cases involved requests by the United States. If the United States wished to export items on the CoCom list, other members, reflecting their interest in trade liberalization, were willing to oblige. At the same time, virtually every denial came as a result of the exercise of the U.S. veto.[110] The upshot of

108. In 1977, for example, Japan applied to CoCom to sell three advanced computers to China after, according to Japanese government officials, the United States submitted a similar request. *Japan Times,* February 6, 1977.

109. OTA, *Technology and East-West Trade,* p. 169.

110. In 1977, for example, of thirty-one exception request vetoes, thirty came at the hands of

Table 8. CoCom approvals (by volume, $ million)

Year	Amount	Year	Amount
1967	11	1973	106
1968	8	1974	119
1969	19	1975	185
1970	62	1976	162
1971	56	1977	214
1972	124		

Source: "Special Report on Multilateral Export Controls," in U.S. Congress, House, Committee on International Economic Policy and Trade, *Export Administration Act: Agenda for Reform,* hearings, 95th Cong., 2d sess., October 4, 1978, p. 52.

this pattern was, predictably, discontent on the part of other CoCom members. For the United States to deny the exceptions of others on strategic grounds while it abstained from export competition was tolerable. Not surprisingly, however, its denial of requests while submitting more for approval than did any other member generated resentment in CoCom.[111]

To other CoCom members, delays were even more disturbing than denials. The United States took much longer than did other CoCom members to process exception requests, and it frequently flouted CoCom's procedural rules, which set limits on the amount of time allowed each member to review the requests of others.[112] The problem was compounded because the United States was the only CoCom member to impose reexport control requirements. As a result, requests submitted to CoCom that contained U.S.-origin components or technology were for some time subject to a double review. The CoCom request was processed by the State Department, whereas the reexport request (essentially the same) was handled by Commerce. Such cases, which constituted about 25 percent of all allied exception requests, were subject to the longest delays.[113] To Western Europe and Japan, the additional reexport review implied that the United States lacked confidence in the CoCom

the United States. See "Special Report on Multilateral Export Controls," in United States Congress, House, Committee on International Relations, Subcommittee on International Economic Policy and Trade, *Export Administration Act: Agenda for Reform,* hearings, 95th Cong., 2d sess., October 4, 1978, p. 57.

111. On at least one occasion, the United States denied an exception request of another CoCom member for an item similar to that previously approved for a U.S. exporter. See Stent Yergin, *East-West Technology Transfer: European Perspectives,* p. 48.

112. GAO, *Export Controls: Need to Clarify Policy and Simplify Administration* (Washington, D.C.: GPO, March 1, 1979), pp. 10–11.

113. Ibid., p. 14. The Commerce Department eventually waived its reexport license requirement for items reviewed in CoCom. Personal correspondence with William Root, April 15, 1987.

process. Equally important, the uncertainty associated with U.S. delays hampered the ability of other members to service Eastern markets. As the resentment of other CoCom members reached its peak, U.S. exporters complained that their exception requests were being held hostage in CoCom until the U.S. government granted approvals to other states.[114]

The expansion of trade by the United States, together with its occasional denial and frequent delay of other members' cases, generated suspicion among other members regarding U.S. motives in CoCom. The United States was perceived as abusing its privileged position in CoCom to advance its economic interests by providing its firms with a competitive advantage.

When U.S. officials objected in CoCom in 1975 to German plans to build a $600 million nuclear power plant in the Soviet Union, West German officials suspected that the underlying U.S. motive was to manipulate the global competition between Kraftwerkunion and the U.S.-based Westinghouse.[115] Similarly, when U.S. officials objected to a proposal by Lucas Aerospace of Great Britain to sell an advanced electronic fuel control system to the Soviets for use in their version of the Concorde, Lucas contended that the real reason for denial had nothing to do with Western security but was intended to assist its U.S. competitors.[116] In discussing a contentious case involving the French firm Thomson, a French Foreign Ministry official scoffed at the suggestion that the United States was genuinely concerned that a sophisticated digital switching system could enhance the command and control capabilities of the Soviet military. Rather, he claimed, the United States was bitter over Thomson's success in marketing a system that AT & T had earlier discarded as lacking commercial promise. The real issue behind the U.S. objection was the global struggle between the United States and France for high-technology market shares; "there are no angels in this game, and in a fight even a hit below the belt is a good one, if you can get away with it."[117]

The Sperry case, described earlier, raised similar suspicions. After denying Sperry's request as a protest against Soviet human rights violations, the Carter administration suggested to other members that it would veto a CoCom exception request to replace the sale. This stance irritated West European officials, who viewed it as a blatant attempt to use CoCom to further U.S. foreign policy objectives rather than to advance legitimate military security concerns. A West German CoCom official captured West European sentiment when he asserted that "there is

114. Statement of Edward Loeffler in *Export Licensing: CoCom List Review Proposals of the United States*, p. 15.
115. OTA, *Technology and East-West Trade*, p. 168.
116. See *Daily Telegraph*, January 6, 1977; *Guardian*, January 6, 1977; *Times* (London), January 7, 1977; and *Guardian*, May 27, 1977.
117. Interview, Ministry of Foreign Affairs, Paris, June 15, 1982.

no place for human rights in CoCom."[118] When the French contracted to replace the Sperry order, they did not even bother to ask for CoCom approval. Moreover, the Carter administration confounded its allies shortly thereafter by reversing its position and presenting its own exception request (albeit at a lower technology level) for the Sperry computer. For allies such as Japan, which had agreed to refrain from pursuing the sale, U.S. behavior in CoCom appeared both confusing and counterproductive.[119]

West European accusations notwithstanding, it is doubtful that U.S. officials were sufficiently Machiavellian or unified to use CoCom systematically to advance U.S. commercial interests. The delays in responding to CoCom exception requests were more likely a result of the same interagency conflicts and cumbersome bureaucratic procedures that frustrated U.S. exporters.[120] Nevertheless, the widespread perception that the United States was motivated more by economic than by strategic considerations demonstrated the cynicism with which some members had come to view CoCom and the U.S. role within it. As I noted in Chapter 1, self-imposed costs enhance the credibility of commitments; the message sent, perhaps inadvertently, by the United States to other members was: we are not fully committed to CoCom, and thus there is no need for you to be.

That the United States did little to allay the concerns of other members reflected the lack of appreciation in the U.S. government, at the highest levels, for CoCom's problems. In 1977 the State Department, in recognition of growing resentment in CoCom, presented a plan to the high-level U.S. Economic Defense Advisory Committee (EDAC) designed to alleviate the problem of exception request delays. State came away with very little. First, although the EDAC agreed in principle that the U.S. CoCom delegate in Paris should be given the authority to decide on the spot exception cases that had "clear precedent," the committee first decided to conduct a study of what constituted clear precedent. By the end of 1979, that study had still not been completed. Even if the U.S. delegate had been granted the proposed authority, he or she would still have had less autonomy than other CoCom delegates, many of whom could decide on most exception requests and even make certain list review changes. Second, the EDAC rejected State's proposal that reexport control review be eliminated for items already covered by CoCom,

118. Interview, Ministry of Economics, Bonn, July 8, 1982.
119. Statement of Assistant Commerce Secretary Frank Weil in *United States Export Control Policy and Extension of the Export Administration Act*, pt. III, May 3, 1979, pp. 27–28, and the statement of Lawrence Brady, Department of Commerce, in *Transfer of Technology to the Soviet Bloc*, p. 62.
120. GAO, *Export Controls: Need to Clarify Policy*, p. 11.

on the grounds that the national controls of certain CoCom members were not as effective as those of the United States.[121]

List Coverage and Interpretation

In light of the proliferation of exceptions and discontent over how they were processed, it is not surprising that the United States and its major allies came into conflict over how strictly the multilateral embargo should be interpreted and applied. These conflicts demonstrated (as did U.S. use of the exceptions veto) that the United States, despite the shift in its national policy, continued to expect other Western states to restrict exports of direct military significance. In this sense, the United States continued to play its traditional CoCom role. Given its own trade expansion, however, it was difficult for the United States to lead credibly in urging its allies to accept economic sacrifices in the interest of collective security. Moreover, in contrast to their behavior during the early 1980s, U.S. officials appeared unwilling to exert pressure (i.e, threaten or actually restrict intra-Western technology transfer) in an effort to ensure the compliance of recalcitrant allies or non-CoCom suppliers with their strategic trade preferences.

Between 1967 and 1972, the United States attempted to resist pressure from Great Britain and France to allow the provision of equipment and technical assistance necessary to enable Poland to mass-produce integrated circuits.[122] Poland desired this capability as a means to increase its output of television sets, desk calculators, and small computers. U.S. objections were based on the fact that the ICs produced could also be used in military equipment and, more important, that the technology provided could serve as a starting point from which Poland, and possibly the Soviet Union, could progress to the production of more advanced circuitry with more important military applications. To allow an exception for Poland would breach the CoCom embargo on a critical technology and possibly set a precedent. By 1972, with the onset of détente and America's own barrage of exception requests, the Nixon administration could no longer credibly resist. Nixon was persuaded by British prime minister Heath, against the advice of both the Defense and Commerce Departments, to remove U.S. objections. Exceptions were granted, and the French won the sale. To add insult to injury, the package delivered to

121. Ibid., pp. 13–15. As noted above, State and Commerce eventually worked out the arrangement on reexport licenses that the EDAC failed to endorse in 1977.
122. GAO, *Details of Certain Controversial Export Licensing Cases*, pp. 7–8, and statement of J. Fred Bucy in United States Congress, Senate, Committee on Governmental Affairs, Permanent Subcommittee on Investigations, *Transfer of Technology to the U.S.S.R. and Eastern Europe*, pt. II, hearings, 95th Cong., 1st sess., May 25, 1977, pp. 21–22.

Poland contained U.S.-origin components and technology, which were reexported without U.S. permission.[123]

Another well-known case was brought to light by the U.S. machine tool manufacturer Cyril Bath.[124] In 1977, its application to sell a metal-stretching machine tool to the Soviet Union was denied by the Commerce Department. Although it was intended for civilian use, U.S. officials decided that the machine in question could be used to enhance the performance of Soviet aircraft production. At the same time, a French firm accepted an order for nine metal-stretching machines, virtually identical to the machine of Cyril Bath. Although the item was contained on the CoCom list, the French government never brought the case before CoCom as an exception request. When questioned, the French CoCom delegate claimed the machines were for automotive use and thus did not require CoCom approval. Ironically, foreign availability was not considered a factor in the U.S. review of Cyril Bath's license request: the item was CoCom-controlled and the alternative supplier was a CoCom member.

A third case involved the Japanese sale to the Soviet Union of a floating dry dock for the servicing and repair of large ships at sea.[125] The Carter administration objected, claiming the dock to be militarily significant (despite the fact that it was omitted from the CoCom list at the time) and considering it likely that it would be diverted to military use. With a capacity of eighty thousand tons, it would be the only dock in the Pacific fleet area large enough to accommodate the Kiev-class aircraft carrier. According to the CIA, no existing Soviet shipyard would have been capable of constructing such a dock "without major facility modifications, associated capital expenditures, and interruptions in present weapons systems."[126] The Japanese firm defended its sale, asserting that because the dock would be used primarily for civilian purposes, it did not require export control approval. The Japanese government evidently concurred, declining to block the sale. In response to U.S. criticism, a spokesman for the firm stated that "we have nothing to do with the use to which this dock will be put, and if such a way of reasoning is applied, even foodstuffs can be used for military purposes . . . it was a floating dock we built gladly, as there are no orders for new ships."[127]

Discontent generated by the exceptions and list interpretation issues was exacerbated by the failure of CoCom governments to streamline the multilateral control list. The maintenance of controls on items that ei-

123. GAO, *Details of Certain Controversial Export Licensing Cases*, p. 8.
124. United States Congress, House, Committee on International Relations, Subcommittee on International Economic Policy and Trade, *Export Licensing: Foreign Availability of Stretch Forming Presses*, hearings, 95th Cong., 1st sess., November 4, 1977.
125. U.S. Embassy, Tokyo, *Daily Summary of the Japanese Press*, November 17, 1978, pp. 11–14.
126. CIA, "Soviet Acquisition of Western Technology" (April 1982), p. 8.
127. *Daily Summary of the Japanese Press*, November 17, 1978, pp. 11, 13.

ther no longer embodied the most advanced technology or were widely available to the East from non-CoCom sources was consistently cited by firms in CoCom member states, and by West European governments, as one of the most significant problems plaguing CoCom.[128] Overcontrol diminished CoCom's effectiveness by making the enforcement of controls more difficult, stimulating the demand for exceptions, and straining the credibility of the embargo in the eyes of member governments and their firms.

Responsibility for addressing the overcontrol problem rested primarily with the United States. Other CoCom members generally favored the liberalization of the control list, yet they could not undertake it without U.S. support, as CoCom rules required unanimity on decontrol decisions. Although willing to decontrol some items, U.S. officials resisted a large-scale streamlining of the list, particularly with regard to what many West European officials considered the most critical items: computers and their peripheral equipment. During the 1978 list review, other CoCom members pressed for the relaxation of control parameters on computers, while the United States proposed even higher levels of restriction. The need for liberalization was suggested by the fact that the majority of CoCom exception requests involved computers.[129] At the same time, the credibility of the U.S. proposals was not enhanced because in actuality, U.S. officials reportedly had already prepared twenty-nine exception requests based on their proposed criteria before the negotiations were even completed.[130] The United States appeared to desire restrictive formal controls not necessarily to keep computers from the East but rather to determine the circumstances under which they might be exported.

The list review ended late in 1979 without agreement on computers, which meant that the technical parameters governing that category of items remained those which had been agreed to in 1975. Rapid technological change assured that those parameters were hopelessly outdated and would remain so until the completion of CoCom's next list review, scheduled not even to begin until 1982.[131] In the interim, firms wishing to export computers were expected to obtain a formal exception from CoCom and/or a U.S. reexport license. Given the problem of delays described above, this solution was hardly a satisfactory one. The incentives for firms to bypass CoCom's formal controls were great. The situation became far worse after 1980, when the United States, for for-

---

128. Interviews, Bonn, Paris, and London, May–July 1982, and the testimony of business officials in *Export Licensing: CoCom List Review Proposals of the United States.*

129. OTA, *Technology and East-West Trade*, pp. 158–59. On the 1978 review, see "Technology Transfer Controls: The Great CoCom Debate," *Business Eastern Europe*, November 3, 1978, pp. 345–47.

130. Statement of Lawrence Brady, *Transfer of Technology to the Soviet Bloc*, p. 88.

131. Computer parameters were finally updated in 1984. See Chap. 8.

eign policy reasons, imposed a policy that allowed no exceptions for exports to the Soviet Union, thereby partially eliminating the "escape valve" that had made the outdated control list tolerable to exporters.

Enforcement

A final critical indicator of CoCom's lack of effectiveness concerns the laxity with which export controls were enforced during the 1970s. A GAO study published in 1979 found that in the United States, inadequate resources were devoted to enforcement efforts, a situation that reflected enforcement's low-priority status in the eyes of U.S. officials.[132] The Compliance Division of the Commerce Department, which had primary responsibility for enforcement, relied on spot checks to discover and seize dual-use technology leaving the country illegally. Its coverage, however, was uneven: 88 percent of the checks were conducted at New York exit points, where only 55 percent of controlled shipments left the country, whereas only 4 percent of the checks were made in California, where 25 percent of controlled commodities exited.[133] Inspections were rarely made at night or on weekends, despite the frequency of flights at those times, because of a lack of overtime pay for inspection personnel. In a 1976 study, GAO investigators found the Compliance Division to be "unable to effectively determine compliance with export control regulations" and to have "limitations on its ability to investigate alleged violations."[134] A Senate investigation completed in 1982 reached a similar conclusion, finding the Compliance Division to be "understaffed, ill-equipped and underqualified"; it recommended transferring responsibility for enforcement to the Customs Service of the Treasury Department.[135]

By the mid-1970s it was also clear that the United States could not rely with confidence on Western governments, through the use of the IC/DV procedure, or on Western firms, through the use of reexport licensing, to protect against the illegal diversion of U.S.-origin technology or components. Among the most serious violations were the unauthorized reexport of U.S. semiconductor technology and manufacturing equipment from France and West Germany to Poland and the Soviet Union, and of sensitive electronics and radar equipment from Sweden to the Soviet Union for an air-traffic-control system that American officials refused to license directly from a U.S. firm.[136] In its 1976 investigation, the GAO

132. GAO, *Export Controls: Need to Clarify Policy*, chap. 5.
133. Ibid., p. 52.
134. GAO, *The Government's Role in East-West Trade: Problems and Issues* (Washington, D.C.: GPO, February 4, 1976), p. 30.
135. Statement of Fred Asselin, minority investigator, in *Transfer of Technology to the Soviet Union and Soviet Bloc Nations*, p. 82.
136. The Swedish firm Datasaab was blacklisted by the U.S. Commerce Department and

found the reexport of U.S.-controlled commodities without U.S. approval to be "the most significant form of illegal diversion to Communist states." The GAO also reported that many U.S. officials had so little confidence in the reexport or IC/DV systems that they believed the only meaningful export control safeguard was the initial decision to export.[137]

In light of trade expansion, the widespread belief that the control list was unnecessarily comprehensive, and the problems involved in securing U.S. reexport licenses and general exception approvals, it is not surprising that West European governments considered enforcement even less of a priority than did the United States.[138] France, for example, did not require end-use statements from its exporters; thus, controlled commodities originating in France could be reshipped legally from any non-CoCom state to controlled destinations. West Germany did require such statements, but West German officials admitted that they could still do little to prevent the diversion of West German exports through non-CoCom countries.[139] At least some West European officials appeared to believe that because the Soviets could obtain virtually anything for which they made a determined effort, major efforts to prevent illegal sales were not worthwhile.

In 1979 the GAO concluded that as CoCom was not based on formal agreement, there was little the United States could do, other than using diplomatic persuasion, to improve other states' enforcement.[140] Given the lack of priority accorded to enforcement by the United States itself, it is difficult to see how U.S. officials, had they been inclined to make the effort, could have been credible or convincing in urging others to improve their performance.

An unclassified version of a CIA study and hearings conducted by the Senate's Permanent Subcommittee on Investigations in 1982 brought to light the weakness of CoCom and the ability of the Soviet Union to exploit it during the 1970s.[141] The most dramatic case involved a syndicate of U.S. and West German firms acting as illegal intermediaries between U.S. electronics firms and Soviet purchasing agents. Between 1977 and 1980, enough equipment was apparently shipped illegally— piece by piece, from Silicon Valley through Western Europe to the Soviet Union—to put together at least one entire integrated circuit manufacturing plant. According to the CIA, the ability of the Soviets to achieve

---

eventually agreed to pay a $1 million fine. See *Wall Street Journal*, April 6, 1984, and April 9, 1984.

137. GAO, *The Government's Role in East-West Trade*, p. 44.
138. GAO, *Export Controls: Need to Clarify Policy*, p. 58.
139. Stent Yergin, *East-West Technology Transfer: European Perspectives*, p. 36.
140. GAO, *Export Controls: Need to Clarify Policy*, p. 59.
141. CIA, "Soviet Acquisition of Western Technology," and *Transfer of Technology to the Soviet Union and Soviet Bloc Nations*.

large-scale integration and to apply it to their most advanced weapons systems can be traced directly to combined legal and illegal acquisitions from the West. Overall, the CIA concluded that the Soviets had effectively combined legal purchases with illegal acquisitions of Western technology and equipment to achieve significant benefits across a wide range of military systems and to save hundreds of millions of dollars in research and development costs.

Investigations prompted in 1987 by the Toshiba Machine-Kongsberg diversion revealed in further detail the nature and extent of illicit trade during the 1970s.[142] Toshiba Machine had been approached by the Soviets in 1974 to sell multiaxis machine tools but had refused, citing CoCom restrictions. In light of French sales of such machines between 1976 and 1979, however, the Japanese firm reconsidered and agreed to meet the Soviet request for even more sophisticated CoCom-controlled machines. Questioned in 1987, French officials claimed the machines their firms provided without CoCom approval were only "slightly" above CoCom limits and that the sales took place when "détente was in vogue" and "CoCom did not work very well."[143] The Norwegian firm Kongsberg agreed to provide computer controllers for the Toshiba sale; these were also CoCom-controlled. Further investigation revealed that Kongsberg had illegally shipped over eighty controllers to the Soviets since 1974, in conjunction with equipment supplied to the Soviets by British, French, West German, and Italian machine tool firms.[144] Overall, the evidence clearly suggested that CoCom violations were widespread and a matter of routine, that the Soviet Union was a clear beneficiary of the regime's weakness, and that the U.S. government was either unwilling or incapable of rectifying the problem.

This section has traced the weakness of CoCom during the 1970s to the failure of the United States to play its traditional leadership role. By the manner in which it expanded its own trade, the United States failed to set a domestic example, thereby creating little incentive for others to minimize exceptions, adhere to a strict interpretation of the control list, or devote priority attention to enforcement. By subjecting exception and reexport requests to significant delays, the United States exacerbated the administrative burden of the control system and raised suspicions regarding the integrity of its own motives in CoCom. By failing to acquiesce in the decontrol of less strategic items, the United States increased the pressure for exceptions and reduced the incentives for enforcement. By allowing foreign policy considerations to become entangled with

142. See Chap. 8. Investigations were carried out by the Norwegian and Japanese governments following the discovery of evidence that between 1982 and 1984, Toshiba Machine and Kongsberg Vaapenfabrik had shipped multiaxis machine tools equipped with computer controllers to the Soviet Union to produce quieter propellers for Soviet submarines.
143. See *Wall Street Journal*, October 21, 1987, and *Washington Post*, September 14, 1987, p. 24.
144. *New York Times*, October 22, 1987, p. D1, and *Washington Post*, October 22, 1987, p. A1.

CoCom controls, the United States failed to maintain the soundness of the control process and thereby diminished its legitimacy as a necessary strategic instrument. Finally, by failing to extract the compliance of non-CoCom suppliers, the United States increased the already high incentives of other CoCom governments and their firms to disregard the CoCom regime.[145]

CONCLUSION

The United States entered détente with the hope that its political, economic, and strategic objectives in East-West trade could be maximized simultaneously. It ultimately failed on all three counts. The strategy of tactical linkage failed to influence Soviet foreign policy in accordance with U.S. preferences. With the exception of agricultural products, U.S. firms failed to establish themselves as a major force in the Soviet market. U.S. national security export controls were compromised by the pursuit of political and economic objectives, and by the end of the decade, CoCom was in a state of disarray, lacking the confidence of its most important members.

In 1976, the U.S. Defense Department began to prepare a major initiative to address the national security dimensions of the problem. It involved an effort to reform the U.S. export control system, revitalize CoCom, and even enhance the competitive position of U.S. firms. At the same time, it reopened two major issues that had been resolved during the 1950s: whether to pursue economic warfare and whether to increase restrictions on U.S. technology transfer to other Western states. The basis of the DoD's initiative and preliminary efforts to implement it are the subject of Chapter 6.

145. Of the major non-CoCom suppliers, only the Swiss maintained official CoCom-comparable controls during the 1970s. The Swedish government abolished its technology transfer controls in 1968 because of its disagreement with the U.S. government over Vietnam. This left reexport controls at the firm level as the principal means for the United States to ensure Swedish compliance. The same was true for Austria, which neglected to maintain official export controls after its occupation was ended in 1955. Despite the inadequacy of reexport restraints, it was not until the 1980s that the United States sought to establish or strengthen government-to-government export control links with Sweden, Austria, and other non-CoCom suppliers in Europe and Asia. See Chap. 8, and Jan Stankovsky and Hendrik Roodbeen, "Export Controls outside CoCom," Paper presented at the conference on "CoCom, Alliance Security and the Future of East-West Economic Relations," Leiden, the Netherlands, November 1989.

CHAPTER SIX

From Products to Technologies:
The Bucy Report and Export Control Reform

By 1974, America's willingness to liberalize technology transfers to the Soviet Union was generating considerable concern in the Department of Defense. DoD officials were disturbed by the idea of Soviet officials and technicians touring U.S. factories and negotiating, for example, with leading military contractors in the American aircraft industry, such as Boeing and Lockheed. The Soviets appeared interested in acquiring from such firms complete manufacturing capabilities, along with the process know-how and quality control techniques characteristic of U.S. defense plants.[1] Even if major deals were not consummated, valuable know-how might still be transferred during the negotiating process. DoD officials were particularly concerned that the U.S. government appeared oblivious to the potential national security risks of unrestricted technology exchanges and seemed unprepared to deal systematically with Soviet efforts to acquire American expertise. Malcolm Currie, director of Defense Research and Engineering, argued in 1974 that the United States lacked a national technology export policy and, in light of détente, that the formulation of one should be regarded as of the highest national priority.[2]

In July 1974, Currie directed the Defense Science Board, an advisory body to the Department of Defense, to establish a task force to undertake a comprehensive review of whether, and how, the U.S. and CoCom export control systems could be strengthened considering the changes taking place in the U.S.-Soviet relationship and in the world economy. J. Fred Bucy, then executive vice-president of Texas Instruments, agreed

1. See "Detente: A Trade Giveaway?," *Business Week*, January 12, 1974, pp. 65–66.
2. Malcolm Currie, "Technology Export Policy," *Aviation Week and Space Technology*, January 21, 1974, p. 9.

to chair the fifteen-member task force, which was comprised of high officials from the Defense Department and from private firms specializing in advanced technology. Four subcommittees were established to study the process of technology transfer in jet engines, airframes, solid-state devices, and scientific instruments. These sectors were selected on the basis of their interest to the DoD and in accordance with the judgment of the task force that they were representative of high-technology industries. The task force deliberated throughout 1975 and in February 1976 submitted its final report—which became known as the Bucy Report—to the director of Defense Research and Engineering.[3]

The Bucy Report proved to be among the most influential documents produced on U.S. export control policy. Its findings and recommendations provided the impetus for an attempted revision of the U.S. control system and, more important, for a conceptual shift in the overall approach of U.S. officials. After 1980 the report provided the framework for alliance negotiations in CoCom, as U.S. officials attempted to persuade other members of the merits of the task force's arguments.

Given its significant role in shaping U.S. export control policy, the Bucy Report warrants a detailed examination. In the first part of this chapter, I analyze the report's findings and recommendations in the broader context of export control strategy. In the second, I examine the Carter administration's attempt to implement those recommendations. Although the impact of the report after four years fell far short of the expectations of its architects, in the aftermath of the Soviet invasion of Afghanistan, the ideas associated with it had a major influence on both U.S. and multilateral export control policy.

THE BUCY REPORT: INTERPRETATION

The major recommendations of the Bucy Report can be summarized briefly. First, strategic export controls should focus explicitly on "design and manufacturing know-how."[4] The task force considered this recommendation to be its most important and asserted that in comparison, all other considerations were secondary.[5] Analyses of each of the four advanced-technology sectors supported the conclusion that the mastery of design and manufacturing know-how was the crucial factor account-

3. Defense Science Board Task Force, *An Analysis of Export Control of Advanced Technology: A DoD Perspective*, hereafter cited as *Bucy Report* (Washington, D.C.: Office of Defense Research and Engineering, 1976). The rationale underlying the Bucy Report is discussed by Robert Parker, deputy director of Research and Engineering, in U.S. Congress, House, Committee on International Trade and Commerce, *Export Licensing of Advanced Technology: A Review*, hearings, 94th Cong., 2d sess., March 11, 15, 24, 30, 1976, pp. 210–12.

4. *Bucy Report*, pp. 1–3.

5. Ibid., p. iii.

ing for America's qualitative industrial and military superiority and thus required the most careful protection.

Second, the report distinguished "active" from "passive" technology transfer mechanisms and recommended that the former be controlled most restrictively. Active mechanisms involve close personal contact, that is to say, frequent and specific communication between donor and receiver. The task force found that the more active the mechanism, the more likely was the receiving state to be able to absorb and utilize transferred know-how effectively. The most active mechanisms identified were turnkey factories, licensing accompanied by an extensive teaching effort, and joint ventures. The most passive mechanisms, far less effective in transferring know-how and thus less of a threat to national security, included trade exhibits, product sales without operating and maintenance information, and commercial literature.[6]

Third, the task force distinguished "revolutionary" from "evolutionary" advances in a given technology. The latter were incremental improvements made routinely, whereas the former were quantum leaps based on conceptual departures from current practice. The Bucy Report recommended that all the key elements of a revolutionary advance should be protected unconditionally in order to prevent an adversary from making significant gains in a short period of time without having to incur the costs of research and development.

Finally, with regard to the implementation of export controls, the task force argued that end-use deterrents should not be relied on to prevent the diversion of strategic technologies; that the United States should exercise greater caution in transferring technology to allied and neutral nations; that CoCom should be strengthened; and that the Defense Department should play a greater role in the U.S. export control process.

Although the Bucy Report was explicit in its emphasis on the need to control technology, it was more ambiguous with respect to the crucial issue of export control strategy. One could interpret the report as calling, at the same time, for a revised and streamlined strategic embargo, for a return to economic warfare, and for a fundamental shift in U.S. policy regarding the transfer of technology to other Western states. Given the ambiguity, and salience, of the report, public and private sector officials with different stakes in the export control issue could each point to it with some justification as providing support for their preferred policies.

Strategic Embargo

The Bucy Report clearly articulated the need to maintain a strategic embargo, but one with a different emphasis from that being practiced by

6. Ibid., p. 6.

the United States. According to the report, the current system stressed the control of products rather than focusing on what was truly critical, namely, design and manufacturing know-how. This emphasis was a key weakness because the export of products, by themselves, did not usually involve the transfer of know-how. The task force found, for example, that reverse engineering, or dismantling and attempting to reproduce an item, was usually an ineffective technique for revealing current production methods.[7] The emphasis on products resulted in a control list that was far too long and a licensing system far too cumbersome.[8] The system overcontrolled what was of marginal importance to national security and undercontrolled what was vital. It needed to be reoriented away from products to focus on the relatively small subset of revolutionary technologies of direct and critical military significance.[9]

This proposed shift in emphasis had two major implications. First, end-use statements or safeguards could no longer be plausibly expected to deter the diversion of exports from civilian to military use.[10] Unlike that of a physical product (e.g., a machine tool placed in an auto factory), the diversion of design and manufacturing skills could be neither detected nor prevented by the United States. As the report noted: "The transfer of know-how is irreversible. Once released, it cannot be taken back, contained, or controlled."[11] Moreover, in the case of technical know-how, the Soviets did not have the usual incentive to forego diversion. When a piece of equipment was physically diverted from an automobile to a missile factory, the civilian end user was forced to sacrifice in the interest of the military. Technical know-how, however, is intangible and could be carried in the mind as well as be embodied in machinery and equipment. If absorbed and diffused effectively, it could be "used" simultaneously by both civilian and military recipients.

The Bucy Report recommended that the intrinsic utility of an item, rather than its stated end use, determine whether its export should be allowed. If an item was judged to be of military significance, the prudent course would be to assume that it would ultimately find its way into military use. Compared to the existing strategic embargo, that contemplated by the task force would be less discretionary and, in this sense, more restrictive. Exception requests would not longer be granted on the grounds that items on the control list would be used for peaceful purposes.

7. Ibid., p. 5.
8. Ibid., p. 16.
9. Ibid., pp. 17–18.
10. The safeguard program went into effect among CoCom members in 1976 and applied primarily to computer systems. It required the Western exporter to make periodic on-site inspections of the equipment and facilities sold to Eastern customers to determine whether diversion to military use had taken place. The firm was required to report its findings periodically to its government. See GAO, *Export Controls: Need to Clarify Policy and Simplify Administration* (Washington, D.C.: GPO, March 1, 1979), pp. 55–56.
11. *Bucy Report*, p. 26.

Second, the list itself could be shortened because restrictions on the export of end products that did not carry design and manufacturing know-how could be relaxed. The task force thus claimed an "implicit liberalism" toward East-West trade in its approach.[12] The only products that required strict control were those accompanied by extensive operating information and maintenance procedures (and thus likely to involve the transfer of know-how) and those considered "keystone" equipment. Keystone equipment was critical in that it completed a process line and allowed it to become fully operational. The most commonly cited example was integrated circuit inspection or test equipment.[13] Such equipment was unique and specialized rather than general and multipurpose; it was the critical missing link in an entire production process. All other products could be exported without case-by-case review, which would assist both exporters, by minimizing delays and uncertainty, and export controllers, by enabling them to focus their efforts more carefully on the relatively few critical cases. A highly selective list, one that focused on the strict control of critical technologies and keystone products and avoided the discretionary and cumbersome case-by-case review process, would provide the United States with the best of both worlds: the protection of technology vital to its national security and the ability of its exporters to compete more effectively for Eastern export markets.

The task force's recommendation to strengthen and streamline the strategic embargo took on additional importance in light of two structural changes, apparent by the mid-1970s, that affected the ability of the U.S. government to dominate access to technology of direct military significance. The first involved a gradual shift from the public to the private sector in the discovery and development of critical technologies. As noted in Chapter 4, in the first two decades following the Second World War, the U.S. government accounted for the majority of U.S. research and development (R&D) spending, focusing its efforts in the defense and space industries. In 1963, for example, it accounted for 90 percent of such expenditures on aircraft and missiles and 65 percent on electrical and communications equipment. Overall, the federal government accounted for 58 percent of industrial R&D, whereas private industry made up 42 percent.[14]

One important consequence of the government's dominant position was that for the most critical technologies, military applications preceded widespread civilian use. An obvious example is nuclear technology, which was developed in national laboratories, applied to weapons design and manufacture, and closely guarded by government agencies in order

12. Ibid., p. 29.
13. Ibid., p. 2. The report appeared to conceive of keystone equipment as a bottleneck item in the production process (see Chap. 2, above).
14. Henry R. Nau, *Technology Transfer and U.S. Foreign Policy* (New York: Praeger, 1976), p. 68.

to prevent its early dissemination. Once the military applications had been exploited, the technology was made available for civilian use; indeed, nuclear power plants for naval vessels made an important contribution to the development of civilian nuclear reactor programs.[15] Similarly, ICs were first extensively used by the military, to improve the guidance systems of Minuteman missiles. The Defense Department provided private industry with R&D funds for the development of manufacturing techniques and pilot production, and it also served as a market for the first, relatively high-cost ICs. In 1963, 100 percent of IC production was sold to the Defense Department.[16] Overall, the government's heavy involvement in the development of militarily critical technologies enhanced its knowledge of, and control over, the potential transfer of those technologies to other states.

Gradually, however, the balance of power shifted, and new technology increasingly came to be developed in the private sector, outside the direct influence and control of the federal government. Between 1963 and 1975, the government's share of industrial R&D decreased from 58 percent to 38 percent, whereas that of the private sector rose from 42 percent to 62 percent.[17] As a consequence, improvements in military capabilities became more and more dependent on technological advances made by private industry for commercial purposes. As Thane Gufstafson has noted, "new generations of weapons rest on a multitude of advanced supporting technologies that cannot be said to be inherently either military or civilian."[18] More important, for many new technologies, widespread civilian use came to *precede* military application. U.S. cruise missiles, for example, are vitally dependent on advances in microelectronics first applied to the consumer goods sector.[19] In 1977, Fred Bucy noted a five-year lag between the commercial and the military applications of microprocessor technology.[20]

The predicament such a pattern has posed for the U.S. government is clear. The overseas transfer of militarily significant technologies, those developed in the private sector without government involvement and, in some cases, even without government knowledge of their existence, could potentially take place prior to their application to U.S. defense

---

15. Alexander H. Flax, "The Influence of the Civilian Sector on Military R&D," in Franklin A. Long and Judith Reppy, eds., *The Genesis of New Weapons Systems* (New York: Pergamon, 1980), p. 118.

16. Ibid., p. 122.

17. Nau, *Technology Transfer and U.S. Foreign Policy*, pp. 65–67.

18. Thane Gufstafson, *Selling the Russians the Rope? Soviet Technology Policy and U.S. Export Controls* (Santa Monica, Calif.: RAND Corporation, 1981), p. 3.

19. Flax, "The Influence of the Civilian Sector on Military R&D," p. 120.

20. Bucy's comments are found in U.S Congress, Senate, Committee on Governmental Affairs, Permanent Subcommittee on Investigations, *Transfer of Technology to the U.S.S.R. and Eastern Europe: Part II*, hearings, 95th Cong., 1st sess., May 25, 1977, pp. 4–12.

systems.[21] It has been at least possible, then, for the Soviet military to obtain emerging U.S.-developed technologies even before the U.S. military has put them to use.

As made explicit in the Bucy Report (and in Chapter 2, above), this commercial development of critical technologies suggests that the control criteria of the strategic embargo should emphasize military *significance* rather than military *use*. Use and significance traditionally tended to coincide, particularly when the U.S. government developed those technologies of greatest significance and immediately put them to military use. For export control purposes, it was possible to rely on use as a surrogate for significance. If the U.S. military used more than a certain percentage of the output of an item, that item was deemed to be of likely military importance to the Soviets and was placed under control.[22] By the 1970s, however, the two were no longer synonymous. Those items most significant militarily were not necessarily those developed and used primarily by the military. For example, by 1972, innovations in microelectronics were no longer driven by the Defense Department, and the DoD consumed only 15 percent of U.S. integrated circuit production, despite the item's clear military importance.[23]

The second structural change involved the relative decline of the United States technologically. As was documented in a 1983 Commerce Department study and elsewhere, the postwar period witnessed a gradual erosion in the competitive position of once dominant U.S. high-technology firms.[24] From 1954 to 1980, the U.S. share of OECD exports declined from 25 percent to 16 percent for all manufactured products and from 35.5 percent to 20 percent for technology-intensive ones.[25] West European and, especially, Japanese competition became intense across a range of high-technology sectors, including semiconductors,

21. An interesting example is that of Kevlar, a substance originally produced by DuPont for use in radial tires and later used as the casing material for the propellant in the Trident I missile. Kevlar's military applications were realized accidentally when defense contractors discovered the substance at an aerospace trade fair. See Deborah Shapley, "Technological Creep and the Arms Race: ICBM Problem a Sleeper," *Science* 201 (September 22, 1978), 1104, cited in Gufstafson, *Selling the Russians the Rope?*, p. 5.

22. See the statement of Graham Allison in *Export Licensing of Advanced Technology: A Review,* p. 18.

23. Flax, "The Influence of the Civilian Sector on Military R&D," p. 122. Similarly, the fact that something was used mainly by the military did not necessarily imply it was of military significance. A study conducted in the early 1970s found that liberalizing the export of large computers, which were traditionally restricted because of their use by the U.S. military and the inability of the Soviets to produce them, would not lead to an important increase in Soviet military capabilities. The Soviets typically substituted larger quantities of less powerful machines, and increased manpower, to perform the tasks for which the U.S. military relied on large computers. See Robert Klitgaard, "Limiting Exports on National Security Grounds," in *Commission on the Organization of the Government for the Conduct of Foreign Policy,* vol. IV, pt. VII (Washington, D.C.: GPO, 1975), pp. 443–75.

24. U.S. Department of Commerce, International Trade Administration, *An Assessment of U.S. Competitiveness in High Technology Industries* (Washington, D.C., February 1983).

25. Ibid., p. 46.

machine tools, robotics, and fiber optics. It was apparent by the mid-1970s that America's CoCom allies possessed the capabilities to produce and export civilian products and technologies with potential military applications on a significant scale.

The implications of this trend were similarly clear. More so than during the 1960s, the ability of U.S. officials to maintain an effective strategic embargo depended on the cooperation of other CoCom states. As America's relative technological position declined, its stake in the multilateral control process increased. CoCom needed to be strengthened. The ability of U.S. officials to promote that objective would be enhanced if they could promise, in exchange for tighter technology controls, a meaningful streamlining and liberalization of the control list with regard to end products.

Economic Warfare

Although it clearly called for a strengthened strategic embargo, the Bucy Report at times appeared to move beyond that and to suggest in addition the need for economic warfare. The report's main concern was the protection of advanced U.S. technology, but it often failed to distinguish technology of direct *military* significance from that of primarily *industrial* significance. For example, the report assigned the highest priority to the control of active transfer mechanisms, and at the top of its list were turnkey plants incorporating advanced technology. As I argued in Chapter 5, the purchase of turnkey factories from the West was at the heart of Soviet efforts to modernize their industrial infrastructure. The sale of such plants was the chief focus of Western export competition. To deny the export of turnkey factories on the basis of their effectiveness as a technology transfer mechanism without distinguishing military from industrial applications would be to move export control strategy into the realm of economic warfare. It would mean, in effect, targeting the Soviet military-industrial complex, or military support system, in addition to items with direct and specific military applications.

Similarly, the task force hinted at a broader embargo when it expressed its concern over the contribution made by Western technology to the Soviet industrial base, and its belief that a more advanced industrial base would enhance the ability of the Soviets to absorb and diffuse technology of direct military significance. As the report put it, "development of a highly capable infrastructure prepares a country to be a receptive host for subsequent revolutionary advances it may acquire."[26] The report also deemed active transfer mechanisms to be most effective when the receiving state possessed an adequate infrastructure capable of pro-

26. *Bucy Report*, p. 12.

viding necessary parts, supplies, and manufacturing equipment.[27] The report recommended asking as one of the criteria for embargo whether a given export provided "a critical manufacturing capability, *supportive* of strategic products or technologies."[28] If the development of the Soviet industrial base was the key to future strategic gains, then it would not be unreasonable to expect the prudent export control official to carry out the logical implications and seek to prevent the export of technology that was broadly supportive of the industrial base, even if it lacked direct and immediate military significance.

This logic, embedded in the report, ran directly counter to that of structural linkage, which suggested that the personal contact accompanying large-scale transfers was desirable because it exposed the Soviets to Western ideas, organizational skills, and management methods. The Bucy task force believed personal contact to be dangerous for precisely that reason. Rather than bringing about the transformation of Soviet society, increased exposure to Western ideas and industrial methods would simply enable the Soviets to exploit strategic technological advances more effectively. Thus, although proponents of structural linkage (and U.S.-Soviet détente) encouraged private and governmental scientific and technological exchanges, the task force cautioned against them and urged that they be brought under stricter control.[29]

Indeed, Western architects of détente and the drafters of the Bucy Report offered diametrically opposed visions of the long-term impact of Western technology on the relationship between the Soviet military and civilian sectors. For supporters of détente, technology transfer would exacerbate what was perceived to be the growing competition between the two sectors because it encouraged Soviet consumers to demand a larger share of available economic resources. For the Bucy Task Force, technology transfer would reinforce the intimate and symbiotic relationship between the two sectors, allowing both to become more efficient and resulting in a greater combined threat to U.S. security. These concerns are best expressed in the report's discussion of computer exports.

> The widespread use of computers, *even in commercial applications*, enhances the "cultural preparedness" of the Soviets to exploit advanced technology. It gives them vital experience in the use of advanced computers and software in the management of large and complex systems. The *mere presence* of large computer installations transfers know-how in software, and develops trained programmers, technicians and other computer personnel. All this can be redirected to strategic applications.[30]

27. Ibid., p. 4.
28. Ibid., p. 34, emphasis added.
29. Ibid., pp. 8, 18.
30. Ibid., p. 25, emphasis added.

Intra-Western Technology Transfer

Perhaps most controversial is that the Bucy Report recommended a new approach to advanced-technology trade between the United States and other non-Communist countries. Throughout the postwar period, U.S. officials had sought to minimize the impact of the U.S. East-West control system on intra-Western trade. They refrained from using trade denial to compel other CoCom members to replicate domestically the more restrictive U.S. export control list. In order to prevent the unauthorized transshipment of U.S.-origin technology through NATO members and Japan, they relied on both CoCom controls and the requirement that firms in those states acquire permission directly from the United States before reexporting an item. In the case of non-CoCom suppliers, as of the 1970s the United States relied almost exclusively on reexport controls. Although neither the United States (because unauthorized diversions sometimes occurred) nor other Western states (because reexport controls were viewed as an extraterritorial violation of their sovereignty) were entirely satisfied with this arrangement, both were willing to tolerate it. Other Western states did not wish to jeopardize their access to U.S. technology. Similarly, the consensus among U.S. officials, from the Truman through the Ford administrations, was that the preservation of the free trade system in the West ultimately brought greater benefits to the United States than would any initiative that compromised this system in the attempt to maximize the denial of strategic technology to the Soviet Union.

The Bucy Task Force, however, was deeply skeptical of this long-standing arrangement. It questioned the ability of the United States to protect its national security while allowing other Western states relatively free access to U.S. technology. The task force judged, in effect, that the benefits of unfettered Western exchange were worth risking in order to afford additional protection to U.S. technology. In the interest of maintaining U.S. strategic lead time over the Soviet Union, it explicitly called for tighter restrictions on intra-Western trade.

The report distinguished CoCom members from neutral states. It recognized the importance of maintaining CoCom, noting that "CoCom is the only linkage among the U.S. and its allies that defines strategic technologies and restricts their export to Communist nations. CoCom must be maintained as a viable agreement."[31] It argued that CoCom should be strengthened and that the United States could contribute by taking a leadership position rather than a "reaction-mode stance."[32] In earlier congressional testimony, Bucy had warned of the possible demise of CoCom, claiming that the allies were watching U.S. behavior and

31. Ibid., p. 19.
32. Ibid.

would follow its lead in requesting exceptions and exporting technology to the East.[33]

The report also pointed to CoCom's limitations and weaknesses. It noted that some members perceived less of a need to maintain strict controls and that CoCom's effectiveness had been diluted by exceptions, ambiguous interpretations of lists, and perhaps even conscious violations of CoCom rules.[34] As a result, U.S. lead time in certain strategic technologies had been compromised. Bucy presumably had IC technology, that of his own industry, in mind. Two years earlier he had been a vocal public critic of the French export of IC design and manufacturing know-how to Poland, which included the illegal reexport of what the report would have characterized as U.S. keystone equipment. In what can only be taken as a vote of no confidence in at least some CoCom allies, the Bucy Report warned that "for the most critical technologies, the U.S. should not release know-how beyond its borders, and then depend on CoCom for absolute control."[35]

More explicitly, the Bucy Report recommended sanctions against CoCom members that failed to protect U.S.-origin strategic technology: "Any CoCom nation that allows such technology to be passed on to Communist countries should be restricted from receiving further strategic know-how."[36] This passage presumably referred to the conscious violation of CoCom rules by a member state. What is unclear is whether the task force also meant it to apply to items the United States considered strategic but the allies refused to add to the CoCom list. The U.S. control list was approximately 33 percent longer than the CoCom list at the time. Moreover, if U.S. officials believed the enforcement efforts of other member governments were unsatisfactory, could it thereby conclude that such countries "allowed" technology to be passed on to the Soviets and thus should be denied U.S. technology? Given the extent of the disagreement between the U.S. and other CoCom members over what should be on the list and how it should be interpreted and enforced, carrying out the proposed sanctions could ultimately lead to the interruption of Western trade across a range of high-technology sectors. And if, as the passage suggests, it was offending *nations*, rather than firms, that were to be sanctioned, the global operations of U.S.-based multinational corporations could also be affected. If France were sanctioned for allowing computer technology to be passed on to the Soviets, then U.S.-based firms might be prevented from transferring know-how

33. U.S. Congress, Senate, Committee on Foreign Relations, Subcommittee on Multinational Corporations, *Multinational Corporations and United States Foreign Policy*, pt. 10, hearings, 93rd Cong., 2d sess., June 17, 19, July 17–19, 22, 1974, pp. 252–55.
34. *Bucy Report*, p. 19.
35. Ibid., p. iv.
36. Ibid., p. 22.

to their French subsidiaries, a situation that would complicate their efforts to maximize returns on a global scale.

With regard to neutral states, the Bucy recommendations were even more restrictive. The task force noted that many such states were in the process of building technology bases and hence were potential "pipelines" for technology transfer to the Communist world.[37] Sweden, Switzerland, Austria, and the Middle East were specifically mentioned; Brazil, India, and the newly industrializing states of the Pacific Rim also fit the category. The report argued that because these governments viewed reexport controls as an infringement on their sovereignty, their enforcement efforts were severely lacking.[38] Some firms complied, but reexport controls alone could do little to prevent the transshipment of U.S. strategic technology.[39] Only the Swiss government maintained a viable export control system. Given the "uncertain control and enforcement environment" in these states, the report recommended that "the U.S. should release to neutral countries only the technology we would be willing to transfer directly to Communist countries."[40] In other words, the risk of unauthorized diversion was sufficiently great that for all practical purposes, neutrals should be treated as if they were Communist states. Again, such a restriction could have a profoundly detrimental effect on the export operations of U.S. firms. Moreover, prohibiting the export of strategic technology to neutral states could lead to restrictions on U.S. trade with its CoCom allies, even if there were absolute agreement within CoCom. Since many CoCom members did not impose reexport controls on their trade with neutrals, it would be imprudent for the United States to transfer technologies to CoCom members that in turn exported to neutral states.

Once the Bucy Report's concern with intra-Western technology transfer is recognized, the implications of its central recommendation—that controls shift from products to technology—can be appreciated more fully. U.S. control regulations did, in fact, cover technology, in the form of restrictions on technical data, which was defined as "information of any kind that can be used, or adapted for use, in the design, production, manufacture, utilization or reconstruction of articles or materials."[41] Exports to the East of technical data related to items on the CoCom list already required a validated license. The export of such data to the West,

37. Ibid., p. 20.
38. After 1968, the Swedish government was no longer willing to cooperate with U.S. export controls, leaving the decision to comply fully in the hands of private firms. See Gunnar Adler-Karlsson, "International Economic Power: The U.S. Strategic Embargo," *Journal of World Trade Law* 6 (September 1972), 506.
39. A National Academy of Sciences study concluded in 1987 that "independent foreign companies are either ignorant or casual in their compliance with U.S. re-export controls." See *Balancing the National Interest* (Washington, D.C.: National Academy Press, 1987), p. 107.
40. *Bucy Report*, p. 22.
41. National Academy of Sciences, *Balancing the National Interest*, p. 87.

however, could take place under general license, so long as the recipient provided written assurances that the unauthorized reexport of the data would not take place. Thus the report's contention that the control system overemphasized product controls and underemphasized technology controls applied more appropriately to intra-Western than to East-West trade. By implication, the proposed trade-off—a relaxation of product controls in exchange for tighter technology controls—meant, in effect, looser *product* controls on trade with the *East* and tighter *technology* controls on trade with the *West*.

The Bucy Report's recommendations thus posed a potentially serious threat to the complex web of relationships, which had developed in the era of Western interdependence, between U.S. firms and those of the rest of the non-Communist world. If a full panoply of intra-Western restrictions were adopted, the extent of interference with the global trading system would depend primarily on how broadly "strategic" technology was defined. As we have seen, the Bucy Report left room for a wide range of interpretations. In the best case for intra-Western trade, the United States would develop a very streamlined strategic embargo, and other CoCom members would accept it and agree to enforce it faithfully. This situation would obviate the need for U.S. sanctions against CoCom members. The problem of dealing with neutrals would still remain but would be mitigated as products of lesser strategic significance were dropped from the control list. In the worst case, U.S. officials would propose broad technology controls without liberalizing product controls. CoCom members and non-CoCom suppliers would resist, and the United States would find itself compromising the liberal trading system in the interest of assuring that U.S.-origin technology did not reach the Soviet Union.

Within the U.S. government, the crucial task of implementing the Bucy Report fell primarily to the Department of Defense. Although Defense was not the lead agency in export control policy, conservative critics of détente lobbied hard for an expansion of its role. The DoD's formal powers were enhanced by the Export Administration Act Amendments of 1974, which gave the secretary of defense the authority to review all license applications for exports to Communist countries and to recommend denial to the president on the grounds of national security.[42] A presidential decision to override the Defense Department in any given case was to be submitted to the Congress, which could, in turn, override the president with a majority vote. Now, the Bucy Task Force was proposing a further expansion of DoD's role. It urged that Defense take the initiative in determining which technologies were critical and in

---

42. The amendments are discussed in OTA, *Technology and East-West Trade* (Washington, D.C.: GPO, 1979), pp. 119–20.

devising strategies to protect them.[43] This proposal meant, in effect, that the Defense Department would be primarily responsible for interpreting the Bucy Report and for translating its ambiguous and controversial directives into an export control strategy.

THE BUCY REPORT: IMPLEMENTATION

The Defense Department: Economic Warfare Rejected

The DoD formally accepted the recommendations of the Bucy Report as guidance for policy in August 1977.[44] In a series of policy statements, agency officials made it clear that the DoD viewed export control policy in the larger context of U.S. defense strategy. That is, export controls were intended to contribute to deterrence. As conceived by the DoD, both conventional and nuclear deterrence required the United States to maintain qualitative (i.e., technological) superiority in order to offset quantitative Soviet advantages in the deployment of military systems. The maintenance of qualitative superiority, in turn, required the protection of American lead time in the application of new technologies to military systems. By the mid-1970s, however, DoD officials were expressing concern that U.S. technological superiority had been eroded by the Soviet Union in a number of key areas.[45] By attempting to combine qualitative parity with already existing quantitative advantages, the Soviets threatened to achieve a posture of overall military superiority and thus to undermine the effectiveness of U.S. deterrence.

To maintain technological superiority and thus deterrence, the DoD concluded that the United States needed to rely on four principal mechanisms.[46] First, it needed to make real increases in technology-intensive military research and development. By 1979, the DoD estimated, the Soviet Union was spending twice as much as the United States on mili-

43. *Bucy Report,* p. 28.
44. See Harold Brown, "Interim DoD Policy Statement on Export Control of United States Technology," Memorandum for the secretaries of the military departments et al. (Washington, D.C.: Office of the Secretary of Defense, August 26, 1977). Within the department, primary responsibility for implementing what became known as the "critical technologies approach" to export control was lodged in the Office of the Undersecretary of Defense for Research and Engineering. See statement of Defense Undersecretary William Perry in U.S. Congress, House, Committee on Foreign Affairs, Subcommittee on International Economic Policy and Trade, *Technology Exports: DoD Organization and Performance,* hearings, 96th Cong., 1st sess., October 30, 1979, p. 18.
45. See the comments of Malcolm Currie, on exiting office as director of Defense Research and Engineering. "Technological Superiority Required," *Aviation Week and Space Technology,* February 2, 1977, p. 7.
46. See the statement of William Perry in U.S. Congress, Senate, Committee on Banking, Housing and Urban Affairs, Subcommittee on International Finance, *Trade and Technology, Part II: East-West Trade and Technology Transfer,* hearings, 96th Cong., 1st sess., November 28, 1979, pp. 25–26.

tary R&D. Second, the United States needed to coordinate defense projects with its NATO allies in order to maximize the benefits of technological collaboration. Third, as militarily critical technologies were increasingly developed in the private sector, the U.S. government needed to support, and exploit, commercial advances in technology. Finally, controls on the export of militarily critical technologies to the Soviet Union needed to be strengthened. Defense Undersecretary William Perry argued that the DoD should pursue these four mechanisms simultaneously, as part of an integrated program, while attempting to minimize the conflicts among them.[47]

The DoD accepted the Bucy Report recommendation for a shift in emphasis from product to technology control. Its working assumption was that there existed a "small subset" of militarily critical technologies, which was identifiable and relatively stable over time.[48] Once those technologies had been identified and controlled, restrictions on the export of many less crucial end products could be relaxed. By taking this position, the DoD was able to support both the expansion of American exports and the protection of national security. As Deputy Assistant Secretary Ellen Frost testified before Congress:

> My first point may come as something of a surprise to some of you. It is simply this: in the Department of Defense we share this administration's assumptions that export controls, as such, are inherently undesirable. We take it for granted that the welfare of the United States and the rest of the world is best served when trade among all nations is unfettered by governmental restraints. We believe that U.S. companies can and should export what they wish to sell abroad, unless such exports interfere with something the Executive branch is directed by the Congress to do—in this case to protect and enhance the military security of the United States.[49]

Similarly, other officials noted that although the DoD had typically been perceived as having a "Neanderthal mentality" with regard to exports, the fact that it accepted the critical technologies approach (CTA) "puts that notion to rest," since the CTA would ultimately result in a net relaxation of controls.[50]

The DoD's rejection of economic warfare was as explicit as was its support for a selective strategic embargo. In response to the question of

---

47. See the statement of Perry in *Technology Exports: DoD Organization and Performance*, p. 4.

48. Statement of Defense Deputy Undersecretary Ruth Davis in U.S. Congress, House, Committee on Foreign Affairs, Subcommittee on International Economic Policy and Trade, *Extension and Revision of the Export Administration Act of 1969*, hearings, 96th Cong., 1st sess., February 15–May 9, 1979, pp. 403, 407.

49. Ibid., p. 154.

50. "DoD Revises Policy on Export," *Aviation Week and Space Technology*, September 12, 1977, pp. 12–13.

whether the United States should deny technology that contributed to the civilian sector of the Soviet economy because it allowed the Soviets to devote more energy to military pursuits, Deputy Undersecretary Ruth Davis asserted: "That is an argument that is stated quite frequently. Of course, it has logical merit to it. However, that is an extreme to which in this country, it would not be wise to go, and an extreme to which the Department of Defense would certainly not advocate going."[51] In the same set of congressional hearings, Ellen Frost was asked whether the DoD believed that the export of oil and gas exploration equipment should be denied to the Soviets because it contributed to their overall military potential.[52] She conceded that increasing Soviet oil production could enhance the growth rate of the Soviet industrial base and permit the Soviets to increase their hard currency earnings. For the purposes of export control, however, these factors were of little consequence to the DoD. The crucial questions were whether an increase in Soviet oil production would *directly* enhance Soviet military capabilities and to what extent such a contribution would be militarily significant. As the Soviet military consumed only a small fraction of total Soviet oil output and would enjoy priority over the civilian sector in event of a shortage, neither the export nor the denial of Western energy technology would have any discernible impact on Soviet military capabilities.[53] Thus the Department of Defense had no objection to its sale.

In rejecting economic warfare, the DoD reflected the preference of the Carter administration for a nonconfrontational political relationship with the Soviet Union and for the expansion of U.S. exports.[54] Moreover, DoD officials viewed economic warfare as unnecessary from the perspective of U.S. defense strategy. In Chapter 2 I argued that the most attractive target for economic warfare was the state perceived to be mobilizing for an imminent, protracted war, typified by the two world wars. In such conflicts industrial capacity is a key military asset, and exports of technology or equipment that strengthen the target's industrial base or could be converted to military output during wartime are properly viewed as a threat to national security. DoD officials, however, did not conceive of the Soviet Union as mobilizing for a long, conventional struggle, and U.S. defense planning did not assign a high priority to preparing for such a contingency. U.S. planning focused on the deterrent effects of nuclear weapons; in the event deterrence failed (e.g., in central Europe), U.S. and NATO war plans anticipated a conventional struggle of limited duration and called for the early use of nuclear weapons. Rapid mobilization and the resolve to escalate would play a

51. *Extension and Revision*, p. 435.
52. Ibid., p. 177.
53. Ibid.
54. Statement of William Perry in *Technology Exports: DoD Organization and Performance*, p. 2.

greater role than would the depth or endurance of U.S. and Soviet industrial capacity.

Rather than anticipating or preparing for a protracted conventional struggle, DoD officials viewed the United States and the Soviet Union as locked in an ongoing technological arms race of indefinite duration.[55] By staying ahead, the United States could assure that neither war nor the end of the race would occur. Export controls were needed solely to prevent the Soviets from reaching parity; they were not viewed as a promising weapon that could be used to hamper the ability of the Soviets to run the race at all. The denial of technology that would contribute significantly to the Soviet economy was not perceived as a strategic opportunity, and the sale of such technology was similarly not seen as posing a strategic risk. As Maurice Mountain of the DoD argued:

> [There is a] natural tendency to regard all things that may somehow contribute to military power as having direct military value. To a degree this view is correct, but it is far too general to be useful. The simplest way to avoid such confusion is to note that, whatever impact a technology may have on the quality of human life, the domestic economy, foreign trade, the balance of payments or even diplomatic relations, its national security significance depends entirely on the extent to which it is or may be applied to a *specifically military purpose*. Thus a technology which has no present or future military application can safely be ignored in the control process. . . . *the necessary link between technology and national security is the production of military weapons systems*.[56]

As the above statement and those of other defense officials suggest, the control or promotion of technology transfer could serve purposes other than that associated with a strategic embargo. The design of a broader strategy, however, was beyond the institutional responsibility of the DoD.

> Ever since 1969, the Department of Defense has made its export control judgements entirely in terms of the narrow definition, that is, the military significance of a given transaction or set of transactions. That is what we are best qualified to determine. We are keenly interested in all developments which alter the nature and extent of Soviet power, whether they are military, political or economic in nature. We know that the Soviet threat does not lie solely in warheads or silos, but also in perceptions of power and influence and the overall foundation on which they rest. But when we speak of export controls in terms of this larger meaning of national security, we must defer to the Department of State and N.S.C.[57]

55. See the comments of Ruth Davis in *Extension and Revision*, p. 435.
56. Maurice Mountain, "Technology Exports and National Security," *Foreign Policy*, no. 32 (Fall 1978), 95, emphasis added.
57. Statement of Ellen Frost in *Extension and Revision*, p. 176.

The Dresser Case

Even if not within the DoD, there was support in the United States before 1980 for a broader interpretation of the Bucy Report and for more comprehensive restrictions on U.S. technology exports to the Soviet Union. Proponents included conservatives in Congress (e.g., Senators Henry Jackson and Jake Garn), some members of the Carter National Security Council, and the principal author of the report, J. Fred Bucy. The conflict between their "hard-line" position and the relatively "liberal" position of the DoD was most starkly revealed in the debate over a controversial technology transfer case. Because the Dresser case was subject to such intense public and congressional scrutiny, it forced U.S. policy-makers to reveal their export control preferences and the assumptions underlying them.

In 1978, after years of difficult negotiations with the Soviets, Dresser Industries of Dallas was awarded a $144 million contract to provide a turnkey factory, including the necessary technical data, to manufacture drill bits to be used in Soviet oil exploration.[58] Having nearly exhausted their most readily accessible petroleum reserves, the Soviets needed to dig very deep wells, in difficult terrain, in order to maintain and expand production. Given the limitations of their technology, however, the Soviets were unprepared to produce the quality drill bits in quantities sufficient to undertake the task on a massive scale. The Dresser factory could help to alleviate this crucial economic bottleneck because it was designed to produce 100,000 quality drill bits annually.

Dresser was required to obtain a validated license for the export of both the technical data (the blueprints that explained the production process) and the computer-controlled electron beam welder, which was an essential part of the process line. As drill bits were not on the CoCom list, the technical data to produce them was not controlled multilaterally. The electron beam welder was on the CoCom list, although the United States had the discretion to export it as an administrative exception, as long as it reported the sale to CoCom.[59] Both applications were forwarded by the Commerce Department to the DoD for technical review. Defense officials found nothing objectionable, and in May 1978, they approved the export of the technical data and, in August, the electron beam welder.

Word of the approval generated a strong negative reaction from conservatives in Congress and on the NSC staff. The former were alarmed

58. See U.S. Congress, Senate, Committee on Governmental Affairs, Permanent Subcommittee on Investigations, *Transfer of Technology and the Dresser Industries Export Licensing Actions*, hereafter cited as *Dresser Case*, hearings, 95th Cong., 2nd sess., October 3, 1978. A succinct summary of the DoD position in the case is provided by Ellen Frost in *Extension and Revision*, pp. 181–83.
59. *Dresser Case*, pp. 14, 140–41.

by the factory's potential contribution to Soviet industrial capabilities and by the fact that the administration was de facto setting U.S. policy on Soviet energy without consulting Congress. At the time, the issue of whether the United States should help the Soviets solve their energy problems, thus reducing Soviet incentives for intervention in the Persian Gulf, or refuse to assist, thus exacerbating Soviet economic difficulties, was subject to lively debate.[60] The NSC staff, for its part, was annoyed that export control officials had approved an important sale to the Soviets without allowing the White House to exploit the potential political leverage the export might have provided. Recall that in early August 1978, oil and gas exploration equipment had been placed on the Commodity Control List as part of the tactical linkage strategy favored by National Security Adviser Brzezinski. Since the NSC staff considered the Dresser technology to be unique, or available only in the United States (a point hotly disputed by Dresser's president, Jack Murphy), the license request provided an ideal opportunity for the exercise of linkage.[61] Sustained pressure from Congress and the NSC was sufficiently powerful politically that President Carter was convinced, shortly after the licenses had been approved, to hold them back from Dresser, pending further investigation.

The Defense Department was directed to reconsider the case. In order to assist it, the DoD took the unusual step of asking the Defense Science Board to prepare an independent assessment of the proposed sale. The Defense Science Board assigned the task to J. Fred Bucy, which meant, in effect, that Bucy was given the opportunity to interpret, in an actual case, the ambiguous guidelines put forth in the report that bore his name.

Bucy's findings on the Dresser case ran directly counter to those of the Defense Department. Specifically, he determined that:

(1) Deep-well drilling is a critical technology, wholly concentrated in the United States.
(2) Rock drill bit manufacture in high volume and its supporting design and application techniques are critical technologies.
(3) Computer-controlled electron beam welding has military significance, in that it can be used to make aircraft parts and can be diverted easily.
(4) Tungsten carbide inserts (the items produced by the Dresser technology in an intermediate step in the production of drill bits) have military significance, in that they can be diverted easily into the production of armor-piercing projectiles.

60. See, for example, the editorial in *Washington Post*, October 11, 1978.
61. Statement of Jack Murphy, *Dresser Case*, pp. 171–212.

Thus, Bucy concluded that neither the export of technical data for drill bit manufacture nor that of the electron beam welder should be approved. The United States could approve the sale of drill bits themselves to the Soviet Union but should carefully monitor the quantity sold.[62]

Bucy's general recommendation—that technology should be restricted, whereas products could be approved—was consistent with the thrust of the 1976 report. His third and fourth findings were conceptually compatible with the narrow interpretation of the Bucy Report preferred by the DoD. By defining deep-well drilling and rock drill bit manufacture as critical technologies without linking them directly to Soviet military capabilities, however, Bucy in effect applied the broader (i.e., economic warfare) interpretation of the report. He contended before Congress that these technologies were significant to national security "in the broadest sense."[63] In addition to differing with the DoD over its evaluation of the individual components, Bucy explicitly criticized the DoD for failing to make an overall assessment of the manufacturing process as a complete system.[64]

In other words, along with transferring the technology to produce drill bits, the United States was inadvertently passing on to the Soviets other know-how as well. For example, by becoming familiar with the Dresser process, the Soviets would improve their skills in metallurgy. The DoD, Bucy believed, should have made an assessment of whether enhancing Soviet metallurgical capabilities was in the national security interest of the United States. More important, by allowing the Dresser sale, the United States would be providing the Soviets with "process know-how"—the knowledge, methods, and skills generally associated with industrial production on a mass scale. Process know-how, Bucy emphasized, should certainly be kept away from the Soviets. Applied as a general rule, the denial of process know-how would be tantamount to prohibiting the export of turnkey factories (and perhaps individual pieces of industrial equipment), regardless of whether they possessed direct military significance.

Despite his stature in the defense community and the strong support for his position in the N.S.C., the military services, and among conservatives in Congress, Bucy's advice was rejected by the Department of Defense. With regard to the beam welder, DoD held firm in its judgment that current and anticipated future foreign availability made it futile for the United States to deny it to the Soviets.[65] Similar beam welders were available in Sweden and Switzerland. Defense officials also emphasized

62. J. Fred Bucy, "An Assessment of Rock Drill Bit Technology," reprinted in *Dresser Case*, pp. 6–10.
63. *Dresser Case*, p. 150.
64. Ibid., p. 148.
65. Ellen Frost, *Extension and Revision*, p. 183.

that to produce anything other than drill bits, the welder would require significant modification. In any case, the Soviets were unlikely to divert it given their pressing need to alleviate the bottleneck in quality drill bit production. With respect to the tungsten carbide inserts, the DoD similarly judged that Soviet armor-penetrating capability would not be significantly enhanced by diversion and, again, that the Soviets would probably not divert given the importance of the drill bit facility. In summary, DoD decided that both items had military utility but were not militarily *critical*.[66]

With regard to the entire Dresser factory, defense officials frankly acknowledged that the whole might be greater than the sum of its parts. The indirect benefits of selling an advanced manufacturing process might be more important to the Soviets than the value of each piece of equipment viewed in isolation. Nevertheless, DoD shied away from this broader concern and also from the broader issue of whether U.S. security was increased by assisting Soviet energy development. Despite persistent prodding by Senators, DoD officials refused to take a position on these questions and focused exclusively and narrowly on the direct military significance of the transfer.[67] As Deputy Secretary of Defense Charles Duncan put it in a memo to Senator Jackson, "Our DoD reassessment of our national military interests per se as distinguished from the broader national strategic interests . . . reconfirms previous evaluations which showed small risk to military security from the transfer of any of the components involved in the Dresser application."[68] Undersecretary Perry frankly acknowledged that the Dresser technology would give the Soviets a distinct *economic* advantage, but as far as DoD was concerned, this gain alone was not grounds for denial.[69] Rock drill bit manufacturing, a critical technology to Bucy and others because of its energy and industrial significance, was not a *militarily* critical technology to the DoD. To recommend its denial on the basis of its indirect strategic significance was economic warfare, and the Defense Department was not prepared to take such a step.

Neither, it turned out, was the Carter administration as a whole. In September 1978, President Carter resolved the debate over Dresser by announcing that he had decided to approve the validated license applications. Carter, in effect, sided with the somewhat unexpected "alliance" of the Defense Department and the business community, against the

66. See the statement of William Perry in *Dresser Case*, p. 93.

67. Comments of Perry and Frost in ibid., pp. 98–99. Defense officials claimed to defer to the State and Energy Departments for a position on the broader strategic (as opposed to narrow military) aspects of the Dresser case.

68. Charles Duncan, "The Dresser Industries' Rock Drill Bit Manufacturing Plant Export Application," Memorandum, reprinted in *Dresser Case*, p. 11.

69. *Dresser Case*, p. 93.

architect of the Bucy Report and the hard-liners in Congress and the NSC.

The Dresser approval served as an impetus to the growing minority in Congress that favored a return to economic warfare. In September 1978, seventy-seven congressmen introduced the Technology Transfer Ban Act of 1978.[70] The bill stated that the United States lacked a coherent national policy on technology transfer to the Communist world; that Soviet behavior demonstrated that American and Soviet views of détente basically differed and belied the expectation that an increase in economic interdependence with the West would moderate Soviet military or political behavior; and that trade with the West was being used by the Communist world to acquire strategic technology.[71] As a response, the bill proposed to restrict the export of goods and technologies that could make *any* contribution to any nation's military *or* economic potential that could prove detrimental to the United States. By proposing a return to the language adopted in the 1962 amendment but abandoned in 1969, the sponsors clearly hoped to move U.S. policy beyond a narrow strategic embargo and to prohibit exports—such as the Dresser drill bit plant—that contributed significantly to the Soviet economy, even if they had no direct military utility.[72]

A return to such a broader embargo, however, could not command the support of a majority in the House. By March 1979, conservatives in Congress were forced to lower their expectations and proposed a somewhat less restrictive measure (H.R. 3216).[73] This bill suggested a distinction between "critical" products and technologies—those indispensable to U.S. military systems and superior to those of controlled nations—and "significant" products and technologies—ones that would make an important contribution to an adversary's military potential, even if they were obsolete by U.S. standards. The former were to be unconditionally denied to the East, whereas the latter would be subject to validated license approval. Unlike the earlier proposal, H.R. 3216 did not call explicitly for the denial of trade of economic significance to the Soviets. It did, however, recommend that the DoD be given primary responsibility for administering export controls. Despite the DoD's position in the Dresser case, conservatives still believed that their interest in tighter restrictions on technology transfer to the Soviets would be better served if Defense, rather than Commerce, were put in charge.

70. The bill is discussed in *Aviation Week and Space Technology*, March 19, 1979, pp. 27–28.
71. OTA, *Technology and East-West Trade*, p. 122.
72. See Jeffrey Golan, "U.S. Technology Transfers to the Soviet Union and the Protection of National Security," in *Law and Policy in International Business*, vol. II (1979), pp. 1080–82.
73. H.R. 3216 is outlined by its sponsors, Lester L. Wolff and Clarence E. Miller, in *Extension and Revision*, pp. 537–41.

H.R. 3216 also died in Congress. As I noted in Chapter 5, the legislation finally adopted in 1979 reflected the generally pro-trade sentiment of Congress and the Carter administration. It did not call for economic warfare, left the main responsibility for controls in the hands of Commerce, and placed restrictions on the president's ability to use foreign policy controls. The EAA of 1979 did, however, endorse the idea of focusing controls on militarily critical technologies, and it directed the DoD to produce an initial list by August 1, 1980.[74]

The Conflict over Intra-Western Controls

As was the case for economic warfare, there was support within the United States during the late 1970s for the most controversial of the Bucy Report's recommendations: a more restrictive approach to U.S. technology transfers to the non-Communist world. In addition to raising national security concerns, the Bucy recommendations struck a responsive chord with those who viewed intra-Western technology transfers as a threat to the national *economic* interests of the United States. By the mid-1970s, a number of scholars and analysts were expressing the concern that U.S. firms were eroding the overall competitive position of the nation's industry by liberally sharing their advanced technology with foreign affiliates.[75] The problem, critics charged, was that in addition to "dated" technology, American firms often transferred state-of-the-art design and manufacturing techniques. For example, General Electric shared its most sophisticated jet engine technology with SNECMA, the French government-controlled firm, in exchange for a continued foothold in the West European market. Similarly, Amdahl, a small offspring of IBM, shared advanced computer design and manufacturing know-how with Japan's Fujitsu, in exchange for successive rounds of finance capital.[76] Although these and similar arrangements benefited the particular U.S. firms involved, in some cases assuring their survival, they potentially worked to the detriment of U.S. industry as a whole by creating additional sources of foreign competition. In the all-important electronics sector, for example, despite the fact that most integrated circuit technology was initially created in the United States, by 1978 the United States found itself in the alarming position of having become a heavy net importer of electronics and communications equipment incorporating that technology.[77]

74. *Export Administration Act of 1979*, PL 96-72, September 29, 1979, sec. 5(d).
75. See, for example, Jack Baranson, "Technology Exports Can Hurt Us," *Foreign Policy*, no. 25 (Winter 1976–77), pp. 180–94; Herbert E. Meyer, "Those Worrisome Technology Exports," *Fortune*, May 22, 1978, pp. 106–9; and Robert G. Gilpin, *U.S. Power and the Multinational Corporation* (New York: Basic Books, 1975).
76. Baranson, "Technology Exports Can Hurt Us," pp. 182–83.
77. Meyer, "Those Worrisome Technology Exports," p. 108.

Although he did not explicitly address it in the report, Bucy viewed the economic implications of technology imports as a matter of national concern.[78] Writing in the *New York Times* in 1976, he noted that design and manufacturing know-how was being transferred to many nations that could "use it to beat us in the international marketplace." He argued further: "The threat is therefore twofold. Exporting design and manufacturing know-how to potential enemies strengthens them militarily. And exporting that same know-how to potential economic competitors— friends or foes—strengthens them to compete against us for world markets."[79] The potential economic dangers of technology transfer provided an additional rationale for those U.S. officials who preferred tighter restrictions on intra-Western trade in order to prevent the leakage of U.S. technology to the East.

Support for the Bucy Report's more restrictive approach to intra-Western transfers was found in both the House and the Senate. H.R. 3216 called for validated license requirements on the transfer of critical goods and technologies to *all* destinations. This new obligation would greatly increase controls on U.S. trade with non-Communist countries because at this time, as noted above, only strategic products required validated licenses for shipment to the West. In addition, H.R. 3216 also stipulated that no license be granted to any Western firm whose *government* failed to provide "adequate assurances" regarding reexport. Representatives Wolff and Miller contended that this provision was necessary because of the ineffective control systems of CoCom and other non-Communist states.[80] They believed that responsibility for protecting U.S. technology needed to be placed squarely on other Western governments rather than solely on the firms that imported the technology.

The existing regulations had, of course, served to mitigate the impact of U.S. East-West trade controls on U.S. diplomatic relations. Western governments could claim they did not recognize U.S. reexport authority, even if their firms felt compelled to comply with the regulations for fear of losing access to U.S. sources of supply. The requirement that governments provide explicit assurances would force the United States and its major trading partners to confront the issue of extraterritoriality directly. Although the possibility of diplomatic conflicts may have troubled the State Department, those in favor of tightening the loopholes in the U.S. embargo appeared to consider it worth the risk and to believe that if Western governments were unwilling to guarantee the protection of U.S. technology, their firms did not deserve the privilege of receiving it.

78. Former British CoCom delegate Roger Carrick has claimed the existence of a hidden agenda of intra-Western technological protectionism in the Bucy Report. *East-West Technology Transfer in Perspective* (Berkeley, Calif.: Institute of International Studies, 1978), p. 84.
79. J. Fred Bucy, "Going, Going, Goooonnnnne," *New York Times*, September 11, 1976.
80. *Extension and Revision*, pp. 538–39.

Sympathy for that position was found in the Senate as well. During the debate over the renewal of the EAA, a minority faction led by Senator Jackson called for the use of U.S. trade sanctions against CoCom allies that made available to the East items controlled by the United States for national security purposes.[81] Reminiscent of the early 1950s, this provision would, in effect, require CoCom members to accept fully what the United States proposed for alliance control. Sanctions would be imposed if U.S. officials failed to persuade the allies in negotiations to accept their proposals. A similar provision was found in H.R. 3216, which attempted to revive the Battle Act by calling on the president to deny military and economic aid to recalcitrant CoCom members.[82] As the State Department noted in its comments on the proposal, not only was it likely to be counterproductive diplomatically, but it also failed to account for the fact that the United States no longer provided significant amounts of aid to its CoCom allies.[83]

From the perspective of U.S. exporters, nothing in the Bucy Report was viewed as potentially more destructive than the possibility of increased intra-Western restrictions. U.S. corporate officials generally received the report with cautious optimism. With regard to exports to the East, they welcomed the intended relaxation of product controls in exchange for tighter restrictions on a selected set of technologies. Some did fear the possibility that the government might ultimately increase its controls on technology without significantly liberalizing product controls.[84] Regardless of whether products or technologies were at issue, it was generally easier in the United States to have items added to the list than to have them removed. In any event, the ultimate impact of the report's recommendations on Eastern trade was of far less concern than its potential impact on the global operations of U.S. firms. As an extensive commentary by the Machinery and Allied Products Institute (MAPI) put it:

> In part because U.S. trade and commercial relationships with communist countries are not substantial in terms of U.S. interests worldwide, U.S. industry is especially disturbed about the implications of the DoD initiatives with respect to commercial relationships with *noncommunist* countries. Any benefits gained from reducing the risk of diversion

81. "Technology Export Control Advances," *Aviation Week and Space Technology*, July 30, 1979, p. 16.
82. *Extension and Revision*, p. 540.
83. See the comments of William Root, director, Office of East-West Trade, Department of State, in U.S. Congress, Committee on Banking, Housing, and Urban Affairs, Subcommittee on International Finance, *U.S. Export Control Policy and the Extension of the Export Administration Act*, pt. III, hearings, 96th Cong., 1st sess., May 3, 1979, pp. 167–69.
84. See the statement of Thomas Christiansen, representing the American Electronics Association, in U.S. Congress, House, Committee on International Relations, Subcommittee on International Economic Policy and Trade, *Export Licensing: CoCom List Review Proposals of the United States*, hearings, 95th Cong., 2d sess., June 14 and 26, 1978, pp. 63–65.

from noncommunist countries must be weighed against the disadvantages to U.S. industry (which would be refused the right to exploit the innovation internationally) and the U.S. foreign trade position, to foreign countries which would be denied the economic benefits of the technology, and to U.S. diplomatic relations with the affected countries.[85]

U.S. exporters suspected that intra-Western trade was, in fact, the primary target of the Bucy Report, despite its emphasis on East-West trade and Soviet military capability. Their suspicion was justified; as noted above, the export of strategic technology (i.e., proprietary technical data) to the East was already under validated license control. As the Defense Department conceded, as far as Eastern exports were concerned, shifting from product to technology control merely constituted a change in emphasis.[86] This was not the case with regard to non-Communist markets, and given their substantially greater stake in such markets, U.S. exporters were understandably disturbed at the prospect of tighter controls. Expanded restrictions on the export of know-how threatened to hamper the flow of information from U.S. corporations to their foreign subsidiaries, affiliates, and licensees. The problems posed by export controls in trade with the East—delays, uncertainty, and unpredictability in the licensing process—might be posed in the more significant realm of intra-Western trade. Moreover, once technology was formally under control to all destinations, the U.S. government might be tempted to employ controls for foreign policy as well as national security purposes. Eventually, foreign firms might be compelled to seek other sources of supply, and foreign governments could be expected to retaliate, thereby reducing the flow of technology into the United States.[87]

Although a minority in Congress supported intra-Western controls and the business community generally opposed them, the position of the DoD was ambivalent. On the one hand, DoD officials recognized that the liberal policy of the United States on technology transfer to friendly and allied countries posed national security risks.[88] Given their pro-trade orientation and their weak control and enforcement policies, other Western states could have U.S. critical technology transshipped through them to the Communist world. In addressing this concern, Defense Secretary Harold Brown recommended that the export of critical technology to all destinations require a validated license, which would enable the govern-

85. Machinery and Allied Products Institute, *U.S. Technology and Export Controls* (MAPI, April 1978), p. 5.
86. Statement of Ruth Davis in *Extension and Revision*, p. 434.
87. MAPI, *U.S. Technology and Export Controls*, pp. 50–51.
88. Maurice Mountain, "The Continuing Complexities of Technology Transfer," in Gary K. Bertsch and John R. McIntyre, eds., *National Security and Technology Transfer: The Strategic Dimensions of East-West Trade* (Boulder, Colo.: Westview Press, 1983), pp. 18–20.

ment to monitor, though not necessarily prohibit, intra-Western trade in critical technology.[89]

At the same time, however, DoD officials were sensitive to the costs of increased intra-Western restrictions and, in particular, to the possibility that such restrictions would jeopardize collaboration between the United States and its NATO allies in defense technology. The Carter administration entered office with a major commitment to improving NATO RSI (rationalization, standardization, and interoperability). In order to maximize the return on the collective NATO investment in research and development, the administration proposed that the United States and Western Europe share critical technology more fully.[90] It pledged to allow the "early release" of U.S. critical technology to West European firms, as a means to encourage coproduction, and also to allow those firms to participate in the development of new U.S. weapons systems. Such initiatives would clearly be difficult to undertake in the face of strict controls on the transfer of technology to U.S. allies. In light of the potential conflict between the two initiatives and the need to make a choice, William Perry acknowledged that his office considered NATO collaboration to be a higher priority than tighter intra-Western export controls.[91]

The intra-Western trade issue generated sufficient controversy for the Carter administration to address it explicitly in a report to the Congress in 1978.[92] To the relief of the American export community, the administration decisively rejected the introduction of a more restrictive approach to U.S. technology trade with the Western world. It contended that such restrictions would have a "stultifying" effect on U.S. innovation and thus on the position of the United States in the global economy.[93] Efforts to retain specific technologies in the United States would not necessarily ensure continued U.S. superiority; rather than "hoarding" technologies, U.S. interests were better served by the maintenance of free trade combined with a vigorous R&D effort at home.[94] The president's report strongly supported the transfer of technology to NATO allies in the interest of improving defense collaboration. And on the crucial issue of technology diversion, the report stated: "Even taking into account the possibility that some leakage may have occurred there is no apparent need for tighter controls vis-à-vis other Western countries. Any

89. Harold Brown, "Interim DoD Policy Statement," p. 2.
90. "DoD Revises Policy on Export." See also Harold Brown, "Interim DoD Policy Statement," p. 3.
91. Personal interview with William Perry, April 23, 1981.
92. *International Transfer of Technology*, Report of the President to the Congress, prepared for U.S. Congress, House, Committee on International Relations, Subcommittee on International Security and Scientific Affairs, 95th Cong., 2d sess., December 1978.
93. Ibid., p. 17.
94. Ibid., p. 26.

effort to impose such controls would have disruptive political and economic effects, and would have the potential for undercutting cooperation with other nations controlling exports."[95] The report argued elsewhere that tighter controls on U.S. exports to other Western countries would have a "disproportionately high political cost in terms of weakening our alliance relationships."[96] For the Carter administration, the potential increase in the protection of U.S. technology afforded by increased vigilance over intra-Western trade was not justified by the political and economic costs, to the United States and within the Western alliance, that such a strategy entailed.

Following the preference of the administration, and the majority in Congress, the EAA of 1979 endorsed neither more restrictive intra-Western controls nor sanctions against uncooperative CoCom members. Rather, in the case of "foreign availability" of items that U.S. officials deemed to be of national security significance, it directed the president to enter negotiations in the attempt to eliminate the alternative source of supply.[97] The legislation did not stipulate further action, however, in the event negotiations failed.

The Military Critical Technologies Effort

Having decided to focus narrowly on the control of technology of direct military significance, the DoD began in 1977 to construct its initial Military Critical Technologies List (MCTL). Its first step was to determine which broad areas of technology were of primary concern to U.S. strategic lead time. By 1979, the following fifteen had been selected:

(1) Computer network technology
(2) Large computer system technology
(3) Software technology
(4) Automated real-time control technology
(5) Composite and defense materials processing and manufacturing technology
(6) Directed energy technology
(7) Large-scale integration (LSI) and very large-scale integration (VLSI) design and manufacturing technology
(8) Military instrumentation technology
(9) Telecommunications technology
(10) Guidance and control technology
(11) Microwave component technology

95. Ibid., p. 11.
96. Ibid., p. 24.
97. *Export Administration Act of 1979*, sect. 5(f)(4).

(12) Military vehicular engine technology

(13) Advanced optics (including fiber optics) technology

(14) Sensor technology

(15) Underseas system technology[98]

Second, the DoD sought to determine which specific technologies within each broad category were of critical military value. Within the composite materials area, for example, laser and plasma arc welding were deemed militarily critical, whereas electron beam welding was not.[99] Third, for each specific technology, arrays of know-how, keystone products and materials, and products that reveal know-how needed to be identified. Finally, for each specific militarily critical item, the DoD planned to make an assessment of Soviet capabilities, the differential between U.S. and Soviet capabilities, and potential alternative sources of supply, in order to determine which items should actually be subject to export controls.

The MCTL project turned out to be far more time-consuming than either the Bucy Task Force or the DoD had anticipated. By the end of 1979, very little progress had really been made. The DoD had not completed the steps outlined above and thus was unprepared to provide the administration with an actual control list that could serve as an alternative to, or something to be merged with, the existing CCL. The EAA of 1979 directed the Defense Department to produce an initial MCTL by August 1, 1980; on that date, the DoD made available only a table of contents, which contained well over one hundred general-technology areas within the broad categories listed above. It conceded that "the specificity with which the technology elements within these areas are identified in many cases needs further refinement for control decisions."[100]

In part, the inability of the DoD to make significant progress can be explained by the ambitious nature of its undertaking. Rather than beginning with the original control list and attempting to modify it, defense analysts chose to begin with a blank slate and attempted to select their critical-technology categories from an initial list of over eight hundred candidates. More important, however, the lack of progress was due to the fact that the entire project was a remarkably low priority for the DoD. Deputy Undersecretary Ruth Davis claimed in 1979 that the implementation process to date had been characterized by "ad hoc actions,

98. Statement of Ruth Davis, *Extension and Revision*, p. 410. By 1984, five categories had been added: chemical technology, nuclear-related technology, cryptologic technology, energy systems technology, and energetic materials technology. See Office of the Undersecretary of Defense, Research and Engineering, *The Military Critical Technologies List* (Washington, D.C., October 1984).

99. *Extension and Revision*, p. 428.

100. *Federal Register* 45 (October 1, 1980), 65015.

one-time studies, and reliance on volunteer-participatory groups from industry and government."[101] She contended that completion of the project was managerially and technically feasible but vitally depended on the extent to which the DoD, together with the Carter administration as a whole, was willing to consider it a high priority and to devote economic and political resources to it. The complacency of the DoD effort earned it a severe rebuke from Fred Bucy, who publicly questioned the department's commitment and ability to carry forward and implement the recommendations made by the Defense Science Board in 1976.[102]

Even more troubling, the preliminary list that did emerge from the DoD was far more comprehensive than had been envisioned. Those familiar with the initial version (the MCTL was classified until November 1984) suggested that it was significantly longer than the existing CCL and thus made a mockery of the DoD's characterization of its work as an effort to identify the "small subset" of truly critical technologies. The initial list, along with supporting information on the military applications of each technology, totaled some five thousand pages.[103] According to RAND analyst Thane Gufstafson: "[The list] contains a virtual roll-call of contemporary techniques, including videodisk recording, polymeric materials, and many dozens of others equally broad. If this collection had automatically become the basis for the official CCL, . . . the entire Department of Commerce would not have been large enough to administer the export control program."[104] The National Academy of Sciences produced a similar assessment, noting that the list lacked prioritization, and that its development had "not been disciplined by considerations of clarity, foreign availability, or enforceability, considerations that should be reflected if it is to be used as an operational control list accessible to licensing officers and exporters."[105]

The process by which the initial list was constructed helps to explain its all-encompassing content. The Office of Defense Research and Engineering received assistance from a multitude of agencies, including the Army, the Navy, the Air Force, the National Security Agency, NASA, Commerce, State, Defense Intelligence, and Central Intelligence.[106] Moreover, executive officials familiar with the process noted that list construction took place under the watchful eye of Senator Jackson and other conservatives; no agency wished to be held responsible for neglect-

101. *Extension and Revision*, pp. 409–10.
102. "Export of Technology to Communist Countries," address to the Armed Forces Communications and Electronics Association Symposium, Washington, D.C., January 11, 1979.
103. See the statement of Oles Lomacky in U.S. Congress, House, Committee on Foreign Affairs, Subcommittee on International Economic Policy and Trade, *Export Administration Amendments Act of 1981*, hearings, 97th Cong., 1st sess., March 26, April 14, 28, and May 13, 1981, pp. 105–29.
104. Gufstafson, *Selling the Russians the Rope?*, p. 4.
105. National Academy of Sciences, *Balancing the National Interest*, p. 129.
106. Statement of Lomacky in *Export Administration Amendments Act of 1981*, p. 106.

ing something of potential significance.[107] The list grew more comprehensive, as officials, particularly those in the military services, erred heavily on the side of caution in their appraisals. The elasticity of the central concepts (e.g., "critical" technology, "revolutionary" advance) facilitated this tendency. Industry officials, whose participation was intended as a counterweight to that of the security-conscious military services, expressed frustration at their apparent lack of influence over the process. Hugh Donahue of Control Data, for example, complained that industry recommendations for the removal of innocuous items were usually ignored; revised versions of the list always seemed to come out with a new increase, rather than decrease, in items to be controlled.[108]

Although many technologies were singled out for control, at the same time there was little evidence of government willingness to follow through on the Bucy recommendation that controls on products be relaxed. The DoD was reluctant to endorse large-scale product decontrol before the completion of the MCTL exercise. Yet even assuming the list were to be completed, there remained reason to doubt that product controls would be relaxed significantly. Consider that in addition to design and manufacturing know-how (i.e., technology), the DoD was committed to maintaining controls on keystone products, products that revealed critical technology, and products that possessed intrinsic military value. As the GAO pointed out, virtually every product on the existing control list could arguably be placed within at least one of these categories. Indeed, if any general characterization were to be valid, it would be that items on the CoCom list possessed intrinsic military value. The GAO's findings reinforced the lingering suspicion of industry officials: by depicting the future shape of controls as a simple trade-off between technology and products, the Defense Department "ran the risk of promising more than it could deliver."[109] Given the record of DoD implementation until 1980, it is difficult to disagree with those skeptics in industry and government who feared that the ultimate outcome of the MCTL effort would be a longer control list, and the maintenance of the complex and time-consuming case-by-case licensing process.

ALLIANCE COORDINATION

For the Bucy Report recommendations and, in particular, the MCTL to be effective export control mechanisms, the United States required

107. Personal interview with William Perry.

108. Donahue's comments were made at "Technology, Trade, and U.S. Security: Balancing Conflicting National Objectives," *A National Issues Forum* (Washington, D.C.: Brookings Institution, September 26, 1984).

109. GAO, *Export Controls: Need to Clarify Policy and Simplify Administration*, p. 63.

the support and cooperation of its major allies. Between 1977 and 1980, the Carter administration made very little progress toward gaining that support. Having failed to complete the critical technologies list, administration officials were unprepared to make specific export control proposals in CoCom. Thus, rather than taking the form of negotiations, discussions of the Bucy Report within the alliance were of a general, exploratory nature.

West European officials greeted the Bucy Report with a mixture of skepticism and indignation. Their skepticism was directed at the proposed shift from product to technology controls. In general, they viewed the idea as conceptually attractive yet ultimately impractical. Technology, defined as know-how not embodied in machinery and equipment, was simply too elusive to control effectively.[110] Blueprints could travel across borders via diplomatic pouch, and "ideas" could be transferred interpersonally. As one West European official put it, effective controls would require the government "to open every letter and put a guard on every scientist who left the country."[111] British officials in particular contended that they lacked the legal authority to control disembodied technology and, given the impossibility of enforcement, saw no reason to attempt to obtain it.[112]

Officials in Britain, France, and West Germany also shared the skepticism of American business concerning the report's promise of eventual product decontrol. They questioned whether the U.S. government could seriously commit itself to relaxing restrictions on items that might be used by the Soviet military, even if their military significance was judged to be less than critical. Political pressure from within the United States would make it difficult for any U.S. administration to carry out the second part of the Bucy plan. For nations with serious trading interests, it would be imprudent to accept tighter controls on technology and then wait, as one U.S. official put it, "for the other shoe to drop."[113]

West European officials were also predictably hostile to the recommendation that end-use statements be disregarded in the control process. An unconditional embargo (i.e., no exceptions) was viewed as unattractive, particularly if product controls remained in effect. West European officials generally perceived a trade-off between the length of the

110. Interview, British Ministry of Defence, London, June 2, 1982. See also OTA, *Technology and East-West Trade*, p. 180.
111. "Export Policy Seen Harmful to CoCom," *Aviation Week and Space Technology*, September 10, 1979, pp. 26–27.
112. Interviews, British Ministry of Defence, and U.S. embassy, London, May 25, 1982. German officials, in contrast, stressed that they possessed the legal authority to control technology, yet they conceded that there existed significant practical problems in the implementation of such controls. Interview, Economics Ministry, Bonn, July 8, 1982.
113. The statement was made by a U.S. State Department official, in reference to West European attitudes. Interview, Department of State, Office of East-West Trade, Washington, D.C., May 12, 1982.

control list and the need for exceptions. Unless the list could be drastically pared, exceptions served, in the words of former British CoCom delegate Roger Carrick, to "lubricate" the embargo and help make it more tolerable to those states with a strong interest in trade.[114] Arguing in defense of end-use statements and safeguards, Carrick and others claimed that the more active the transfer mechanism, the easier it became for Western governments to monitor the end user in Eastern Europe. Moreover, they felt East European importers were naturally reluctant to divert Western technology, for fear of jeopardizing their sources of supply. Overall, the growing suspicion within the United States regarding technology transfer and interpersonal contact between East and West was conspicuously lacking among officials in the capitals of Western Europe.[115]

The portion of the Bucy Report that dealt with reexport controls inspired considerable indignation in Western Europe. Officials resented the report's questioning of their commitment to export controls, particularly in light of the liberal exceptions and technology transfer policy of the United States during the 1970s. They reacted even more strongly to the recommendation that U.S. sanctions be used against uncooperative CoCom member governments. As Congress debated the merits of sanctions, West European officials publicly warned that such a strategy could bring dire consequences to CoCom. One West European CoCom official stated bluntly that "the Soviets might be able to do that with Comecon, but we are allies, not satellites."[116]

Although they were unprepared during the 1978–1979 list review to make specific proposals regarding the control of critical technologies, U.S. officials did put forth a number of general proposals, in order to pave the way for future negotiations. U.S. delegates proposed to amend CoCom strategic criteria to include references to technology as well as products and also to include technology controls as an integral part of the CoCom list rather than merely as a footnote to product controls.[117] Other CoCom members accepted both U.S. recommendations. Given that they did not necessarily accept the U.S. definition of technology, which included disembodied know-how, it does not appear that any substantive concessions were thereby made.

A third U.S. proposal—that when products at the lower end of the performance scale were decontrolled by CoCom, the technology related

114. Carrick, *East-West Technology Transfer in Perspective*, pp. 38–40.
115. See Angela Stent Yergin, *East-West Technology Transfer: European Perspectives*, Washington Papers (Beverly Hills, Calif.: Sage Publications, 1980), p. 8.
116. "Export Policy Seen Harmful to CoCom," p. 27. Comecon was the economic organization linking members of the Soviet bloc. The formal name is CMEA—Council for Mutual Economic Assistance.
117. Statement of Frank Weil, Department of Commerce, in *U.S. Export Control Policy and the Extension of the Export Administration Act*, p. 40.

to those products would remain under control—was rejected by other CoCom members. The implicit trade-off was that the United States would be more sympathetic to product decontrol if its allies embraced the priority of protecting technology. For the allies, however, nothing was to be gained through such a deal. Most products at the lower end of the performance scale were subject to CoCom's administrative exception rule—they could be shipped without prior CoCom (i.e., U.S.) approval. Eliminating the inconvenience of having to report such sales to CoCom after the fact in exchange for adding restrictions on the technology related to lower performance products was hardly an attractive trade-off for other CoCom members.

In summary, in the alliance, as well as in the United States, the critical technologies approach was essentially a non-starter. In Western Europe, officials were content to react passively to the Carter administration's failure to generate specific proposals, because doing so relieved them of the burden of a possible confrontation over the reform of CoCom. In the United States, critics of the existing system were left to bemoan the fact that although nearly four years had passed since the drafting of the Bucy Report, both the Unites States and CoCom continued to misdirect their export control efforts, to the apparent benefit of the Soviet Union.

The Soviet invasion of Afghanistan provided an impetus, for both the Carter and the Reagan administrations, to treat technology transfer and the reform of CoCom as major issues of national security concern. At the same time, however, it helped to generate support for the two most controversial agendas of the Bucy Report: economic warfare and intra-Western controls.

CHAPTER SEVEN

Afghanistan, Poland, and the Pipeline:
The Renewal and Rejection of Economic Warfare

The invasion of Afghanistan in December 1979 marked a turning point in U.S. relations with the Soviet Union. Calling it "the gravest threat to peace since World War II," President Carter considered the invasion to be of global significance, claiming it jeopardized the security of Iran, Pakistan, and Persian Gulf oil deemed vital by the West. It represented, for the first time in the postwar era, the direct application of Soviet military force outside its tacitly acknowledged East European sphere of influence. The Carter administration felt compelled to respond decisively: the president stated that "aggression unopposed becomes a contagious disease" and that "no country committed to world peace and stability can continue to do business with the Soviet Union."[1]

In the United States, détente became increasingly controversial by the late 1970s. The invasion reinforced in the government and the public the growing belief that the Soviet Union had violated the spirit of détente by extending its influence in the third world, abusing human rights, and engaging in a sustained military buildup.[2] The Carter administration's response signaled its willingness to abandon détente and return to a more confrontational relationship. By requesting the Senate to defer further consideration of SALT II, Carter placed in abeyance the arms control negotiations that had been a cornerstone of the détente process. By sharply curtailing scientific, cultural, and economic exchanges, he removed what had become important symbols of a positive bilateral relationship. By proclaiming the Carter Doctrine, he communi-

---

1. *New York Times*, January 5, 1980, p. 6.
2. On the demise of détente, see Robert Legvold, "Containment without Confrontation," *Foreign Policy*, no. 40 (Fall 1980), 74–98, and John Lewis Gaddis, *Strategies of Containment: A Critical Appraisal of Postwar American National Security Policy* (New York: Oxford University Press, 1982), chaps. 9 and 10.

cated the renewed willingness of the United States to confront the Soviets or their proxies in the third world, with direct military force if necessary, to protect U.S. interests. By sending Defense Secretary Brown to Peking to discuss closer defense ties and the possible sale of U.S. military equipment, he completed the administration's tilt toward China. The above initiatives were accompanied by the escalation of hostile rhetoric and an attempt to isolate the Soviet Union diplomatically.

As it did in the past, the shift in U.S.-Soviet political relations helped to inspire a change in U.S. economic strategy. After the invasion of Afghanistan, it was no longer politically credible for U.S. government officials to argue that trade expansion would have a moderating influence on Soviet behavior. Similarly, in an atmosphere of confrontation, it was politically unacceptable to justify trade with the Soviet Union on the grounds of U.S. economic interests. The invasion placed U.S. corporate officials, some of whom had labored for decades to establish the political respectability of exports to the Soviet Union, in the unenviable position of having to defend publicly their belief that such trade was not incompatible with patriotism and American security.[3] That task was made all the more difficult by the discovery that Soviet forces entered Afghanistan aboard trucks produced, with the help of U.S. and Western technology, at the Kama River plant.[4] The contribution of U.S. trade to Soviet military activity was thus dramatically displayed. In the highly charged atmosphere of U.S. domestic politics, the issue of the plant's ultimate military significance—the Soviets could presumably have invaded without the help of Kama trucks—was overwhelmed by the political implications of clear evidence of its military use.

The imposition of U.S. sanctions in response to the Soviet invasion marked the resolution of the debate that had raged for the first three years of the Carter administration between advocates of trade promotion and trade denial. The sanctions, which followed by only a few months the passage of the Export Administration Act of 1979, repre-

3. See, for example, the statement of the National Machine Tool Builders Association in U.S. Congress, Senate, Committee on Banking, Housing, and Urban Affairs, Subcommittee on International Finance, *U.S. Embargo of Food and Technology to the Soviet Union*, hearings, 96th Cong., 2d sess., January 22 and March 24, 1980, pp. 244–49.

4. The discovery before Afghanistan was invaded that part of the output of the Kama plant was being put to military use sparked a dispute in the Commerce Department involving Lawrence Brady, deputy director of the Office of Export Administration. Brady urged his superiors to revoke the Kama licenses and went public when they refused to do so, claiming that their refusal symbolized the larger failure of Commerce to protect U.S. security in the interest of promoting East-West trade. Commerce officials did revoke the licenses after the Soviet invasion. On the controversy, see the statements of Assistant Commerce Secretary Stanley Marcuss and of Lawrence Brady in U.S. Congress, Senate, Committee on Banking, Housing, and Urban Affairs, Subcommittee on International Finance, *Trade and Technology, Part II: East-West Trade and Technology Transfer*, hearings, 96th Cong., 1st sess., November 28, 1979, pp. 59–70 and 82–86, and statement of Undersecretary of Defense William Perry in U.S. Congress, Senate, Committee on Governmental Affairs, Permanent Subcommittee on Investigations, *Transfer of Technology to the Soviet Bloc*, hearings, 96th Cong., 2nd sess., February 20, 1980, pp. 52–53.

sented a decisive victory for hard-liners on the NSC staff over the State and Commerce Departments. Equally important, they signaled a shift in the position of the Department of Defense. Before the invasion of Afghanistan, high officials in the DoD viewed a strategic embargo as sufficient to protect U.S. security. Following the Soviet invasion, however, they adopted the broader interpretation of the Bucy Report and supported an expansion of controls into the realm of economic warfare.

The purpose of this chapter is twofold. First, I establish and explore the logic of the U.S. return to economic warfare between 1980 and 1984. Both political and strategic considerations underlay the shift. To U.S. officials, particularly those in the Reagan administration, economic warfare complemented both a confrontational political approach and a defense strategy that deemed it necessary to prepare for a long conventional war against a hostile yet economically vulnerable adversary. Economic warfare, however, was not simply an appropriate response to a potential military *threat*. U.S. officials also viewed it as affording an attractive strategic *opportunity*. Given the Soviet economic predicament, a broad denial strategy could help to exacerbate the burdens of the Soviet empire and enable the United States to gain decisive advantages in the ongoing geopolitical competition between the superpowers for influence.

Other Western governments, particularly those of France and West Germany, resisted economic warfare. In their view, to pursue that strategy would jeopardize the Eastern economic links that West European firms had developed during the 1970s; it would also threaten the cordial East-West political relationship that both provided the impetus to positive economic relations and was reinforced by them. West European leaders were primarily concerned with the regional objective of maintaining a stable, nonconfrontational relationship with the Soviets in Europe. In the absence of the perception of a direct military threat, they viewed economic warfare as involving economic sacrifice and political risk, with little to be gained in return.

My second purpose in this chapter is to examine the effort and explain the failure of the United States to gain alliance support for economic warfare. When U.S. officials failed to persuade their West European counterparts in CoCom and elsewhere of the merits of the strategy, they turned to coercion. The result was an alliance confrontation, the costs of which forced the United States to retreat in the alliance context and in national policy as well. The conflict of 1982 over economic warfare thus concluded as had that of 1954, with West European governments prevailing. The manner in which the United States conducted the 1982 conflict, however, had detrimental consequences for both the competitive position of U.S. firms and the government's ability to insulate the Western trading system from the effects of East-West export controls.

AFGHANISTAN AND U.S. SANCTIONS

Although the various economic measures taken by U.S. officials in 1980 are usually subsumed under the heading of "Afghanistan sanctions," different measures were intended to serve different objectives.[5] The most familiar and controversial sanction, the grain embargo, was clearly conceived as an instrument of tactical linkage. It was intended to punish the Soviets for their foreign policy and, more important, to send a signal of U.S. resolve regarding future acts of Soviet aggression.[6] Grain was a bottleneck item in the Soviet economy. Given their dismal 1979 harvest, the Soviets planned to purchase twenty-five million tons of grain from the United States, the largest amount for a given year in the history of U.S-Soviet agricultural trade.[7]

Similar motives informed the decisions taken in February to embargo phosphates and to boycott the Moscow Olympics. The former was based on the assumption that if grain were to be denied, so too should be the supply of fertilizer necessary to grow grain. Occidental Petroleum bore most of the burden of the phosphate embargo, losing almost $400 million in expected 1980 sales to the Soviets.[8] The Olympics boycott, announced after an initial U.S. attempt to have the games moved from Moscow, was largely a symbolic gesture, although U.S. firms were prohibited from selling any goods to the Soviets in connection with the games.

Each of the above measures was viewed by the Carter administration as conditional. It was prepared in principle to lift all of them pending desirable changes in Soviet behavior. The sanctions affecting advanced technology, however, were different in origin and intent. First, they were the product of a more deliberate and systematic review, an interagency reassessment of the relationship between East-West technology transfer and U.S. interests. Although these measures were inspired by the Afghanistan invasion, they were, in the words of one U.S. official, more

5. The measures taken are discussed in Homer E. Moyer, Jr., and Linda Mabry, "Export Controls as Instruments of Foreign Policy: The History, Legal Issues, and Policy Lessons of Three Recent Cases," in *Law and Policy in International Business* 15 (1983), 1–171. See also U.S. Congress, House, Committee on Foreign Affairs, Subcommittee on Europe and the Middle East, *An Assessment of the Afghanistan Sanctions: Implications for Trade and Diplomacy in the 1980s*, report, 97th Cong., 1st sess., 1981.

6. For analyses of U.S. motives, see David Baldwin, *Economic Statecraft* (Princeton: Princeton University Press, 1985), pp. 261–68, and Kim Richard Nossal, "Sanctions as International Punishment," *International Organization* 43 (Spring 1989), 301–22.

7. Moyer and Mabry, "Export Controls as Instruments of Foreign Policy," p. 49. In accordance with the 1975 Long-Term Grain Agreement, the Soviets were guaranteed the right to purchase at least eight million tons annually. The additional seventeen million they requested at the end of 1979 could be exported at the discretion of the president. The embargo announced by Carter in early January 1980 covered only the latter; the initial eight million tons of grain were allowed in order that the United States not be accused of breaking the agreement.

8. Ibid., pp. 37–41.

than a "spasmodic reaction" to it.[9] Second, the technology sanctions were intended to remain in effect despite short-term changes in Soviet behavior. They were driven more by national security than by foreign policy considerations, although in one critical case, that of the no-exceptions policy, the distinction between the two motivations was not readily apparent.

Under orders from the president, on January 11, 1980 the Commerce Department froze all existing validated licenses for export to the Soviet Union and ceased to process new requests, pending the completion of a high-level review of export control policy. An interagency committee was assigned to reconsider the national security implications of U.S. technology trade with the Soviets; it was to focus on the extent to which particular categories of items contributed to Soviet military potential, were available from alternative suppliers, and were probable candidates for CoCom control.[10] After approximately two months of regular meetings, the committee completed its deliberations, and in March the administration put forth a new set of export control guidelines.[11] Although they received far less publicity than did the grain embargo and other foreign policy sanctions, their impact on U.S. policy and alliance coordination nevertheless proved to be profound and enduring.

The new guidelines contained three important initiatives.[12] First, the United States adopted a "no-exceptions" policy with regard to the Soviet Union. The government would no longer consider or submit to CoCom requests from U.S. firms to export items on the multilateral control list.[13] Given the exceptions request, and approval, record of the United States during the 1970s, this represented a significant change in policy. It reflected the emerging consensus among U.S. officials that Soviet end-use assurances could not be trusted to protect against the diversion of U.S. exports to military use.[14] In accordance with the new policy, eight

9. Statement of William Perry in *Transfer of Technology to the Soviet Bloc*, p. 28.
10. For accounts of the interagency review process, see the statement of William Perry, ibid., pp. 23–30; the statement of Homer Moyer, Jr., general counsel, Department of Commerce, in ibid., pp. 136–40; and the statement of Commerce Secretary Philip Klutznick in *U.S. Embargo of Food and Technology to the Soviet Union*, pp. 29–39.
11. *New York Times*, March 19, 1980, p. D1.
12. The best statements and discussions of the revised controls are provided by Eric Hirschhorn, deputy assistant secretary of commerce, and William Perry, in U.S. Congress, Senate, Committee on Governmental Affairs, Permanent Subcommittee on Investigations, *Proposed Legislation to Establish an Office of Strategic Trade*, hearings, 96th Cong., 2d sess., September 24–25, 1980, pp. 74–144.
13. The no-exceptions policy did not cover administrative exceptions, which involved items that could be exported to the East at the national discretion of member governments. Governments were required to report such exports to CoCom after the fact. Ibid., pp. 78–80, 102. Also exempted from the new rules were exports related to health and safety; those which protected Western access to vital commodities; and those involving spare parts or servicing for previously exported equipment. See Department of Commerce, Office of Export Administration, *Export Administration Annual Report: FY 1980* (1981), p. 20.
14. Statement of William Perry in *Proposed Legislation*, p. 139.

U.S. license applications for CoCom-controlled items, which probably would have been approved before the invasion of Afghanistan, were denied by the Department of Commerce. The shipments covered by the applications were worth an estimated $1 billion and included equipment used in the development of microwave semiconductors as well as technical data and equipment for a telecommunications plant.[15]

The no-exceptions policy applied *only* to the Soviet Union. Although DoD officials preferred this restriction to be applied uniformly to Warsaw Pact members, officials in State and Commerce argued that East European states should not be punished for Afghanistan and should be given incentives to act independently of the Soviet Union. An obvious problem with differentiated controls was that prohibited technology could be transshipped through East European states to the Soviet Union. In response, the interagency committee determined that exceptions would be granted to East European states only for items "clearly appropriate to their end-use" and for which diversion to military use, or to the Soviet Union, was "readily detectable."[16] Nevertheless, by targeting only the Soviet Union, the administration signaled that its no-exceptions policy was driven as much by foreign policy considerations (i.e., the desire to punish the Soviets) as by national security concerns (i.e., the desire to strengthen the strategic embargo). As I will discuss in Chapter 8, this entangling of national security and foreign policy considerations in U.S. policy came to pose a significant constraint on the ability of the United States to strengthen CoCom.

The second major initiative consisted of steps designed to make the application of national security controls more restrictive. The Carter administration extended the scope of U.S. controls for certain categories of exports, mainly computers and numerically controlled machine tools, by lowering the threshold of technological sophistication at which validated license requirements went into effect. It also added to the list militarily significant items that had previously eluded control, such as electronic-grade polysilicon, certain types of lasers, and machinery and equipment for the manufacture of integrated circuits.[17] The administration committed itself to treating end-use assurances from importers, and claims of widespread foreign availability by exporters, with more skepticism than it had in the past.

Evidence that controls were applied more restrictively in practice is provided by the results of the review of suspended licenses, which took place after the interagency committee completed its work. Of over 1,000

15. Statement of Perry in *U.S. Embargo of Food and Technology to the Soviet Union*, p. 91.
16. Perry, in ibid., pp. 94–95, 103.
17. *Federal Register* 45 (May 5, 1980), 29568–69.

licenses that had been frozen in January 1980, only 281 were ultimately reinstated.[18]

The third major initiative moved U.S. policy beyond a strategic embargo and into the realm of economic warfare. As part of its more restrictive approach to technology transfer, the administration decided to control the export of "process know-how" to "militarily relevant industrial sectors" in the Soviet Union.[19] The process know-how proposal did not receive wide publicity and does not appear to have been fully formed in administration thinking in 1980. Nonetheless, official briefings to the press, testimony before Congress, and interviews I conducted confirmed that the controls were designed to target the leading sectors of the Soviet economy. Among those industries singled out by the administration as "helpful to the military establishment" were energy, aerospace, chemicals, machine tools, metallurgy, motor vehicles, and shipbuilding.[20] The administration reserved the right to add other industries found to be "defense-supporting." It proposed to deny process know-how to these sectors in the Soviet economy, regardless of whether the technology in question had direct military significance. The controls were directed solely at the Soviet Union, since "building up certain sectors of the *Soviet economy proper* could be more detrimental to the U.S. than building up a similar sector in Hungary, Poland or Rumania."[21]

The process know-how proposal was a direct outgrowth of the Bucy Report. By adopting it, the administration in effect endorsed the broader (i.e., economic warfare) interpretation of the report, as outlined in Chapter 6. The principal targets of the new controls were active transfer mechanisms, and, in particular, turnkey factories.[22] Under the new standards, the export of U.S. technology to the Kama River truck plant that occurred during the early 1970s would not have been approved. Similarly, in a direct echo of the Bucy Report, officials stated that energy equipment would be permitted for export to the Soviets on a case-by-case basis, whereas the technology required to produce that equipment would be denied.[23] The controversial technical data package sold by Dresser Industries to the Soviets in 1978 fell victim to the new restrictions; in an ironic final twist to that case, Dresser's license was revoked in

18. Moyer and Mabry, "Export Controls as Instruments of Foreign Policy," p. 53, and *Export Administration Annual Report: FY 1980*, p. 27.

19. The process know-how control idea is discussed by W. Graham Claytor, deputy secretary of defense, and William Perry in *U.S. Embargo of Food and Technology to the Soviet Union*, p. 93, and by Perry and Eric Hirschhorn in *Proposed Legislation*, pp. 101, 119, 137, 143.

20. Statement of Hirschhorn in *Proposed Legislation*, p. 101.

21. Statement of V. Garber, Department of Defense, in U.S. Congress, Senate, Committee on Armed Services, Subcommittee on General Procurement, *Soviet Defense Expenditures and Related Programs*, hearings, 96th Cong., 2d sess., November 1, 8, 1979, and February 4, 1980, p. 187, emphasis added.

22. *Wall Street Journal*, March 19, 1980, p. 22.

23. *New York Times*, March 19, 1980, p. D1.

December 1980.[24] Fred Bucy's interpretation of the 1976 report, which had been rejected in 1978, was now embraced by the Carter administration as official policy guidance.

Administration officials recognized that the new controls took them beyond the strategic embargo's exclusive concern with exports of direct military significance. In explaining the relationship between process know-how controls and the militarily critical technologies effort, Commerce Undersecretary Herzstein stated:

> There is no direct linkage between the two. The critical technologies list that is mandated by the E.A.A. of 1979 will affect our consideration of applications to export goods and technologies at the upper reaches of our technology capabilities. *These will be technologies that have direct military significance.*

> The policy concerning process know-how concerns goods and technology that *may or may not currently require validated licenses* before they can be exported to a communist country. The objective of the policy currently being developed would be to look beyond those individual items and *consider the potential military relevance of the factory into which the products would be incorporated.*[25]

Similarly, Undersecretary Perry described the two principal export control objectives of the DoD as follows: "(1) to maintain the U.S. lead-time in the application of technology to military capabilities[, and] (2) to avoid the facilitation of [Soviet] industrial modernization related to significantly greater defense productivity."[26] The distinction between the "old," yet strengthened, strategic embargo and the "new" economic warfare was thus clear. Perry went on to argue that the first objective was well served by the CCL and, increasingly, the Military Critical Technologies List, whereas the second required more attention, both from the United States and in CoCom. In something of an understatement, Perry noted that "some" past export approvals had been detrimental to the second objective. More to the point, one might argue that the fundamental assumptions underlying the entire Western economic approach to détente ran directly counter to the second objective.

Within the U.S. government, the initiative for process know-how controls came from the DoD. Given its approach before the Soviet invasion of Afghanistan, it is important to consider why the department shifted its position. Clearly, the DoD was not immune to the political pressure for tighter controls that this invasion generated. For example, before the invasion, department officials judged the diversion of Kama trucks to

---

24. *New York Times*, December 4, 1980, p. D1.
25. *Proposed Legislation*, p. 101, emphasis added.
26. Statement of William Perry, ibid., p. 143.

the Soviet military to be insignificant; afterward, they considered Kama's military contribution to be "substantial."[27] More important, by 1980, DoD officials were coming to accept that Western technology transfer had made an important *indirect* contribution to the buildup of Soviet military power during détente.[28] Indeed, they came to the view that such contributions—to the modernization and expansion of the Soviet industrial base—were ultimately as important to the Soviets as were direct contributions made by the diversion of technology to specific military use. The West, in effect, was providing the Soviets with the economic and industrial wherewithal to maintain and increase their military potential over the long run. The invasion of Afghanistan served as the catalyst that propelled these concerns to the political forefront; controls on process know-how were an initial attempt to address the problem thus posed to Western security.

The Response of Other CoCom Members

Relative to that of U.S. officials, the reaction of British, French, and West German officials to the invasion of Afghanistan was subdued. They viewed it as neither a major geopolitical turning point nor grounds to abandon détente. West German chancellor Schmidt frankly asserted that his government "would not permit ten years of detente to be destroyed" by Afghanistan and asked rhetorically whether Kabul was more important to the West than Berlin.[29] In order to emphasize its remoteness from their central concern of European stability, some West European officials even initially referred to Afghanistan as an "East-South" issue. Although the Carter administration strove to isolate the Soviets, French and West German leaders believed that more, rather than less, communication was desirable. Both Schmidt and President Giscard d'Estaing conducted summit meetings with Soviet leader Brezhnev during 1980.[30] Eventually, West European leaders came to condemn the invasion with harsh rhetoric; in the interest of preserving the benefits of détente in Europe, however, their substantive response remained relatively cautious. Not surprisingly, with the partial exception of Great Britain, they were reluctant to join the United States in the imposition of economic sanctions.

With regard to the grain embargo, in response to high-level U.S.

27. Compare the remarks of Perry in *Trade and Technology*, p. 39, to his comments three months later in the post-Afghanistan hearings, in *Transfer of Technology to the Soviet Bloc*, p. 36.

28. See the statement of Jack Vorona, Defense Intelligence Agency, in *Soviet Defense Expenditures and Related Programs*, pp. 69–79, and the statement of Claytor and Perry, *U.S. Embargo of Food and Technology to the Soviet Union*, p. 93.

29. See "Grain Becomes a Weapon," *Time*, January 21, 1980, p. 14.

30. Giscard and Brezhnev met in Warsaw on May 19; Schmidt journeyed to Moscow on June 30.

diplomatic pressure, the European Community (EC), Australia, and Canada pledged not to replace American grain and to keep their own sales to the Soviets at normal levels.[31] Canada abandoned its pledge in November, after U.S. officials concluded a grain agreement with China, a traditional importer of Canadian grain. More important, Argentina deliberately exploited the U.S. embargo by significantly increasing its sales to the Soviets, even at the expense of denying supplies to some of its normal customers. In the final analysis, the Soviets managed to replace most of the grain denied by the United States, though at a somewhat higher cost.

The Carter administration also urged its allies to deny official export credits and guarantees to the Soviets. Such a sanction was essentially cost-free for the United States because of the restraints already imposed by the Jackson-Vanik and Stevenson Amendments. The British were receptive to the idea and in January announced that their 1975 credit agreement with the Soviets, due to expire in February 1980, would not be renewed.[32] Whether this constituted a major statement is arguable, however; the Soviets had chosen to use only about half the credit made available since 1975, and the Thatcher government was already contemplating nonrenewal, as Conservatives generally viewed the subsidized loans as a bad deal for Britain economically. In any case, the French and West Germans, with more at stake, were categorically opposed to credit restrictions. Their prospects for large-scale machinery and equipment exports to the Soviets hinged on the availability of credit.[33] As a collective sanction, EC members would go only as far as to commit themselves to abide by the interest rate rules established by the OECD Consensus. In effect, this commitment merely constituted a pledge to stop cheating on an agreement that had already been made.

Western Europe and Japan were somewhat more receptive to U.S. initiatives in CoCom. On the one hand, they acquiesced in the no-exceptions policy and proved willing to add some, though not all, items proposed by U.S. officials to the multilateral control list. On the other hand, they categorically rejected the process know-how proposal. In terms of strategy, the allies proved resistant to economic warfare for fear of risking their political and economic stake in Soviet trade. At the same time, they were willing to accept some improvements in CoCom's strategic embargo, despite uncertainty over whether U.S. proposals were motivated by strategic concerns or merely by the desire to use CoCom to punish the Soviets.

The allies' acquiescence in the no-exceptions policy is best explained

31. Robert Paarlberg, *Food Trade and Foreign Policy* (Ithaca: Cornell University Press, 1985), chap. 6.
32. *Times* (London), January 25, 1980.
33. *Financial Times*, March 3, 1980.

by economic and political considerations. To individual CoCom members, the economic costs of adopting it were modest. By 1979, almost half the requests for general exceptions were made by the United States, with the remainder divided among fourteen states. More important, since the restriction applied only to the Soviet Union, exceptions could still be requested for Eastern Europe. In 1979, of approximately fifteen hundred CoCom exception requests, only 25 percent were for the Soviet Union.[34] That administrative exceptions were left unaffected made the proposal even more economically palatable. Other CoCom members resisted U.S. pressure to restrict administrative exceptions, believing that national discretion on low-technology items was essential in maintaining business support for, or toleration of, the CoCom embargo.[35]

In light of the proposal's modest economic cost and given the significance attached to it by the United States, it made political sense for other CoCom members to comply. U.S. officials emphasized at the outset of the post-Afghanistan CoCom negotiations that at the very least they expected cooperation on the exceptions issue.[36] Moreover, given CoCom rules, the United States could have imposed its will unilaterally by exercising its veto over any and all CoCom exception requests. Thus, other members were, in effect, forced to acquiesce, even though they never formally accepted a no-exceptions proposal.[37] Their alternative was open defiance of the United States, either by refusing to submit cases to CoCom or by submitting them and subsequently ignoring the U.S. veto. Defiance would probably lead to a confrontation in CoCom or in the alliance more generally. Other member governments sought to avoid such a result in the immediate aftermath of Afghanistan; thus, they grudgingly tolerated the policy, despite their skepticism regarding the strategic significance of the Kama diversion of Western technology and their belief that Soviet end-use assurances were generally reliable.[38]

The allies' response to U.S. proposals for additions to the CoCom list demonstrated both their willingness in principle to bolster the strategic embargo and their sentiment that Afghanistan did not justify allowing security concerns to be given overriding priority in the export control process. Other members agreed to add fiber optics, polysilicon, and IC equipment to the list after being convinced, by evidence presented in

34. See "High-Tech Sales to U.S.S.R. Further Reduced," *Science* 208 (April 4, 1980).
35. Interview, British Department of Trade, London, May 25, 1982.
36. *Business Week*, February 25, 1980, p. 59.
37. Personal correspondence with William A. Root, former director of the Office of East-West Trade, Department of State, April 15, 1987.
38. Interview, Economics Ministry, Bonn, July 7, 1982. During 1980, only five formal requests were made for exceptions to the Soviet Union. Four were withdrawn, and the United States vetoed the other. Statement of William Perry in *Proposed Legislation*, p. 119.

CoCom by the United States, of their direct military significance.[39] On computers, however, U.S. proposals struck others as an attempt to use Afghanistan as a pretext to force a resolution of previously disputed issues in CoCom, in favor of the U.S. position.[40] The 1978–1979 list review had ended in a stalemate on computers, with the United States favoring more restrictive, and other CoCom members less restrictive, control parameters. After the invasion of Afghanistan, U.S. officials put forth a computer proposal even more restrictive than that which had been unacceptable in CoCom in 1979. To other CoCom members, the invasion did not justify acceptance of the prior U.S. proposal, much less something more restrictive.

Whether accepted or rejected, the specific list addition proposals were viewed by the allies as worthy of serious consideration in CoCom. The same cannot be said of the process know-how proposal made by the United States in March 1980.[41] Various West European officials described it as inappropriate for CoCom because it was not concerned with products and technologies of direct military significance. Some emphasized that CoCom was not in the business of retarding Soviet industrial productivity or preventing improvements in the management of the Soviet economy.[42]

Clearly, resistance to process know-how controls was more than simply a question of what was or was not an appropriate concern for CoCom. The proposal struck directly at the heart of Western Europe's and Japan's economic stakes in trade with the Soviet Union. Western export competition during détente consisted primarily of a struggle to sell advanced machinery and equipment, often in the form of turnkey plants, to the Soviets. To deny such sales to the chemical, energy, machine tool, motor vehicle, and other "militarily relevant" sectors would significantly hamper West European and Japanese capital goods producers in the competition for the Soviet market. Many such firms traditionally relied on plant exports to achieve economies of scale; for smaller firms, one major sale could provide employment and profits over several years. As Peter Kenen noted: "Output and investment in the capital goods industries determines the health of the European economy, those industries themselves depend heavily on export sales, and those sales are bunched

---

39. Statement of Perry in *Proposed Legislation*, and in *U.S. Embargo of Food and Technology to the Soviet Union*, p. 93. Interviews with West European officials provided support for Perry's claim, as does the comment of William Root that other CoCom members usually support consistent and well thought out national security control proposals advanced by the United States. See "Trade Controls That Work," *Foreign Policy*, no. 56 (Fall 1984), 61–80.

40. Interview, ICL House, London, June 2, 1982.

41. See *Financial Times*, March 24, 1980, and *Frankfurter Allgemeine Zeitung*, March 22, 1980.

42. Interviews, British Ministry of Defense, London, June 2, 1982; Economics Ministry, Bonn, July 7, 1982; and British embassy, Washington, May 6, 1982.

in large, long-term contracts. Every single contract counts."[43] By 1980 the Soviet market was of greater importance: Western demand for complete factories was relatively stagnant, and third world industrializers were forced to cut back because of debt accumulation.

A U.S. initiative related to the process know-how proposal, namely, that all exports valued at over $100 million be reviewed by CoCom, was similarly doomed to failure. Under such a stipulation, most turnkey plants (and other large-scale ventures, such as the Siberian gas pipeline) would require multilateral approval, regardless of their direct military significance. Led by France and West Germany, other members rejected this proposal, claiming the value of a sale to be inappropriate criteria for CoCom review.[44] Some West European officials found the idea objectionable, even with a strong presumption for approval, noting that the mere existence of a CoCom review might jeopardize proprietary commercial information. There was simply too much at stake for the details of proposed large-scale export packages to be aired in front of potential competitors in a multilateral forum.

To shift CoCom's mission to that of a broader industrial embargo required the acquiescence of West European leaders at the highest levels. Similarly, in 1950 it took Secretary of State Acheson to convince his French and British counterparts of the strategic necessity of economic warfare. The Carter administration made an initial attempt but was unable to convince America's allies that the invasion of Afghanistan posed a threat to Western security similar in degree to that represented by the invasion of Korea.

The invasion did not prevent other CoCom members from continuing their export to the Soviets of turnkey factories and other large-scale projects. The two most controversial cases of 1980 concerned major contracts U.S firms were forced to abandon after the imposition of U.S. sanctions. Creusot-Loire of France agreed to provide the Soviets with a turnkey steel plant, worth over $300 million, after Armco and its Japanese collaborator, Nippon Steel, were compelled to withdraw.[45] Similarly, a contract to build an aluminum factory worth roughly the same amount was awarded to the Klöcker group of West Germany after Alcoa informed the Soviets that it could no longer participate.[46] The Carter administration protested in both cases, claiming the allies had broken their pledge not to pick up orders that U.S. firms were forced to abandon. The West German and the French governments countered that the

43. Statement of Peter Kenen in U.S. Congress, House, Committee on Foreign Affairs, Subcommittee on Europe and the Middle East, *U.S.-West European Relations in 1980*, hearings, 96th Cong., 2d sess., June 25, July 22, September 9, 15, 22, 1980, p. 111.

44. Interviews, Economics Ministry, Bonn, July 7, 1982, and Department of Trade, London, May 25, 1982.

45. *Financial Times*, September 17, 1980, p. 1, and September 19, 1980, p. 1.

46. *Financial Times*, August 20, 1980, and September 16, 1980.

Table 9. Western exports to the Soviet Union, 1979–1983 ($ millions)

Country	1979	1980	1981	1982	1983
U.K.	891	1,058	871	620	676
F.R.G.	3,619	4,373	3,394	3,870	4,418
France	2,005	2,465	1,865	1,559	2,240
Japan	2,443	2,796	3,253	3,893	2,833
Italy	1,222	1,267	1,285	1,499	1,884
U.S.	3,616	1,513	2,431	2,613	2,003

Source: IMF, *Direction of Trade Yearbook, 1983.*

newly signed contracts were significantly different from those originally agreed to by Armco and Alcoa. Whatever the validity of this argument, U.S. officials clearly believed that France and West Germany had violated at least the spirit of the agreement; French and West German officials appeared equally strong in their conviction that business was business and to pass up such attractive export opportunities would be foolish.

Overall, France increased its exports by approximately 23 percent and West Germany by about 20 percent during 1980. U.S. sales declined dramatically, from $3.6 billion in 1979 to $1.5 billion in 1980, which dropped the United States from first to fourth place in the export competition among Western states (see Table 9). More important, the grain embargo and the freeze on high-technology licensing enhanced the reputation of U.S. exporters as unreliable suppliers to the Soviet market.

TRADE AND WESTERN SECURITY: THE REAGAN ADMINISTRATION

The Reagan administration came to office determined to treat East-West economic relations as a profound and immediate national security concern. The consensus among top officials was that Western trade, although it had done nothing to moderate Soviet behavior, had enhanced the ability of the Soviets to achieve military parity during the 1970s and thus to pose a greater threat to Western security. By the end of 1981, the administration viewed the introduction of a broad strategy of economic denial as an appropriate response, both for the United States and, it hoped, for the Western alliance.

One major aspect of Reagan administration strategy, which built on Carter's 1980 initiatives, involved a full-scale effort to strengthen the strategic embargo. This included an expansion of the control list; the launching of a Customs Service program to bolster the enforcement of controls; a tightening of restrictions on scientific communication; and a public relations campaign intended to increase awareness in the business community of the national security risks of technology diversion. The administration simultaneously devoted considerable attention to the improvement of multilateral controls in CoCom. In this effort it was somewhat successful, although its initiatives and methods at times proved controversial. This controversy, however, paled in comparison to that generated by the second major aspect of administration policy—a determined effort to conduct economic warfare against the Soviet Union. Reagan strategists conceived of economic warfare in far more ambitious terms than did those in the Carter administration in 1980, both with regard to what it could accomplish and in their efforts to obtain alliance support.

Economic Warfare as Grand Strategy

To Reagan administration officials, economic warfare was attractive in that it complemented their preferred political approach to the Soviet Union and also fit well with their conception of how best to further U.S. security militarily. Politically, the administration sought to distance itself from the rhetoric and policies of détente. The president publicly questioned the desirability of negotiating with the Soviets and made it clear that an emphasis on arms control would be replaced by a program to rebuild American military strength, in response to what was depicted as a decade of unilateral disarmament. Reagan and other high officials also took to the rhetorical offensive, extolling the virtues of democracy, castigating the Soviet Union as an "evil empire," and challenging the political legitimacy of Soviet rule, both domestically and with regard to Eastern Europe.[47] Economic warfare was compatible with this political approach. Trade denial reinforced the administration's attempt to isolate the Soviet Union politically and to cast doubt on its status as a responsible member of the international community. Given this overall approach, the president's decision in April 1981 to lift the grain embargo should properly be viewed, as it was both within and outside the administration at the time, as an anomaly.[48]

With respect to U.S. military security, economic warfare was attractive for two reasons. First, it appeared to be a necessary defensive measure,

47. See, for example, the infamous "evil empire" speech of the president, to the National Association of Evangelicals, reported in *New York Times*, March 9, 1983.
48. The contradiction between the grain decision and overall Reagan strategy may not be as great as commonly perceived. Henry Nau has argued that during 1981, administration policy reflected a "prudent" approach (i.e., something between détente and denial), within which the lifting of the grain embargo could be justified. See Nau, *The Myth of America's Decline* (New York: Oxford University Press, 1990), pp. 297–301. The contradiction became more glaring after 1981, when U.S. policy shifted decisively to economic warfare.

an appropriate response to a strategic *threat*. The administration endorsed and popularized the idea that Western technology, by contributing to Soviet military efforts, had helped to alter the strategic balance in
a manner disadvantageous to the United States. Although they acknowledged its direct impact, administration officials, following the lead of the
Carter DoD in 1980, placed even greater emphasis on the *indirect* contribution of Western technology. Commerce Assistant Secretary Brady
argued that "we cannot ignore that transfers of dual-use technology
financed with export credits have conferred on the Warsaw Pact countries the production and technical know-how and equipment to accelerate the modernization of key sectors of their industrial base."[49] Similarly,
in its annual dramatization of the threat, *Soviet Military Power*, the Defense Department emphasized the indirect (as well as direct) manner in
which Western technology contributed to the Soviet Union by

(1) yielding a direct, near-term military advantage through transfers
    leading to a Soviet technological breakthrough, filling a gap or
    overcoming a bottleneck in a mature Soviet technology;
(2) providing an indirect, long-term military advantage by helping to
    overcome technological lags in the Soviet industrial infrastructure;
(3) contributing to the overall growth of the Soviet economy by en-
    hancing productivity;
(4) releasing funds for military production.[50]

Given that the Soviet leadership had placed a high political priority on
"the introduction of new manufacturing technologies in all sectors of the
military-industrial complex," the prudent response for the Reagan administration was to accept and broaden the process know-how controls
introduced in 1980.[51] Consider the following statements made by officials from the three key departments in reply to a congressional inquiry
regarding what needed to be controlled for national security reasons:

State: "Our concern extends particularly to the military impact of
transfers to Soviet *defense priority industries*."

Commerce: "The administration is determined to tighten strategic
trade controls on technology and goods which can upgrade Soviet
production in areas relevant to their military strength."

Defense: "Our policy of restricting exports to the Soviet Union of
technology, materials and equipment, which could contribute signifi-

49. Statement of Lawrence Brady in U.S. Congress, Senate, Committee on Foreign Relations, Subcommittee on International Economic Policy, *East-West Economic Relations*, hearings,
97th Cong., 1st sess., September 16, 1981, pp. 6–7.
50. U.S. Department of Defense, *Soviet Military Power* (Washington, D.C.: GPO, 1983), p. 75.
51. U.S. Department of Defense, *Soviet Military Power* (Washington, D.C.: GPO, 1984), p. 93.

cantly, *directly or indirectly*, to Soviet *defense industries* . . . is essential to U.S. security."[52]

Defense also suggested this course to be the appropriate direction for America's multilateral strategy by noting the real weakness of CoCom to be its allowance of legal trade in such sectors as motor vehicles and shipbuilding, which improved the Soviet industrial base.[53]

By endorsing economic warfare, however, the Reagan administration went well beyond the "long-term threat" perspective accepted by Carter in 1980. It also viewed economic warfare as an attractive strategic *opportunity*. Top administration officials viewed the Soviet economy at least to be severely constrained and perhaps to be on the verge of collapse.[54] (Given the developments of the late 1980s, these views proved remarkably prescient.) Over the next decade, the Soviets would be plagued by the serious problems of slow growth, declining productivity, and labor shortages. They would also face shortages of hard currency due to the general drop in world oil prices combined with the gradual decline of Soviet oil production. Hard currency problems would be exacerbated by Soviet willingness to bear the burdens of empire. Economic subsidies to Eastern Europe, particularly Poland, were significant and likely to increase over time as a result of the region's economic and debt problems. The Soviets were also propping up a wide range of economically-dependent client states in the third world, including Cuba (which alone absorbed more than $3 billion annually), Libya, North Vietnam, Cambodia, Laos, Ethiopia, Angola, Afghanistan, South Yemen, and, perhaps eventually, Nicaragua. A NATO study released in 1983 concluded that economic and military aid to the above states constituted a major and continuing drain on Soviet resources.[55] Signs of Soviet difficulties appeared in 1982: their trade deficit was increasing, their hard currency reserves in Western banks were decreasing, they began to sell gold, and they requested that West European banks accept a delay in their repayments of short-term debt.[56]

To U.S. officials, the denial of technology would exacerbate the Soviet industrial slowdown. The denial of credits would remove an important short- and medium-term source of hard currency. The abandonment of

52. See *East-West Commercial Policy: A Congressional Dialogue with the Reagan Administration*, a Study prepared for the use of the Joint Economic Committee, U.S. Congress, 97th Cong., 2d sess., January 16, 1982, pp. 3, 8, 25, emphasis added.
53. DoD, *Annual Report to the Congress*, Fiscal Year 1983 (Washington, D.C.: GPO, February 8, 1982), p. II-31.
54. A detailed summary is found in *New York Times*, March 14, 1982, p. A1. See also Bruce Jentleson, *Pipeline Politics: The Complex Political Economy of East-West Energy Trade* (Ithaca: Cornell University Press, 1986), p. 177.
55. *New York Times*, January 23, 1983, p. 7. See also Charles Wolf et al., *The Costs of the Soviet Empire* (Santa Monica, Calif.: RAND Corporation, September, 1983).
56. *New York Times*, March 14, 1982, p. A1.

the Siberian pipeline deal, which officials considered the cornerstone of their economic warfare strategy, would prevent the Soviets from offsetting the decline in oil exports with an increase in gas exports and would thus deny the Soviets an annual hard currency windfall on the order of $8 billion.

Overall, eliminating the Western contribution would force the Soviets to confront a series of difficult trade-offs among consumption, investment, military spending, and foreign commitments.[57] As the CIA soberly noted, "the competition among rival claimants for resources will become intense."[58] For the Soviets, it was indeed a no-win situation. Consumption could be cut significantly only if the Soviet regime were willing to risk political instability. Investment had to be maintained if the economy was at least to stand still. In the eyes of Reagan strategists, the Soviets were left in the unenviable position of cutting military spending and/or foreign commitments at the expense of their position in the long-term geopolitical struggle with the United States.

Thus, an economic offensive could play a major role in the achievement of U.S. security goals. It could replace arms control in furthering the objective of reducing Soviet military outlays. The administration wished to accelerate the arms race; economic warfare would make it difficult, if not impossible, for the Soviets to compete. The economic dimension of the arms race was suggested in the Pentagon's controversial Five-Year Defense Guidance, which was leaked to the press in 1982. It argued that the United States should seek to develop weapons systems that are "difficult for the Soviets to counter, impose disproportionate costs, open up new areas of major military competition, and obsolesce [sic] previous Soviet investment."[59] The arms race, in effect, should be conducted in a manner that exposed Soviet economic vulnerabilities, and economic warfare would help to accomplish that end.

At least some administration officials believed economic warfare might bring changes in the internal political structure of the Soviet Union. Without Western economic assistance, if the Soviets were determined to maintain both domestic consumption and investment goals as well as military and foreign commitments, they would be forced eventually to undertake major economic reforms. Economic liberalization, however, would probably lead to the decentralization of political power. (This link was also made by advocates of détente, who held, however, that trade promotion rather than trade denial would create the condi-

57. See statement of Edward A. Hewett in U.S. Congress, Senate, Committee on Foreign Relations, *Economic Relations with the Soviet Union*, hearings, 97th Cong., 2d sess., August 12–13, 1982, pp. 128–31.
58. CIA, "Soviet Acquisition of Western Technology" (Washington, D.C., April 1982), p. 11.
59. Richard Halloran, "Pentagon Draws Up First Strategy for Fighting a Long Nuclear War," *New York Times*, May 31, 1982, p. A1.

tions for domestic economic reform.) As Defense Secretary Weinberger noted:

> Without constant infusions of advanced technology from the West, the Soviet industrial base would experience a cumulative obsolescence, which would eventually also constrain the military industries. The Soviet leaders know full well by now that their central planning system is fatally flawed. *But their system cannot be reformed without liberalizing Soviet society as a whole.* Hence without advanced technology from the West, the Soviet leadership would be forced to choose between its military-industrial priorities and the preservation of a tightly controlled political system. By allowing access to a wide range of advanced technologies, we enable the Soviet leadership to evade that dilemma.[60]

The line of attack suggested by Weinberger's statement found its way into official U.S. policy. National Security Decision Directive 75 called for, as an explicit objective of the United States, the use of economic pressure to influence the internal politics of the Soviet Union.[61]

Underlying the two arguments for economic warfare (defensive response to threat, offensive response to opportunity) were two different, yet compatible conceptions of the relationship between the Soviet economy and military. On the one hand, the administration viewed the Soviet economy as a "war economy," one devoted overwhelmingly to the long-term pursuit of supporting and servicing the military sector. In a statement reminiscent of NSC-68, the Defense Department asserted that "underlying Soviet military power is a vast and complex industrial, mobilization and logistics support system designed to focus the resources of the Soviet state on the capability to wage war."[62] Defense and CIA officials similarly argued that the level of sophistication of the industrial base determined, in the long run, the ability of the Soviets to make quantitative and qualitative improvements in their military capabilities.[63] The implications for the transfer of technology were clear: a contribution to the industrial base was, ultimately, a contribution to Soviet military power.

Although they saw the Soviet economy and military as being in close collaboration over the long run, administration officials also considered them to be locked in a competitive struggle for resources in the short

---

60. Weinberger is quoted in *The Economist*, May 22, 1982, p. 69, emphasis added, and in Jentleson, *Pipeline Politics*, pp. 177–78. The changes of the late 1980s suggest that the defense secretary was, in one sense, correct: the Soviets proved incapable of both undertaking meaningful reform and maintaining political control domestically and in Eastern Europe. The choice between "military industrial priorities and the preservation of a tightly controlled political system," however, was *not* ultimately forced by Western economic warfare.

61. *Los Angeles Times*, March 16, 1983, p. 1.

62. *Soviet Military Power*, 1984, Preface, p. 1.

63. CIA, "Soviet Acquisition of Western Technology", p. 3.

run. In the context of this second conception, the "resource-releasing" effects of trade took on greater significance. Trade that acted to increase the availability of resources served to alleviate the sectoral competition. Thus, in the short run, the denial of Western credits and grain sales could be thought to be as important as the denial of sophisticated industrial technology.

As the above discussion suggests, during its first term the Reagan administration, to a greater extent than any since perhaps the Truman administration, viewed East-West economic denial as an essential component of U.S. security and defense strategy. The interest of the Defense Department in the issue, as reflected by the attention devoted to it in annual posture statements, the Five-Year Defense Guidance, and speeches by high officials, was unprecedented. High Defense officials such as Weinberger and Assistant Secretary for International Security Affairs Richard Perle played dominant roles in the export control policy process. Economic confrontation with the Soviets became a peacetime surrogate for military struggle.

Economic warfare also came to complement particular military initiatives. In conventional-force planning, for example, the Defense Department saw fit to abandon the "fallacy" that a future war with the Soviets would be short in duration.[64] It contended that the United States must revise its assumptions and prepare itself, as the Soviets presumably had, to engage in a protracted conventional conflict. By abandoning the assumption that either side would necessarily resort to nuclear weapons at an early stage in a conflict, the DoD accentuated the military significance of a powerful industrial base, which in turn strengthened the rationale for economic warfare. Improving the fitness of the U.S. industrial base and assuring that technology transfer did not enhance the Soviet industrial base both took on the status of important military priorities. Defense officials argued in 1982, for example, that International Harvester should not be permitted to sell an agricultural combine factory to the Soviets because it would probably be converted during wartime to the production of tanks and armored personnel carriers.[65] As I argued in Chapter 2, assumptions made regarding the likelihood of war, and the type of war anticipated, are relevant in determining the potential attractiveness of economic warfare as a strategic weapon.

Another facet of Reagan conventional defense strategy was an expressed commitment to "horizontal escalation."[66] This plan called for

---

64. See DoD, *Annual Report to the Congress,* Fiscal Year 1983, p. I-16. See also Paul Bracken, "Mobilization in the Nuclear Age," *International Security* 3 (Winter 1978–79), 74–93.

65. See "The Harvester Case," *Wall Street Journal,* January 21, 1982.

66. DoD, *Annual Report to the Congress,* pp. I-15–16. For discussion, see Joshua M. Epstein, "Horizontal Escalation: Sour Notes on a Recurrent Theme," *International Security* 8 (Winter 1983–84), 19–31.

the United States to respond to Soviet aggression in Europe by taking to the offensive and counterattacking globally at the points of greatest Soviet vulnerability. The targets most frequently cited were Soviet naval bases (hence the rationale for two additional U.S. carrier battle groups) and Soviet client states in Eastern Europe, Cuba, and Vietnam. By globalizing a conflict initiated by the Soviets, U.S. officials planned to precipitate the collapse of the Soviet empire. As Weinberger noted, U.S. war-fighting strategy should take account of the fact that "the Soviet empire is not a voluntary association of democratic nations."[67]

Economic warfare can be viewed as an attempt to achieve the ambitious objectives of horizontal escalation without engaging in military conflict. By exacerbating the financial burdens of empire, the United States sought to force the Soviets to rethink their global commitments, particularly in Eastern Europe. This rationale, in large part, underlay Weinberger's contention in 1982 that the West should formally declare Poland in default on its loans, thus leaving the solution to Poland's profound economic problems fully in the hands of the Soviets. As on the military battlefield, the United States should not merely react defensively in the economic competition. By raising the stakes and exploiting its natural advantages over the Soviets (here, economic power being the functional equivalent of naval power), in the best of all possible worlds the United States could help to bring about the demise of the Soviet empire peacefully.

Administration officials did not embrace the above arguments with the same level of enthusiasm. Although all appeared to agree on the desirability of economic warfare, they differed on precisely how far to push the Soviets. Officials in the DoD and the CIA generally took the most confrontational position; for example, they advocated a total embargo on industrial technology and equipment.[68] A second group, led by Secretary of State Haig, sought not to push the Soviets so hard as to provoke an unpredictable and dangerous reaction. The Soviets should be left with some room for maneuver, and the United States should allow some nonstrategic trade conditionally, in the hope of influencing Soviet behavior. In 1981, for example, Haig recommended, although Weinberger and CIA director William Casey were opposed to the idea, that the president allow Caterpillar Tractor to sell $50 million worth of pipelayers to the Soviets.[69]

The real difference between the two groups, however, concerned how the United States should deal with its Western allies. Multilateralists, led by Haig, placed highest value on the maintenance of harmonious al-

67. DoD, *Annual Report to the Congress*, p. I-16.
68. *New York Times*, July 19, 1981, p. A1.
69. Ibid., and see *East-West Commercial Policy: A Congressional Dialogue with the Reagan Administration*, p. 23.

liance relations and argued that economic warfare should be conducted only to the extent the allies were willing to participate. Unilateralists, led by Perle and Weinberger, believed the allies should be forced to participate, if necessary. They should be made to choose between loyalty to the United States and to their East-West trade commitments. If they chose the latter, the United States, in turn, should rethink its own alliance commitments. Weinberger suggested as much in 1982, warning that "isolation is never far from the surface in the U.S."[70]

Economic Warfare as Grand Strategy:
West European Perspectives

West European officials found economic warfare objectionable on economic and political grounds and held strategic assumptions diametrically opposed to those held by top U.S. officials.

The West European position during the early 1980s centered on "compartmentalization," a concept many officials used to characterize the relationship between the Soviet economy and military.[71] For the purposes of trade policy, the Soviet military could be conceived of as detached or isolated from the rest of the economy. Because the military sector was more advanced, it tended not to rely on the civilian economy for technological developments. Because it was the overriding sectoral priority, it did not, in any meaningful sense, compete with other sectors for the allocation of scarce resources.

This conception of the military/economy relationship called into question both the threat and the opportunity rationales for the Reagan administration's economic warfare strategy. If the economy and military sector were not symbiotic, there was no justification for fearing the long-term contribution of advanced technology to the civilian sector. Building up the Soviet economy was not necessarily a threat to Western security. A French Foreign Ministry official claimed that the Soviet military sector was "inwardly oriented"; it relied on the indigenous R&D community for technological advances. As the military dominated available Soviet resources (e.g., scientists, engineers, skilled labor, and scarce inputs), the civilian sector was forced to be outwardly oriented, drawing on Western technology to generate improvements in productivity. The source of innovation and the level of development of the two sectors was markedly different. Moreover, as the Soviets were "bad at diffusion, there is little movement between the two sectors."[72] The military was self-sufficient; its qualitative advances were not dependent on the technological devel-

70. *Business Week*, February 22, 1982, pp. 60–72.
71. Interviews, Foreign Ministry, Paris, June 15, 1982; Economics Ministry, Bonn, July 7, 1982; and the British embassy, Washington, May 6, 1982.
72. Interview, Paris, June 15, 1982.

opment of the economy as a whole and thus were not affected by the contribution of Western technology to the civilian sector. West German Economics Ministry officials questioned the appropriateness of referring to a "military-industrial complex" in the Soviet Union.[73] The United States, they contended, was guilty of "mirror imaging." The fact that military and civilian developments proceeded apace in the United States, with commercial innovations producing military spinoffs and vice versa, was not reason to conclude that the same dynamics existed in the Soviet Union.

With regard to economic warfare as a strategic opportunity, West European officials disagreed with U.S. officials' bleak assessment of the Soviet economic predicament. Moreover, they stressed that any economic hardship inflicted on the Soviets had to be weighed against the losses that trade denial would bring to the West and, in particular, to their own trade-dependent economies.

Even if the West could place effective economic pressure on the Soviets, compartmentalization dictated that the Soviet military was unlikely to bear the brunt of resource deprivation. This outcome was especially probable if deprivation resulted from outside economic pressure; Soviet leaders could then point to external aggression as justification both for maintaining (or even increasing) the level of military commitment and for imposing hardship on the consumer. West European officials further feared that, provoked by economic warfare, the Soviets might become a more belligerent and dangerous international actor than would otherwise have been the case. As French president Mitterrand stated, economic warfare was a serious step, one traditionally complemented or followed by military confrontation.[74]

It should come as no surprise that the United States—less dependent on trade, more willing to abandon détente, and more concerned with containing the Soviets globally and matching their military spending at the margin—viewed the Soviet economy and military as being closely integrated. At the same time, Western Europe—more trade dependent, more sensitive to the need to preserve détente, and engaged in neither an arms race nor a geopolitical competition with the Soviet Union—considered the two to be more in isolation. Perceptions reinforce, and are reinforced by, the interests that ultimately drive export control policy. Given the clear conflict of interests and the belief on both sides that the issue was one of vital national security significance, the stage was set for an eventual alliance confrontation.

---

73. Interview, Bonn, July 7, 1982.
74. *Washington Post*, June 4, 1982, p. 1.

ALLIANCE COORDINATION: CoCom, CREDITS, AND THE PIPELINE

The Reagan administration first raised the export control issue formally at the Ottawa Summit of July 1981. In preparation for that meeting, top administration officials clashed over how other CoCom members should be approached.[75] Weinberger favored pressing them to both abandon the pipeline deal and place broad restrictions on export credits and technology transfer. He argued that Reagan could use his personal prestige to convince the allies to cooperate. Haig was more cautious, contending that the allies were not prepared to adopt such policies and that forcing the issue could spark a trans-Atlantic conflict. Ultimately, the president sided with the State Department and decided to raise objections in principle to the pipeline project but not to attempt to impede it directly.[76] The administration used the summit primarily to emphasize the seriousness with which it viewed trade with the Soviet Union, a message to which other Western leaders were less than fully receptive. In the final communiqué, the seven Western states pledged, in noncommittal fashion, to ensure that "in the field of East-West relations, our economic policies continue to be compatible with our political and security objectives."[77]

More important for the United States was that the allies agreed to consult one another on the improvement of the multilateral export control system. U.S. officials took this pledge as grounds for convening a sub-cabinet-level CoCom meeting, which took place in January 1982. The administration advertised the meeting as "the first broad reconsideration of our technology control system in nearly thirty years."[78]

From the perspective of the United States, the results of the January meeting were mixed. Member states did agree that CoCom needed to be revitalized.[79] They pledged an effort to "harmonize" their licensing procedures, in order to avoid the disputes over list interpretation that had marked the 1970s, and to place the exporters from each member on an equal competitive footing. They also agreed that enforcement should be improved and affirmed their willingness to abide by the no-exceptions policy, at least until the conclusion of the 1982–1984 list review.

75. *New York Times*, July 19, 1981, p. A1.
76. Nau, *The Myth of America's Decline*, p. 301.
77. The communiqué is reprinted in *Department of State Bulletin* (August 1981), pp. 8–9. The quote is found in sec. 36.
78. See Caspar Weinberger, "Technology Transfers to the Soviet Union," *Wall Street Journal*, January 12, 1982.
79. See the statement of James Buckley, undersecretary of state, in U.S. Congress, Senate, Committee on Governmental Affairs, Permanent Subcommittee on Investigations, *Transfer of U.S. High Technology to the Soviet Union and Soviet Bloc Nations*, hearings, 97th Cong., 2d sess., May 4–6, 11–12, 1982, p. 158.

The Reagan administration sought more from the meeting than a reaffirmation of the strategic embargo and promises to strengthen it. Its intent was signaled by Commerce Assistant Secretary Brady on the eve of the meeting: "We are pledged to limit the direct *and indirect* contributions made by our resources to the Soviet military buildup, and we are pledged to substantially reduce Soviet leverage over the economies of the Western world."[80] At the meeting, the U.S. delegation proposed that CoCom's control criteria be revised to include the denial of exports of turnkey plants and technology to defense priority sectors.[81] It also revived the Carter administration's suggestion for an automatic review of any East-West trade deal valued at over $100 million. In short, the Reagan administration took up where the Carter team had left off and used the January meeting to attempt to convince West European leaders to shift alliance strategy in the direction of economic warfare.

Other CoCom members again resisted. Their response was suggested by the leader of the U.S. delegation, Undersecretary of State James Buckley, who tactfully reported to Congress that CoCom members "exercise considerable care to avoid controls whose principal impact would be economic rather than military."[82] A British Defence Ministry official subsequently described the process know-how initiative as "far too broad to be useful" and feared that if it were ever adopted, CoCom would be flooded with exception requests.[83] Rather than focusing on broad sectors such as shipbuilding or communications, he suggested that the United States would be better served by concentrating on such specific items as protective coatings for submarines. In other words, tightening the strategic embargo would probably bring greater payoffs than an ambitious, yet ill-conceived attempt to blunt the overall development of the Soviet industrial base.

The Reagan administration's broad control proposals thus suffered the same fate as did the Carter initiatives of 1980. Other CoCom members made it clear at the January meeting that CoCom was not an appropriate forum for economic warfare, for discussing the political risks inherent in East-West energy trade, or for employing foreign policy sanctions against the Soviet Union.[84] Their commitment to participate in a revival of CoCom did not imply a willingness to broaden its purpose.

The Polish Crisis and Sanctions

The debate over economic warfare moved beyond CoCom and gradually emerged publicly. On December 12, 1981, the Polish military leader-

80. *New York Times*, January 14, 1982, p. 1, emphasis added.
81. Interviews, Foreign Ministry, Paris, June 15, 1982; Ministry of Defense, London, June 2, 1982; and Foreign Ministry, Bonn, July 6, 1982. See also *New York Times*, January 20, 1982.
82. Statement of James Buckley, *Transfer of U.S. High Technology*, p. 160.
83. *Interview*, Ministry of Defence, London, June 2, 1982.
84. See *Financial Times*, January 19, 1982.

ship suspended the activity of the independent trade union Solidarity and imposed martial law. The Reagan administration responded by invoking sanctions against Poland and, as it was convinced of Soviet complicity in the crackdown, against the Soviet Union as well. As had the Carter administration after the invasion of Afghanistan, U.S. officials initially acted without coordinating their response with other Western states. As a result, the administration faced the problem of securing alliance support for its foreign policy sanctions while it continued to press for compliance with regard to economic warfare. In the months that followed, the two strategies became increasingly entangled. The Polish situation gradually became less a foreign policy crisis that demanded a concerted alliance response and more a rationale used by U.S. officials to justify increasing pressure on other CoCom members to practice economic warfare.

The Reagan sanctions imposed against Poland in December 1981 included the suspension of $100 million in agricultural assistance, the withdrawal of Poland's right to fish in U.S. waters, and the refusal to extend Poland's credit line at the Export-Import Bank.[85] The Soviet sanctions that followed shortly thereafter were largely symbolic but included two of substantive economic and political significance. First, the administration added energy transmission and refining items to its list of commodities controlled for foreign policy purposes. Previously, only energy production and exploration items had been controlled. This extension of controls covered contracts signed by Caterpillar in 1981 for pipelayers and spare parts valued at $400 million and contracts signed by General Electric, in connection with the Siberian pipeline, worth $175 million.[86] Second, the administration again indefinitely suspended the processing of all validated licensing requests. In effect, this meant an embargo on the export or reexport to the Soviet Union of any material, equipment, or technology on the U.S. Commodity Control List. In announcing the suspension, the Commerce Department neither set a deadline nor stated the conditions that would lead to its lifting.[87]

In part, these sanctions reflected public pressure on the president to do "something" in response to the Soviet role in Polish repression. At the same time, they served the purpose of those in the administration who believed U.S. export controls should have been made even more restrictive in 1981. As noted above, Weinberger and Casey objected to the Caterpillar deal and also opposed allowing General Electric to participate in the pipeline project. The Polish crisis resolved that debate in the administration in favor of a maximally restrictive stance on industrial exports.

85. See Moyer and Mabry, "Export Controls as Instruments of Foreign Policy," pp. 60–92.
86. Jentleson, *Pipeline Politics*, pp. 204–5.
87. Moyer and Mabry, "Export Controls as Instruments of Foreign Policy," p. 101.

The governments of Western Europe, particularly West Germany, responded to the imposition of martial law more cautiously. Chancellor Schmidt initially declined to implicate the Soviets and did not deem it necessary to abandon his coincidental visit to East Germany.[88] The British and French governments proved more eager to condemn the repression, yet they were as reluctant as the West Germans to follow the United States in imposing economic sanctions. All were sensitive to the economic costs, particularly because a grain embargo was conspicuously absent from the Reagan administration's list of punitive economic measures. Thus, there existed the probable recurrence of a familiar pattern: U.S. unilateral sanctions, followed by West European support in words but not in deeds, followed by U.S. pressure and the outbreak of an alliance conflict. Ironically, after the invasion of Afghanistan, the allies had pledged to avoid such a conflict, and they agreed to prepare a contingency plan to be implemented in the event of Soviet intervention in Poland. During 1981 the United States, Great Britain, France, and West Germany met secretly to develop such a plan, but the West European governments proved reluctant to commit in advance to a specific set of sanctions.[89] The allies were even more unprepared for Soviet intervention short of outright invasion; they had not agreed on a course of action in response to internal Polish repression that was aided and abetted by the Soviets.

U.S. officials, particularly Secretary of State Haig, were determined that the Polish crisis not be overshadowed by an alliance crisis. A trade-off emerged at an emergency meeting of NATO foreign ministers in early January. The United States would refrain from pressuring its allies to endorse or adopt its sanctions; in return, the allies pledged not to undermine U.S. measures and also to devise and implement their own sanctions in the near future.[90] Haig stressed that in considering the issue, each NATO government had a different set of problems and assets with which to deal.[91] A confrontation was temporarily avoided.

By accepting in principle the need for future action, the allies afforded the United States the opportunity to maintain pressure for more meaningful economic restrictions. For U.S. officials, the quest in the alliance for effective Polish sanctions became indistinguishable from the broader attempt to retard the Soviet economy over the long run. The Polish situation was depicted as another part of the larger pattern of Soviet

88. Schmidt's reluctance to follow the lead of the United States was sharply criticized in much of the U.S. press. See, for example, the editorial entitled "Questions for Mr. Schmidt," *Wall Street Journal*, January 4, 1982.

89. See Nau, *The Myth of America's Decline*, p. 305, and also *Wall Street Journal*, December 16, 1980; *New York Times*, March 29, 1981; and *Economist*, April 11, 1981.

90. The text of the NATO Declaration on Poland is reported in *New York Times*, January 12, 1982, p. A8.

91. *New York Times*, January 7, 1982, p. A1.

behavior—including Afghanistan and the military buildup—that justified a fundamental shift in alliance strategy. In the communiqué put out
after the NATO meeting, the allies pledged to "reflect on longer-term
East-West economic relations . . . in light of the changed situation."[92] To
the Reagan administration, the token restrictions on Soviet imports
agreed to by the European Community in March were clearly not sufficient; U.S. officials set their sights higher, on the Siberian pipeline.[93]

The Pipeline

The details of what Soviet officials routinely referred to as "the East-
West deal of the century" are by now familiar.[94] West European firms
contracted to supply wide-diameter (fifty-six–inch) steel pipe, turbines,
compressors, and related technology to the Soviets for the construction
of a three thousand–mile pipeline from the Urengoi gas fields of Sibe-
ria, through Eastern Europe, to a linkup with the West European gas
grid. Natural gas would be supplied to West Germany, France, Italy,
Belgium, and Ireland. The Soviet equipment purchases were financed
in large part by export credits subsidized by participating Western governments. Equipment contracts awarded to each country's firms were
tied, approximately, to the amount of gas that country agreed to purchase. Western credits would be paid off by the Soviets in the form of gas
deliveries, made after the pipeline was completed (which was estimated
to be 1984) and put to use. The deal was generally valued at $10–15
billion, and Soviet gas was to be pumped to the West for twenty years.

The initial concern of the United States was that the pipeline would
make the allies, particularly West Germany, dangerously dependent on
Soviet gas supplies and thus susceptible to Soviet political influence. By
1990, the Soviet share of West German natural gas consumption would
nearly double, from 16 to 30 percent. For Western Europe as a whole,
the increase would be from 10 to 16–18 percent. In addition to consumers, the primary users of Soviet gas included such key industries as
steel, chemicals, and petrochemicals, whose health was vital to West
European economic stability. In the event of a gas cutoff, firms in
such industries might not be able to shift quickly to alternative energy
sources. Moreover, unlike oil, there was no spot market for natural gas,
and stockpiling was extremely expensive. These realities, U.S. officials
feared, would prompt West European leaders to accommodate Soviet
foreign policy. Indeed, the administration suspected that reluctance to

92. NATO Declaration on Poland.
93. Members of the European Community only agreed to reduce their imports of Soviet
goods by $140 million, or approximately 1.4 percent. *New York Times*, March 12, 1982, p. A6.
94. Details of the various aspects of the pipeline deal can be found in Jentleson, *Pipeline
Politics*, chap. 6, and U.S. Department of State, Bureau of Public Affairs, *Soviet-West European
Natural Gas Pipeline*, Current Policy, no. 331 (October 1981).

jeopardize trade ties already constrained West European foreign policies, including their unwillingness to back fully the U.S. position on the Afghanistan invasion and Polish martial law. Dependence on gas imports, combined with an existing and growing reliance on Soviet export markets, even more pointedly raised the specter of "Finlandization." To U.S. officials, the very growth of East-West interdependence, initially designed to constrain Soviet foreign policy and divide the Eastern alliance, now threatened to pacify Western Europe and disrupt the cohesion of NATO.

To West European officials, the Reagan administration both exaggerated the dangers and underestimated the benefits of the pipeline. West German leaders argued that although their imports of Soviet gas would increase significantly, overall dependence on Soviet energy supplies would be 5 percent at most by 1990. To guard against potential blackmail, the French planned a six-month strategic stockpile of Algerian gas, while the West Germans arranged for emergency supplies from the Netherlands. West European officials also pointed to the existence of mutual dependence: 30 percent of Soviet machinery imports came from the West, and thus Soviet leaders were unlikely to risk the economic relationship for political gain.[95]

Moreover, the pipeline would assist the West Europeans in fulfilling an agreed alliance objective—the diversification of sources of energy supply. Since the OPEC shock of 1973–1974, the Western allies had committed themselves to avoid overreliance on Persian Gulf oil. The Iranian upheaval of 1979 and the ensuing Iran-Iraq war further emphasized the desirability of this objective. Although exchanging dependence on Middle Eastern oil for that on Soviet gas was not exactly what the United States had in mind, West European officials contended that the Soviets had proved themselves to be reliable suppliers. In addition, helping to develop Soviet gas reserves would increase the availability and lower the price of energy globally and would reduce the incentive for Soviet intervention in the Persian Gulf.

In the midst of a recession plaguing traditional heavy industries, the export benefits of the pipeline were also considerable. The primary contract won by West Germany's Mannesmann and France's Creusot-Loire, to deliver and install twenty-two compressor stations, was valued at $1 billion.[96] Mannesmann was also the lead supplier of wide-diameter pipe, for which the total Soviet order was worth over $2 billion. Italy's Nuovo Pignone won a $900 million order for nineteen compressor stations and fifty-seven turbines, and AEG-Kanis and Britain's John Brown landed orders for forty-seven turbines worth $300 million and twenty-

95. *New York Times*, February 14, 1982, p. E5, and August 6, 1981.
96. *New York Times*, September 30, 1981, and *Economist*, October 31, 1981.

one turbines worth $190 million, respectively. For the West German firms in particular, the deal was a godsend. Mannesmann's profits had steadily declined for four years, and its plant at Mülheim was entirely dependent on Soviet orders for its survival. AEG-Telefunken (of which AEG-Kanis was a subsidiary) had not earned a profit since 1976; company officials estimated that the Soviet contracts would provide twenty to twenty-five thousand jobs over a two-year period.[97]

While continuing its efforts to persuade the allies that the strategic risks of the pipeline outweighed its economic benefits, the Reagan administration raised the possibility of compensation and proposed U.S. coal as an alternative to Soviet gas.[98] Exploiting America's huge reserves and refining the coal in Western Europe had the advantage of keeping the expansion of Western Europe's energy supply within the alliance. One key problem was that the United States could not easily increase its coal exports significantly; that would require, among other things, a considerable expansion of U.S. port capacity. Whether the Congress would actually commit funds to such a project, and in what time frame, was open to question. Moreover, coal was not a perfect substitute for gas and would create environmental problems.

West European officials did not reject the idea outright; they claimed to be open to any discussion of energy alternatives. They stressed, however, that U.S. coal could only complement, rather than substitute for, Soviet gas. Some West Europeans reacted to the U.S. offer with cynicism, viewing it as primarily motivated by a desire to assist the U.S. coal industry, whose position in world markets was under attack from foreign suppliers.[99] More important, West European officials questioned whether the United States could be counted on as a reliable supplier. They pointed to the soybean embargo of 1973, which was based on short-supply concerns; the restrictions on U.S. oil exports adopted in 1976 as a response to OPEC price increases; and the Carter administration's restrictions on nuclear materials exports, part of its nonproliferation initiatives.[100] Such concerns prompted acerbic comments, such as those by Chancellor Schmidt: "You have not given us a single gallon of oil, and you cannot do it, or of gas. So we have to diversify."[101] Even if the United States could and would commit itself to making available sufficient energy supplies, its offer would still be, at best, partial compensation. Equip-

97. Jentleson, *Pipeline Politics*, pp. 188–89.
98. Ibid., pp. 185–88. In addition to U.S. coal, U.S. officials also proposed—without the support of the Norwegian government—that the allies increase their reliance on Norwegian gas. Ibid., p. 187.
99. See, for example, "USSR Sanctions Make U.S. Laugh all the Way to the Bank," *New Statesman*, January 15, 1982. In May 1982, the United States sent a coal-selling mission to Western Europe, which was intended to capture markets from the struggling Polish industry. *Financial Times*, May 20, 1982.
100. Interview, U.S. Embassy, Bonn, July 6, 1982.
101. See William Safire, "Helmut's Pipeline," *New York Times*, February 19, 1982, p. A31.

ment exports were as important to the West Europeans as were gas imports, and the United States was in no position to serve as an alternative market for the former. At the time of the pipeline dispute, the Reagan administration was actually working to *restrict* West European steel exports from entering the U.S. market.[102] It was thus asking, in effect, that already hard-pressed West European steel firms sell neither to the Soviets nor to the United States. Simultaneously, the Reagan administration continued to allow American farmers to sell grain to the Soviets and, immediately after the Ottawa Summit, granted Caterpillar Tractor a license to sell two hundred pipelayers. U.S. assurances that grain was nonstrategic and that the pipelayers would not be used on the disputed pipeline failed to address the main West European concern: that U.S. policy forced them to bear disproportionately the economic burden of containing the Soviets.

By late 1981, a second argument against the pipeline was gaining prominence, especially among DoD officials. The pipeline threatened not only to enhance Soviet leverage but also to destroy both the opportunity and the threat components of the administration's economic warfare strategy. According to the CIA, natural gas exports from the pipeline would earn the Soviets $8 billion annually.[103] Thus, after paying off its creditors, the Soviets would enjoy a flow of hard currency, over two decades, sufficient to alleviate much of the internal struggle over resource allocation that was anticipated by U.S. officials. The opportunity to squeeze the Soviet economy and "send economic shock waves from Eastern Europe to Cuba to Vietnam" would be wasted, as gas exports and earnings offset the decline in oil output and exports, which even the Soviets admitted would take place by 1985.[104]

Hard currency earnings would be used by the Soviets, administration officials feared, to purchase advanced Western technology to "support their industrial base and military machine."[105] This concern reveals an important link between CoCom and the pipeline in administration thinking. If CoCom members had agreed to target Soviet defense priority sectors, the availability of hard currency would be somewhat less of a threat. The failure of the Reagan administration to gain acceptance at the January meeting for a broader industrial embargo made it all the more imperative that the pipeline be stopped, because the Soviets were

102. On June 11, 1982, the U.S. Commerce Department made a preliminary ruling to impose countervailing duties on British, French, Italian, and Belgian steel. See *Financial Times*, June 21, 1982, p. 3.
103. *New York Times*, March 14, 1982, p. A1.
104. See Lawrence Brady's comments in *Business Week*, November 21, 1983, p. 23, and "Soviet Loss Seen from Oil Cuts," *New York Times*, March 31, 1983.
105. The statement was made before Congress by Stephen Bryen of the Defense Department. *New York Times*, February 10, 1982, p. D1.

permitted under the criteria of the strategic embargo to purchase industrial equipment and technology from Western Europe and Japan.

West European officials were predictably unresponsive to the so-called hard currency argument. Schmidt went even further, reassuring the United States that the Soviets would not use their earnings to buy Western technology; rather, they would use it to purchase American grain. Schmidt's public rebuke prompted U.S. conservatives to redouble their efforts to persuade the president to impose a grain embargo, not only to restore U.S. credibility, but also to force the West Germans, in the words of columnist William Safire, "to decide whose side they're on."[106]

U.S. officials naturally seized on the imposition of martial law in Poland as a third reason why the allies should abandon the pipeline. Although West European leaders rejected a direct request to that effect, U.S. officials did have the means to apply indirect pressure. The three European turbine contractors—John Brown, AEG-Kanis, and Nuovo Pignone—all built their turbines under license from, and with parts (rotors and blades) supplied by, General Electric (GE). The December 1981 sanctions banned GE from further participation in the project, which meant, in effect, it could no longer supply its affiliates. Some of the turbine parts had been shipped to Western Europe before December; thus West European production could continue for a short while. Eventually, the West Europeans would be forced either to cease production of the turbines or to manufacture themselves the parts once supplied by GE. The key actor in a so-called European solution was Alsthom Atlantique, a French concern that produced rotors and blades, under license from GE, and had a contract with the Soviets to provide spare parts for the pipeline turbines.[107]

For the allies to adopt the European solution was a serious step given the commitment they made at the NATO foreign ministers' meeting not to take actions that would undermine U.S. sanctions. West European leaders made it clear, however, that such a step would be taken should it become necessary to carry out the contracts already signed with the Soviets. The West German government announced, for example, that it interpreted the NATO agreement as applying only to cases where U.S. firms were primary suppliers, which suggested its willingness to allow West German firms to fulfill their orders.[108]

By February 1982, some officials in the Reagan administration had grown impatient and were urging the president to force the issue by formally extending the U.S. sanctions to cover the West European affili-

---

106. See Safire, "Helmut's Pipeline."
107. For a detailed discussion, see the statement of Hewett, *Economic Relations with the Soviet Union*, pp. 44–45.
108. *New York Times*, January 14, 1982, p. A7.

ates and subsidiaries of American firms. Within the administration there ensued a debate over whether the United States had the legal authority to apply its controls extraterritorially, particularly because the contracts had been signed before the enactment of new U.S. restrictions. Behind this legal dispute, of course, was the more important political issue of whether to risk a confrontation in the alliance by attempting to coerce the West Europeans. In late February, Reagan decided against an extension of the sanctions but left open the option of taking that step in the near future.[109]

A sanctions extension was not the only coercive measure advocated in the United States at the time. As noted above, defense officials urged that the administration formally declare the Polish government in default on its Western loans. Such a step, they reasoned, would discourage all East-West economic relationships, resulting in the drying up of Western credits to the East and perhaps even a West European reconsideration of the pipeline deal.[110] The State and Treasury Departments opposed a formal declaration of default because they feared the adverse impact on heavily exposed West European banks and, more important, on the stability of the international monetary system as a whole. The president decided against coercion, again leaving open the possibility of future action.

There was also discussion in the Congress of withdrawing U.S. troop support in Western Europe unless the allies abandoned the pipeline. Senator Ted Stevens, head of the Appropriations Subcommittee on Defense, floated the possibility, contending that "if the allies feel so secure in their relationship with the Russians, then I think it's time for us to reexamine the number of troops we have in Europe."[111] The pipeline, the European peace movement, and West European reticence on Afghanistan, Poland, and defense spending all combined to raise to political salience the issue of whether U.S. commitments to the alliance should be reassessed.

Clearly, the pipeline dispute threatened to have broader alliance ramifications. For multilateralists in both the United States and Western Europe, it became imperative to search for a solution that would both placate U.S. hard-liners and enable West European governments to act in accordance with their own conceptions of their vital interests. Secretary Haig sought such a compromise in March 1982 by shifting the focus of U.S. pressure on the allies from the pipeline to export credits.

109. *New York Times*, February 27, 1982.
110. See Benjamin J. Cohen, *In Whose Interest?* (New Haven: Yale University Press, 1986), pp. 177–203.
111. Stevens is quoted in *New York Times*, March 3, 1982, p. A9.

Export Credits

Restrictions on Western export credits were attractive to the Reagan administration as both a sanction and an instrument of economic warfare. By offering long-term loans, Western governments enabled the Soviets to maintain their imports of Western technology and to postpone difficult resource allocation decisions. By offering them at below-market rates, those governments were actually providing subsidies to the Soviet economy, which suggested that the Western taxpayer, in effect, was subsidizing the Soviet military buildup. In the pipeline deal, official credits to purchase the equipment were extended to the Soviets at 7.75 percent, at least four points below prevailing market rates.[112] To U.S. officials, credits made the deal all the more unattractive; to the West Europeans, credits made it possible.

Even if the pipeline could not be stopped, U.S. officials believed a broad Western agreement to increase the cost and reduce the availability of credits would reduce the flow of hard currency to the Soviets. The Reagan administration contended that how the West managed its credit dealings with the Soviets should be considered a "critical strategic issue" for the rest of the century.[113] In March 1982, a high-level U.S. delegation visited Western Europe seeking both an increase in the interest rate charged to the Soviets and a decrease in the *volume* of loans made available. The official rationale for the request was to punish the Soviets for Poland; less stated yet equally important was the desire to find an instrument of economic warfare that could command multilateral support.

West European governments refused to single out and punish the Soviets by raising interest rates or reducing the volume of available credit. Credits, as Helmut Schmidt asserted, would not be used as a political or strategic weapon.[114] They were willing to contemplate a general increase in interest rates, however, and thus also in those that applied to the Soviet Union, as part of a broader revision of Western credit policy, presuming there existed an *economic* rationale for doing so. The wide gap between market rates and the "consensus" rates agreed to by Western states in accordance with the OECD Arrangement on Officially Supported Export Credits suggested such a rationale, and negotiations within the OECD that were designed to bridge the gap were already underway.[115] Hence, the U.S. desire to raise Soviet rates, for strategic

112. Often, however, the lower interest rate was partially offset by the willingness of the Soviets to pay higher prices for the equipment purchased, which they considered to be more politically acceptable. Interview, Ministry of External Trade, Paris, June 16, 1982.
113. The statement, by Undersecretary of State Lawrence Eagleburger, is quoted in *New York Times*, March 31, 1982, p. A8.
114. *New York Times*, March 16, 1982, p. A9, and *Financial Times*, June 1, 1982, p. 3.
115. An agreement was reached within the OECD in May 1982; it had the effect of signifi-

and political reasons, could coincide with West European willingness to consider raising rates across the board, for economic reasons.

The existence of some degree of mutual interest on credits and the shared desire to avoid an alliance showdown set the stage for a possible compromise at the June 1982 economic summit at Versailles. West European officials, particularly the French, rejected U.S. demands for explicit credit restrictions against the Soviets. As stated in the communiqué, however, they did agree on the need to exercise "commercial prudence in limiting export credits" in their dealings with the Soviet Union and Eastern Europe.[116] U.S. negotiators were willing to portray this agreement as a major achievement, claiming they had effectively convinced the allies of the need for a concerted, cautious Western approach to economic relations with the East. To West European officials, the agreed statement could both satisfy the United States and still not be incompatible with their own preference that credit policy be dictated by economic, rather than political or strategic, considerations.

As Henry Nau, a participant at the summit, has noted, the entire East-West trade discussion at Versailles was plagued by confusion, both within the U.S. delegation and between the U.S. and its allies.[117] The substantive issues were never clearly resolved. West European leaders left convinced that they had struck a tacit bargain with the Reagan administration. In exchange for a show of solidarity at the summit and a general pledge to treat future economic relations with the Soviets with "prudence," they believed the administration had agreed to drop its objections to the pipeline and to allow U.S. firms to supply their West European affiliates in order that the deal could progress. At the same time, Reagan administration officials appeared to believe that the allies had committed themselves to substantive restrictions on economic relations with the Soviets. In the aftermath of the summit, both sides were dismayed. President Mitterrand and Chancellor Schmidt, seeking to clarify their interpretations of the communiqué publicly, assured their domestic audiences (and presumably the Soviets) that they had neither agreed to undertake economic warfare nor made an explicit commitment to reduce the credit flow to the USSR[118] (Mitterrand had already incensed the United States by claiming at Versailles that France's ability to cooperate on credits was limited by a secret protocol which had been made with

---

cantly raising interest rates for official credits to the Soviet Union from as low as 7.8 percent to between 12.15 and 12.49 percent. The agreement did not pertain to the volume of credits. See Beverly Crawford, "Western Control of East-West Trade Finance: The Role of U.S. Power and the International Regime," in Gary K. Bertsch, ed., *Controlling East-West Trade and Technology Transfer* (Durham: Duke University Press, 1988), pp. 305–8.

116. The text of the communiqué is reprinted in *New York Times*, June 7, 1982, p. D6.

117. Nau, *The Myth of America's Decline*, p. 310.

118. See *Financial Times*, June 7, 1982, p. 2; *Journal of Commerce*, June 9, 1982, p. 3A; and an interview with Mitterrand reported in *Washington Post*, June 15, 1982, p. 1.

the Russians.)[119] The Reagan administration viewed this claim as violating the spirit of the accord at best and as an explicit repudiation of it at worst. Such statements exacerbated the domestic pressure already placed on the administration by critics who claimed that the United States had "sold out" to the allies at Versailles by accepting the "lowest common denominator" of alliance consensus.[120] On June 18, 1982, with the support of Defense Secretary Weinberger and National Security Adviser Clark, President Reagan resorted to coercion and announced an extension of the December 1982 U.S. pipeline sanctions to cover the activities of West European affiliates and subsidiaries of U.S. firms.

EXTRATERRITORIALITY AND ALLIANCE CONFRONTATION

The stated rationale for the extension of sanctions was the continuation of martial law in Poland. Administration officials claimed they had reserved the right to "close the loophole" left by the December 1981 measures if the situation in Poland did not improve. Yet given that the United States had deeper reasons to oppose the pipeline and that Poland was virtually ignored in the subsequent alliance dispute, this consideration seems at best to have been of secondary importance. The sanctions extension was directed as much at U.S. allies as at the Soviet Union or Poland. It was at an effort to "get the attention of the allies" and demonstrate how seriously the administration considered the East-West trade issue.[121] It also represented a shift in tactics in pursuit of alliance economic warfare. If the allies could not be persuaded of the strategy's merits (and the failure at Versailles provided conclusive evidence), then they should be coerced, regardless of the detrimental impact on alliance harmony. Secretary of State Haig, the most ardent critic of this position, was purposely excluded from the June National Security Council meeting at which the decision was made; he subsequently resigned.[122]

West European governments were furious. The decision took them by surprise, and they felt betrayed by the Reagan administration. West German officials bitterly complained that they had taken great pains to portray Reagan to their people as a man of peace, and in response he

119. *Financial Times*, June 30, 1982. According to French officials, the secret protocol was to last until 1984. It determined the interest rates at which the Soviets would be granted French credits and stipulated that there would be no limits on the volume of credits extended. Interview, Paris, June 16, 1982.

120. See the commentary by George Will in *Newsweek*, June 21, 1982, p. 76, and by William Safire in *New York Times*, June 4, 1982, p. 31.

121. Henry Nau's account emphasizes the administration's frustration (especially the president's) at the apparent unwillingness of other Western leaders to take U.S. concerns over East-West trade seriously. See *The Myth of America's Decline*, p. 312.

122. Alexander Haig, *Caveat: Realism, Reagan, and Foreign Policy* (New York: Macmillan, 1984), pp. 303–16. See also Jentleson, *Pipeline Politics*, p. 195.

had humiliated them.[123] The decision was immediately denounced in West European capitals; government leaders, including Britain's prime minister Thatcher, made it clear that the United States would be defied and the pipeline would be built. The European Community issued a formal protest to the United States that explicitly called for the withdrawal of the measures, contending they were "contrary to international law" and constituted "unacceptable interference on the commercial policy of the European Community."[124]

The Japanese government also reacted angrily, as the decision affected a joint Japanese-Soviet petroleum exploration project off Sakhalin Island. To complete their part of the $200 million deal, Japanese firms required $2 million of indispensable equipment from Schlumberger, a U.S.-based supplier. Because the project had been underway since the mid-1970s and Japan had already committed significant resources to it, Japanese officials, including Prime Minister Suzuki, lobbied hard for an exception to the U.S. sanctions. The June decision signaled the Reagan administration's unwillingness to comply, despite Japanese arguments that they, unlike the Europeans, had faithfully supported U.S. sanctions after the invasion of Afghanistan, regardless of significant economic sacrifice.[125]

To West European governments, the extraterritorial dimensions of the June decision were most objectionable. Indeed, in extending the pipeline sanctions, the United States exercised its already controversial extraterritorial authority in unprecedented fashion.[126] First, it reopened an old wound by asserting the right to regulate the activities of U.S. firms incorporated abroad, by defining them legally as "persons subject to U.S. jurisdiction." Before 1977, the United States exercised such authority only in wartime or national emergency, under the auspices of the Trading with the Enemy Act (TWEA).[127] When that act was terminated

123. Interview, Economics Ministry, Bonn, July 7, 1982.
124. See "European Community: Notes and Comments on the Amendments of 22 June 1982 to the Export Administration Act," presented to the U.S. Department of State on 12 August 1982, reprinted in A. V. Lowe, *Extraterritorial Jurisdiction* (Cambridge: Grotius Publications, 1983), pp. 198, 211.
125. On U.S. sanctions and the Sakhalin licenses, see *Financial Times*, June 11, 1982; *Financial Times*, July 9, 1982; and *Far Eastern Economic Review*, July 23, 1982, pp. 43–50.
126. The June extension generated a vast literature on the legal and political implications of U.S. extraterritorial controls. See, for example, Kenneth W. Abbott, "Defining the Extraterritorial Reach of U.S. Export Controls: Congress as Catalyst," *Cornell International Law Journal* 17 (Winter 1984), 79–158; John Elliott, "Extraterritorial Trade Controls—Law, Policy, and Business," in Martha L. Landwehr, ed., *Private Investors Abroad—Problems and Solutions in International Business in 1983* (New York: Matthew Bender, 1983), pp. 3–35; Duane D. Morse and John S. Powers, "U.S. Export Controls and Foreign Entities: The Unanswered Questions of Pipeline Diplomacy," *Virginia Journal of International Law* 23 (1983), pp. 537–65; and Janet Lunine, "High Technology Warfare: The Export Administration Act Amendments of 1985 and the Problem of Foreign Reexport," *New York University Journal of International Law and Politics* 18 (Winter 1986), 663–702.
127. See Ellicott, "Extraterritorial Trade Controls," pp. 2–8.

in 1977, however, the claimed authority over foreign subsidiaries was quietly transferred to the Export Administration Act. This transfer constituted a significant expansion of authority, because the powers delegated under the EAA were not confined to wartime or emergencies. Broad controls over foreign subsidiaries could now be exercised routinely and against targets (e.g., the Soviet Union) to which they had not normally applied. The June extension constituted the first important application of this expanded authority. West European governments, who had contested the U.S. claim to regulate foreign subsidiaries under the TWEA as contrary to international law, found its more routine application as a foreign policy instrument even more disagreeable. The June extension was particularly irritating because U.S. officials claimed, and U.S. courts upheld the position, that foreign firms incorporated in the United States should be subject to U.S. laws.

Second, the extension constituted a very public exercise of U.S. reexport control authority, which West European governments, unlike West European firms, never recognized as legitimate. Recall that conflict was generally avoided because U.S. authority was exercised quietly, with Western governments leaving the decision whether to comply up to their firms. In this case, however, the challenge was so public, on such a major issue, with such high political and economic stakes, that governments could hardly afford not to become involved. More important, the Reagan administration pushed even further beyond the limits of West European tolerance by applying its reexport controls *retroactively*. This action, too, was unprecedented. At the time the pipeline contracts were signed, West European firms had acted fully in compliance with U.S. regulations.[128] The United States was now asserting, in effect, that foreign users of U.S.-origin equipment or technology had to observe not only the U.S. regulations that existed at the time of transaction but also any other restrictions that U.S. officials might adopt, for whatever reason, at any time in the future. The U.S. contention that technology constituted a "continuous export" did not impress allied governments. Rather, West European officials complained that the attempt to exercise already questionable authority retroactively "called into question the basic principles of the world trading system," such as contract sanctity.[129]

The June extension placed in an unenviable position those participating West European firms whose governments had an attitude of defiance. If these firms chose to recognize U.S. reexport control authority and comply with the sanctions, they would lose their badly needed equipment orders and would also be subject to Soviet penalties for breaking the terms of the contracts. In addition, complying with the

128. Morse and Powers, "U.S. Export Controls and Foreign Entities," p. 540.
129. *Financial Times*, July 15, 1982.

United States might eliminate the firms from competition for future Soviet orders. This consideration was an important one because, along with the export pipeline, the Soviets required additional turbines and compressors for five domestic pipelines to be built under the current five-year plan.[130] Another $10 billion in contracts was at stake, and those firms involved in the export pipeline contract would likely have an inside advantage in that competition.

If they chose not to comply, however, the firms would be subject to U.S. retaliation. The Export Administration Act gave the Reagan administration the right to issue "denial orders," which would prohibit U.S.-based firms from exporting to West European firms that were found to be in violation of U.S. laws. The latter would be forced to choose, in effect, between the Soviet market and their U.S. sources of supply. Given their greater overall stake in trade with the United States, most of the firms involved probably would have chosen to sacrifice the pipeline deal.[131]

West European governments did not allow their firms the luxury of an unconstrained choice. Either by invoking domestic legislation or through political pressure, they forced the companies to defy the United States and fulfill the pipeline contracts.[132]

The Reagan administration, as it had threatened, responded by imposing retaliatory sanctions. For Western Europe, the price of refusing to engage in economic warfare against the Soviet Union was a trade war with the United States. The Department of Commerce first placed temporary denial orders against Creusot-Loire and Dresser France, immediately after they had followed the French government's demand that they ship completed compressors to the Soviet Union. This action posed enormous problems for Dresser France, which relied on its parent firm, Dresser U.S., for its design and manufacturing know-how.[133] Dresser

130. *International Herald Tribune*, June 30, 1982, p. 1.

131. See *Economist*, June 26, 1982, p. 18.

132. The British invoked, for the first time, the Protection of Trading Interests Act of 1980, which gave the government the authority to prohibit British companies from complying with foreign directives if the government judged that such compliance would be harmful to the trading interests of Great Britain. See "United Kingdom: Protection of Trading Interests Act 1980" and "The Protection of Trading Interests" (U.S. Reexport Control Order, 1982), reprinted in Lowe, *Extraterritorial Jurisdiction*, pp. 186–96. The French government issued a "requisition order" to force its firms to ship pipeline equipment; if they refused, company managers would be subject to fines and up to one year in prison, and the government itself would seize and ship the items in question. The Italian government simply ordered Nuovo Pignone to ignore the sanctions and honor its contracts. The West German government claimed no legal authority to force the issue but sent formal requests to the firms involved urging them to adopt the general European strategy of defiance. On France, see *New York Times*, July 23, 1982, p. 1. On Italy, see *Financial Times*, July 8, 1982, p. 1, and *New York Times*, July 25, 1982, p. 1. On Germany, see *Financial Times*, August 26, 1982, p. 14.

133. On sanctions against French firms, see *New York Times*, August 27, 1982, p. 1. On the problems of Dresser, see *Washington Post*, August 25, 1982, p. 1, and Ed Luter, Dresser vice-president, "What They're Doing to Us at Dresser," *Washington Post*, August 29, 1982, p. C8.

France possessed the equipment and know-how necessary to complete the pipeline order but would be forced to shut down within a few months if the denial order remained in effect.

Denial orders were subsequently issued against John Brown, Nuovo Pignone, AEG-Kanis, and Mannesmann following each of their first shipments of pipeline equipment.[134] In order to limit the damage to the British firm, which was a highly diversified concern with extensive trade links to the United States, the sanctions were made to apply only to energy technology and equipment, not to all trade. This more accommodating measure, however, was accompanied by an effort to extend the sanctions to other firms. In September, denial orders were issued against thirteen French affiliates of Creusot-Loire involved in the pipeline deal. In October, the Reagan administration sought to cast the net even further by requesting U.S.-based energy equipment suppliers to furnish extensive information on all their European affiliates that might be involved, however modestly, in the pipeline deal.[135]

With neither side prepared to back down, the danger of an all-out trade war became increasingly evident. Members of the European Community retaliated against the United States in the GATT: they formally demanded that the latter reform its DISC (Domestic International Sales Corporation) system on the grounds that it constituted an unfair export subsidy. The complaint broke a private agreement between the two sides that EC members would not register a public complaint about the DISC system.[136] Sir Roy Denman, the EC trade negotiator, publicly warned of possible further retaliation, including lawsuits against the U.S. government, withdrawal of EC support in the GATT for a major U.S. initiative to liberalize trade in services, and, most important, the raising of import barriers to U.S. agricultural exports.[137] The members of the EC had the potential to inflict significant harm on the United States, given that U.S. exports to the EC totaled $52 billion annually, producing a U.S. export surplus of $18 billion. Their incentive to retaliate was heightened by a concurrent U.S. decision to raise trade barriers against West European steel and by the Reagan administration's continued refusal to lower U.S.

---

134. On John Brown, see *Washington Post*, September 1, 1982, p. A1, and *New York Times*, September 10, 1982, p. D1. For Prime Minister Thatcher's reaction, see *Washington Post*, September 2, 1982, p. A33. On the sanctions against Nuovo Pignone, which inspired a march by workers on the U.S. consulate in Florence, see *Journal of Commerce*, October 19, 1982, p. 23, and October 20, 1982, p. 23. On AEG-Kanis, see *Washington Post*, October 1, 1982, p. 18.

135. *New York Times*, September 8, 1982, p. D1, and October 1, 1982, p. D1. In addition to penalizing subsidiaries of Creusot-Loire, the United States imposed sanctions against two British firms that, on order from the British government, fulfilled small contracts for fire protection equipment and radio antennae. *Financial Times*, September 11, 1982, p. 26, and *New York Times*, September 11, 1982, p. 35.

136. *Financial Times*, July 1, 1982.

137. An interview with Denman is reported in *International Herald Tribune*, July 2, 1982. See also *Business Week*, July 19, 1982, pp. 50–51.

interest rates, which West European leaders believed to be a major cause of their protracted recessions.

The pipeline crisis also threatened other areas of mutual alliance concern. First, West European support for a revitalized strategic embargo in CoCom was jeopardized. The all-important 1982 list review was to begin in November; it is difficult to imagine that West European governments would be responsive to U.S. proposals for stricter controls while West European firms were being blacklisted by the U.S. government. Indeed, close advisers to French president Mitterrand reportedly urged him to remove France from CoCom in retaliation for the American sanctions.[138] Fortunately for CoCom, Mitterrand did not accept this advice. Second, already fragile public support in Western Europe for the deployment of U.S. intermediate-range nuclear missiles was put under additional strain. The crisis reinforced in the West European public the growing suspicion that the United States was the more aggressive and dangerous global power, an attitude suggested by placards at West German peace demonstrations that read "Better to have gas from the East, than rockets from the West."[139]

Trans-Atlantic political relations had dropped to perhaps their lowest point in the postwar era. The sentiment was best captured by French foreign minister Claude Cheysson, who spoke of a "progressive divorce" taking place between the United States and Western Europe: "We no longer speak the same language. There is a remarkable incomprehension, and that is grave. The United States seems totally indifferent to our problems. It is the major ally and the world's biggest country and we don't even talk anymore."[140] The obvious beneficiary of this state of affairs was the Soviet Union. What began as a U.S. attempt to punish and weaken the Soviet Union was rapidly evolving into a major Soviet political triumph.

The November Resolution

By September, the futility of the sanctions extension was evident. Although construction of the pipeline might be delayed, it would not be stopped, and the CIA estimated that the Soviets would be able to begin their gas deliveries on schedule.[141] The longer the sanctions remained in effect, the greater the political cost to the Western alliance. Thus, Secretary of State George Shultz sought a means to resolve the dispute that would minimize political embarrassment to the United States. Shultz and President Reagan quietly revealed that they would be receptive to drop-

138. *Washington Post*, November 29, 1982, p. A13.
139. *Financial Times*, June 28, 1982, p. 3.
140. *New York Times*, July 23, 1982, p. A1.
141. *Washington Post*, September 1, 1982, p. A1.

ping the sanctions if they could obtain a set of "stronger" measures to replace them.

A mutually acceptable agreement was reached in November 1982. President Reagan simultaneously announced the lifting of the pipeline sanctions and the launching of a series of detailed studies to be conducted by alliance members. Overall assessments of Western economic relations with the East would be made under the auspices of NATO and the OECD. CoCom would undertake a study of strategic trade controls, as a means to strengthen their effectiveness. The International Energy Agency (IEA) would evaluate the costs and benefits of East-West energy trade; no new deals would be signed while the study was underway. Finally, to ensure that the Soviet economy was not subsidized preferentially, the OECD would explore possible means to harmonize alliance credit policies. The results of the studies were to be reported at the 1983 Williamsburg Economic Summit. U.S. officials claimed a major triumph, contending that the sanctions had served their purpose: they had jolted the allies into thinking seriously about the strategic dimensions of trade with the Soviet Union and into accepting the need to adopt collective policies compatible with the long-term security interests of the West.[142]

Although that assertion contained an element of truth, the main purpose of the agreement was to provide the Reagan administration with the opportunity to make a face-saving retreat.[143] The French government initially proved unwilling to grant even that, as Mitterrand announced immediately after Reagan's statement that France was not a party to the agreement. By December, France and the United States resolved their differences, and the French announced their support of the alliance studies.[144]

The November resolution marked the failure of U.S. coercion and the end of U.S. active opposition to the pipeline. Both the December 1981 and the June 1982 pipeline sanctions were removed; therefore, U.S. firms were again free to participate in the project and also to compete for future contracts. Furthermore, although it was initially unrelated

142. Analysis and details of the November resolution can be found in *New York Times*, November 14 and 15, 1982, p. A1, and *Washington Post*, November 14 and 15, 1982, p. A1.

143. This assessment is shared by Jentleson, *Pipeline Politics*, p. 197, who argues that the United States did "more conceding than leveraging." A similar conclusion is reached by Beverly Crawford and Stefanie Lenway; see "Decision Modes and International Regime Change: Western Collaboration on East-West Trade," *World Politics* 37 (April 1985), 397. David Baldwin argues that how one assesses the dispute hinges critically on whether one adopts a "hard- or soft-line" perspective on U.S. relations with the Soviet Union. He concludes that the lessons of the dispute are ambiguous, since "it may never be clear just who was trying to influence whom with respect to what." See *Economic Statecraft* (Princeton: Princeton University Press, 1985), p. 289.

144. Nau, *The Myth of America's Decline*, pp. 315–16, and *New York Times*, December 15, 1982, p. A1. Mitterrand objected to the apparent linkage between the lifting of the sanctions and the commencement of the alliance studies. The French position throughout the crisis was that the U.S. measures were unjustified and in violation of international law; thus, they should be removed without any quid pro quo.

to the pipeline, the Reagan administration also lifted its suspension of the processing of validated license requests for export to the Soviet Union.[145] Thus, remarkably, with no reference to the status of the Polish situation, the United States simply abandoned all of the substantive measures it had imposed on the Soviet Union in December 1981 as a response to the Polish crisis. This action casts further doubt on the administration's persistent claim that its opposition to the pipeline and its refusal to allow industrial trade with the Soviet Union were motivated by concern over the plight of the Polish people. Rather, it lends credibility to the argument that Poland had served as a convenient rationale for the pursuit of economic warfare.

By removing the December as well as the June sanctions, the administration helped to alleviate its conflict with U.S. firms and their supporters in Congress. Immediately following the November resolution, 250 U.S. corporate executives visited Moscow, with the encouragement of the administration, in the hope of renewing their badly damaged position in the Soviet market.[146]

Initial results of the alliance studies were available by the start of the Williamsburg Summit.[147] Overall, they reflected the willingness of America's allies to reconsider the relationship between East-West trade and alliance security yet reaffirmed their reluctance to move in the direction of economic warfare. The NATO study produced a general agreement that the strategic dimensions of East-West trade should be evaluated more systematically, but West European states proved unwilling to upgrade the NATO Economic Secretariat and give it a greater say in East-West trade.[148] On credits, the OECD study led to an increase in interest rates, but the allies refused to endorse either an automatic adjustment mechanism for rates or limits on maturities or the volume of credit available. The more general OECD study provided an assessment of the economic advantages and disadvantages of East-West trade and a commitment to produce future studies. An American proposal for the exchange of information on major East-West industrial projects, echoing earlier proposals in CoCom, was rejected, however. The IEA study acknowledged the risks of high levels of energy dependence on a single supplier, but the allies could not agree on a specific percentage limit on imports from the Soviet Union by any single member. The allies did

145. *New York Times*, November 16, 1982, p. D6, and *Federal Register* 47 (November 18, 1982), 51858–59.

146. Domestic constraints on the Reagan administration's ability to sustain the pipeline sanctions are emphasized in Jentleson, *Pipeline Politics*, pp. 203–8.

147. For analysis of the studies, see Nau, *The Myth of America's Decline*, pp. 316–20; Angela Stent, "Technology Transfers in East-West Trade: The Western Alliance Studies," *AEI Foreign Policy and Defense Review* 5 (1985), 44–52; and John P. Hardt and Donna Gold, *East-West Commercial Issues: The Western Alliance Studies*, Congressional Research Service, November 5, 1984.

148. French foreign minister Claude Cheysson explicitly rejected the idea of an "economic NATO." See his interview in *Washington Post*, March 21, 1983.

agree to abandon plans for a second pipeline; that decision, however, was driven more by market considerations than by political or strategic factors.[149]

Finally, the CoCom study afforded the Reagan administration one last chance to make its case for economic warfare. As part of that study, the United States convinced its allies to consider—once again—whether industrial equipment and processes of *indirect* military significance should be controlled by CoCom.[150] At issue generally was whether CoCom's criteria should shift from "direct military significance" to the broader conception of "strategic relevance" preferred by the United States. Specifically, U.S. negotiators proposed twenty-two categories of energy production equipment and technology for multilateral control.[151] These so-called other high-technology items were to be shifted from foreign policy control (their current U.S. status) to national security control. Having failed to stop the pipeline, the administration was hoping to influence the future development of the Soviet energy sector by recasting it as a national security export control issue, one appropriate for consideration in CoCom.

Members of CoCom made good on their pledge to consider an extension of the control criteria but ultimately refused to endorse it. Only two of the twenty-two proposed categories were considered sufficiently strategic in its narrow sense to be added to the CoCom list. Lacking support in the alliance, the administration acquiesced in the export of American equipment. A $40 million license for submersible pumps, held in abeyance until the conclusion of the CoCom discussions, was granted to the Hughes Tool Company in March 1984.[152]

CONCLUSION

The East-West trade experience of 1980–1983 highlighted both the potency and limits of the power of U.S. executive officials. First, U.S. officials demonstrated that they still possessed and could utilize the extraordinary export control powers delegated to them in the crisis circumstances of the early cold war.[153] They quickly imposed foreign policy sanctions in response to Soviet provocations, with little regard for the

149. Stent, "Technology Transfers in East-West Trade," p. 49, and Jentleson, *Pipeline Politics*, pp. 200–203.
150. Hardt and Gold, *East-West Commercial Issues: The Western Alliance Studies*, p. 9.
151. Nau, *The Myth of America's Decline*, p. 313.
152. *New York Times*, March 7, 1984, p. D1, and Jentleson, *Pipeline Politics*, pp. 209–10.
153. Recent work by William Long has emphasized the enduring power of the Executive in the East-West trade area. See Long, "The Executive, Congress, and Interest Groups in U.S. Export Control Policy: The National Organization of Power," in Bertsch, ed., *Controlling East-West Trade and Technology Transfer*, pp. 27–67, and Long, *U.S. Export Control Policy: Executive Autonomy vs. Congressional Reform* (New York: Columbia University Press, 1990).

fact that U.S. firms were forced to cancel existing or proposed contracts worth hundreds of millions of dollars. They broadened U.S. export control criteria, again with little debate outside the Executive. They twice froze the entire export licensing process as it pertained to the Soviet Union, once for national security and once for foreign policy reasons. Through the extraterritorial assertion of authority, U.S. officials controlled the trade of American multinationals incorporated abroad, even that of wholly foreign-owned firms which were affiliates or business partners of U.S. concerns. Even in lifting the pipeline sanctions, U.S. officials sought to preserve and extend their extraterritorial power.[154] In short, U.S. officials dramatically demonstrated that for the state which depicts itself as most committed to the principles of global economic liberalism, trade—intra-Western as well as East-West—was still a privilege rather than a right.

At the same time, however, U.S. officials proved unable to obtain alliance support for their preferred policies. They failed to persuade their major allies to impose punitive sanctions, broaden CoCom's control criteria, limit the volume of export credits, or terminate the pipeline. Similarly, they could not provide sufficient compensation to achieve the latter objective. And although they could force West European firms to rethink their pipeline contracts, they proved unable to coerce other Western governments. The attempt to do so created an alliance crisis perhaps equaled only by the Suez episode of 1956. The Reagan administration learned by hard experience what the Truman and Eisenhower administrations appeared to intuit—that efforts to force compliance with economic warfare were likely to be prohibitively costly in terms of America's broader alliance interest.

The Western alliance survived the pipeline. An all-out trade war never erupted; the 1982–1984 list review in CoCom concluded with a substantial agreement; NATO's plan to deploy missiles in Europe if arms control talks failed was effectively carried out. If the pipeline did have enduring costs, they were most probably incurred in the relationship between East-West trade and the broader Western trading system.

First, as a result of the pipeline sanctions, U.S. firms gained reputations as unreliable suppliers in intra-Western as well as East-West trade. West European firms, in some cases with the encouragement of their governments, began to reconsider the desirability of their traditional reliance on U.S. technology and equipment. Some turned to other suppliers; others pursued the costly alternative of developing local substitutes. New licensing agreements similar to that between GE and its

---

154. In withdrawing the pipeline sanctions, the Commerce Department retained language that had been added to U.S. export regulations in June, which suggested that reexport controls could legitimately be applied retroactively. See Morse and Powers, "U.S. Export Controls and Foreign Entities," p. 562.

West European affiliates were subsequently entered into only with caution and reluctance.[155] The tendency toward "de-Americanization" became most pronounced in the energy sector, yet it was found elsewhere as well.

Second, the pipeline sanctions politicized the reexport control issue in the Western alliance. Before the pipeline situation, this assertion of U.S. authority was quietly disputed but generally tolerated by Western governments and firms. Afterward, it simply could not remain uncontested. Indeed, in registering their August 1982 protest, West European governments attacked not only the pipeline sanctions in particular but also the very idea of U.S. reexport controls. They termed it "reprehensible" that U.S. regulations forced non-U.S. firms "voluntarily" to act, in effect, as instruments of U.S. policy.[156] The implications of this and subsequent public challenges were clear: despite the resolution of the pipeline dispute, the routine use of U.S. reexport authority had prominently emerged as an issue of alliance controversy. Other Western governments could no longer object in principle, yet acquiesce in practice.

The pipeline sanctions helped to undermine the long-standing commitment of U.S. officials to insulate the Western trading system, and the position of U.S. firms within it, from the effects of East-West export controls. The key to maintaining the insularity of the two realms is an effective CoCom. U.S. efforts to strengthen CoCom, in the context of the intra-Western trade issue, are the subject of Chapter 8.

155. See Ellicott, "Extraterritorial Trade Controls," p. 31; Patrick B. Fazzone, "Business Effects of the Extraterritorial Reach of U.S. Export Control Laws," *Journal of International Law and Politics* 15 (1983), 576; and National Academy of Sciences, *Balancing the National Interest* (Washington, D.C.: National Academy Press, 1987).
156. See "European Community: Notes and Comments," p. 204.

CHAPTER EIGHT

U.S. Leadership and the Struggle
to Strengthen CoCom, 1981–1989

Throughout the two terms of the Reagan administration, the strengthening of CoCom's strategic embargo was a national security priority for the United States. Administration officials ultimately proved more successful in this undertaking than in their efforts to obtain alliance support for economic warfare. By the latter half of the 1980s, it was evident that the multilateral strategic embargo was stronger than it had been during the 1970s. It was equally clear, however, that over the course of the decade, CoCom remained a source of serious controversy among Western alliance members, and its progress in strengthening controls was limited and subject to strain.

In this chapter I argue that the combination of a strengthened, yet troubled CoCom can be attributed primarily to inconsistent U.S. leadership. During the 1980s, U.S. officials undertook some leadership responsibilities effectively as they sought to revitalize CoCom. They set a domestic example, developed initiatives to expand the control list, and launched an ambitious effort to obtain the compliance of non-CoCom suppliers. At the same time, and despite the high priority attached to CoCom, U.S. officials continued to neglect other leadership tasks. They compromised the integrity of the control process and failed to take adequate steps to minimize the economic and administrative burdens of controls. They also violated what might be considered the diplomatic norms of CoCom. The ability of U.S. officials to carry out the full range of leadership responsibilities effectively was constrained by the lack of cohesion in the U.S. policy process, by the overall insensitivity of the U.S. government to the economic costs of controls, and by the ongoing tendency of U.S. officials to entangle national security and foreign policy considerations in the control process.

The discussion is divided into three parts. The first substantiates the

266

above argument by tracing developments in CoCom from 1981 through 1985. It examines U.S. efforts to gain alliance support for an expansion of the control list, improvements in enforcement, and changes in the institutional character of CoCom. It then analyzes the impact on CoCom of America's ongoing interagency struggle over export control policy, of its failure to decontrol less strategic items, and of its tendency to politicize the control process.

Second, the chapter examines the expansion of U.S. restrictions on the transfer of technology to other non-Communist states. For the first time in the postwar era, U.S. officials chose to blur in practice the boundary between East-West and intra-Western controls. Following the logic of the Bucy Report, U.S. officials employed intra-Western restrictions both as a defensive response to the weakness of the export control systems of other governments and as a lever designed to strengthen those systems.

Strains within CoCom and within the alliance over U.S. intra-Western restrictions peaked by 1985. In January 1988, a deal was struck among CoCom governments at high levels similar to that struck in 1954: the United States committed to reduce the CoCom list and liberalize intra-Western trade in exchange for tighter enforcement by other CoCom members. The third part of this chapter explains both the emergence of the January 1988 agreement and the initial effort to implement it in CoCom in 1988 and 1989. By the end of 1989, the need for effective implementation was all the more pressing because structural changes in the international security system posed the most significant postwar challenge to date to the viability of CoCom.

CoCom and U.S. Leadership, 1981–1985

During its first term, the Reagan administration undertook an ambitious defense buildup that called for the development and application of a wide range of advanced technologies to military systems and purposes. Administration officials viewed the tightening of national security export controls as a necessary means to protect those new investments. It was urgent to do so because U.S. officials believed, with the support of intelligence data, that the Soviet Union had organized a systematic effort to acquire militarily significant Western technology and had done so with regularity and success during the 1970s.

The administration took its case to the public, to increase awareness of the problem and gain support for a more restrictive approach.[1] In April 1982 the CIA released a sanitized version of a classified report that

1. See, for example, Caspar Weinberger, "Technology Transfers to the Soviet Union," *Wall Street Journal*, January 12, 1982.

provided information on Soviet methods, targets, and acquisitions.[2] The Senate's Permanent Subcommittee on Investigations elaborated on those findings during extensive hearings.[3] The CIA report was subsequently updated on the basis of the so-called Farewell papers obtained by French intelligence from a high KGB official in 1981 and 1982.[4] A British scholar who was granted access to some of the Farewell papers largely confirmed the CIA's assessment, finding that the Soviet acquisition process was highly organized and covered a wide range of technologies; that the Soviet military had built-in arrangements for the systematic use of documents and hardware obtained; and that the military's adaptation and use of Western technology was different in style from and more successful than was its assimilation by the civilian sector.[5]

In recognition that the effective denial of strategic technology required multilateral support, the administration focused its efforts on what it termed the revitalization of CoCom. U.S. officials used the high-level meeting of January 1982 to give greater political prominence to CoCom and to lay the foundation for three related initiatives—the expansion of the control list, improvements in enforcement, and the strengthening of CoCom's institutional structure.

As I discussed in Chapter 7, following the invasion of Afghanistan, the United States tightened its national controls over technology transfer and broadened its control criteria to economic warfare. Although other CoCom members rejected the broader criteria at the January 1982 meeting, they did prove willing to consider a U.S.-initiated extension of the control list, provided the items in question were of direct military significance. The CoCom review that began in 1982 was protracted and at times contentious, yet it did ultimately produce a considerable expansion in list coverage. According to the U.S. Defense Department, of over one hundred items U.S. officials proposed for list addition in 1982, fifty-eight had been accepted before the final round for CoCom control. Among those subject to new or tighter controls were floating drydocks (which plugged the gap exploited by Japanese firms in 1978), spacecraft and space launch vehicles, superconductive materials, robots and robot controllers, equipment and technology for the production of super-alloys, and industrial marine gas turbine engine technology. CoCom

2. CIA, "Soviet Acquisition of Western Technology" (Washington, D.C., April 1982).

3. U.S. Congress, Senate, Committee on Governmental Affairs, Permanent Subcommittee on Investigations, *Transfer of U.S. High Technology to the Soviet Union and Soviet Bloc Nations*, hearings, 97th Cong., 2d sess., May 4–6, 11–12, 1982.

4. U.S. Department of Defense, "Soviet Acquisition of Militarily Significant Western Technology: An Update" (Washington, D.C., September 1985).

5. See Philip Hanson, "New Light of Soviet Industrial Espionage," *Radio Liberty* RL 36/86 (January 20, 1986), and "Soviet Industrial Espionage," *Bulletin of the Atomic Scientists* (April 1987), pp. 25–29. While corroborating much of the CIA and Pentagon analysis, Hanson does contend that the 1985 update overplayed the evidence because it failed to acknowledge that much of what the Soviets acquired they judged to be of marginal value.

members even agreed in December 1983 to implement new controls on ten priority items immediately, rather than to wait, as per the usual procedure, until the entire list review had been completed.[6]

The final round of negotiations in 1984 proved the most difficult, involving computers, telecommunications equipment, and software. The latter two had not been subject to prior control as separate categories, whereas computers had been at the center of CoCom debate since the mid-1970s. Since that time no agreement had been reached to update the control parameters. The U.S. Defense Department urged comprehensive controls, even on personal computers, which Assistant Secretary of Defense Richard Perle claimed were used by the United States— and could be by the Soviets—to target nuclear weapons in the European theater. Western Europe and Japan resisted, pointing to widespread foreign availability. Even the normally supportive *Economist* editorialized that "an attempt to control a product that can be bought by anybody at a thousand different places is doomed."[7]

A compromise was finally struck in July 1984, involving the above three categories and completing the list review.[8] In overall terms, the United States exchanged some degree of liberalization on computers for tighter restrictions on software and telecommunications equipment and technology. U.S. officials agreed to liberalize the most readily available microcomputers; those most powerful and those "ruggedized" for harsh industrial or military use were maintained under control. Super minicomputers were placed under control in exchange for significant liberalization on mainframes.[9] In an echo of the Bucy Report, CoCom members agreed to maintain controls on technology for the development or production of unembargoed computers and to restrict the integration of computers into other equipment unless both the computer and the equipment were unembargoed.[10]

Controls were placed on software used for the design, development, or production of items on the CoCom list. In addition, specified software

---

6. See Caspar Weinberger, *The Technology Transfer Program*, A Report to the 98th Congress, 2d Session (Washington, D.C.: U.S. Department of Defense, February 1984), pp. 11–12. A useful summary of the list additions is found in *Security Export Control*, British Business Supplement, June 14, 1985, pp. iii–v. See also *Aviation Week and Space Technology*, September 12, 1983, p. 17.

7. Perle's statement is found in U.S. Congress, Senate, Committee on Governmental Affairs, Permanent Subcommittee on Investigations, *Transfer of Technology*, hearings, 98th Cong., 2d sess., April 2, 3, 11–12, 1984, pp. 158–64. See also *Economist*, April 21, 1984, p. 13.

8. For discussion of the compromise, see *Financial Times*, July 16, 1984, p. 1; *Wall Street Journal*, July 17, 1984, p. 35; *New York Times*, July 17, 1984, p. D7; and especially *Aviation Week and Space Technology*, July 23, 1984, p. 21.

9. The threshold at which computer systems would in most cases be denied to the East was raised significantly from a processing data rate (PDR) of thirty-two bits per second to a PDR of forty-eight bits per second. Computers with a PDR of two to twenty-eight bits per second became licensable at national discretion; those in the twenty-eight to forty-eight range required a CoCom general exception but were eligible for favorable consideration. See *Security Export Control*, p. iv.

10. Willie Schatz and Paul Tate, "CoCom Makes Good," *Datamation*, September 1, 1984, p. 70.

within six categories was subject to embargo, items such as networking and signal processing software and that used in image processing, artificial intelligence, and computer-aided design and manufacturing (CAD/CAM). U.S. officials were happy to obtain formal recognition of software's strategic significance, although they were forced to retreat from their earlier sweeping control proposals.[11]

In the telecommunications sector, a ban was placed on the export of all large switches until September 1988, at which point some would become eligible for licensing at national discretion. Technology used for the development of switching equipment was also embargoed, with no presumption of liberalization after September 1988. The concern in this area was the enhancement of Warsaw Pact command and control capabilities, which, like those of NATO and the United States, relied to a considerable extent on the sophistication of the public telephone network.[12]

Competition for sales to the East had become especially intense in telecommunications, and the CoCom agreement entailed significant economic sacrifices for West European firms. Because of pressure from commercial interests, Britain and France initially resisted the CoCom compromise.[13] The British firms Plessey and GEC had landed a £50–100 million contract to sell their System X to Bulgaria, its first significant overseas sale. The CoCom agreement forced them to abandon the deal. The British government warned, however, that its firms would reconsider if non-CoCom suppliers moved to replace the sale. U.S. officials, taking responsibility for the preservation of the agreement, forced the major non-CoCom supplier, Sweden's Ericcson, to observe CoCom's new rules informally.[14]

Similarly, West Germany's Standard Electrik Lorenz (SEL) was awarded a letter of intent from Hungary for a digital exchange system and was promised follow-up contracts in the Hungarian market. The West German government tested CoCom by agreeing to submit an exception request for the sale. The United States vetoed it, despite the fact that, according to SEL, West German politicians at every level intervened on its behalf.[15] Finally France, which in the early 1980s had sold a modified package of digital switching equipment to the Soviets, agreed to observe the letter, if perhaps not fully the spirit, of the new agreement. French officials claimed that bidding on Eastern contracts was permitted, pro-

---

11. Ibid., p. 69. The six categories were development, programming, diagnostic, maintenance, operating systems, and applications software. *Security Export Control*, p. v.

12. NATO announced at about the same time the CoCom agreement was reached that it planned to spend $2 billion to upgrade its own public telephone network to enhance command and control. *Financial Times*, July 25, 1984, p. 11.

13. *Financial Times*, July 16, 1984, p. 1.

14. *Financial Times*, March 24, 1984, p. 1; July 17, 1984, p. 1; and August 4, 1984, p. 1.

15. *Financial Times*, October 24, 1984, p. 1, and December 17, 1984, p. 1.

vided that actual delivery did not take place before September 1988.[16]

The second major U.S. initiative concerned enforcement. As had been the case for list expansion, U.S. national efforts preceded and provided the impetus for multilateral coordination. The cornerstone of the U.S. program was Operation Exodus, which was launched in 1981 under the auspices of the Customs Service.[17] Operation Exodus represented a new priority for Customs and helped to transform what had been a relatively lethargic U.S. effort during the 1970s into a well-funded, at times overzealous campaign to obstruct the outflow of U.S. critical technology.[18] The Commerce Department, traditionally responsible for enforcement, also improved its capabilities by abolishing the understaffed Compliance Division of the Office of Export Administration and creating a new agency devoted exclusively to enforcement. Commerce staff devoted to the task increased from 39 in 1980 to 185 in 1988, while its budget jumped from $1.7 million to $8.3 million.[19]

U.S. officials pursued the bolstering of enforcement bilaterally and multilaterally. At a high-level CoCom meeting in April 1983, member governments agreed to work collectively to tighten safeguards, improve the IC/DV system, and target technologies and third countries particularly susceptible to diversion.[20] Bilaterally, U.S. officials sought to increase awareness of the diversion problem and convince allied governments to devote additional resources and greater scrutiny to licensing and enforcement. Other CoCom members were receptive in varying degrees. Among the major exporters, Great Britain and France were most responsive, whereas West Germany and Japan lagged behind.

Following the example of and responding to pressure by the United States, Great Britain created in 1983 a team of investigators and technology specialists to monitor illegal technology transfers to the Soviet bloc. It established, in effect, a British version of Operation Exodus called Project Arrow, which earned the praise of Richard Perle: "Initially they had not been prepared to go very far. But the British have taken some positive steps in the areas of enforcement and intelligence. Some of the best cooperation we have had has been British."[21] British-based compu-

---

16. *Financial Times*, August 10, 1984, p. 1.

17. See the statement of William Von Raab, customs commissioner, in *Transfer of U.S. High Technology*, pp. 193–217.

18. U.S. Department of the Treasury, U.S. Customs Service, *Operation Exodus* (Washington, D.C.: Customs Publication no. 600, October 1985). Between 1981 and 1985, Exodus was responsible for over forty-three hundred seizures valued at $300 million.

19. *Business America*, February 29, 1988, p. 22. The relationship between Commerce and Customs was competitive and in many ways counterproductive, as each sought to take the lead responsibility for enforcement. See Linda Melvern et al., *Technobandits: How the Soviets Are Stealing America's High-Tech Future* (Boston: Houghton-Mifflin, 1984), chaps. 5 and 6.

20. DoD, *Technology Transfer Program*, p. 24.

21. Perle is quoted in Melvern et al., *Technobandits*, p. 137. See also *Financial Times*, November 1, 1983, p. 12.

ter smugglers, accustomed to a slap on the wrist or small fine upon detection, began to face stiff fines and jail sentences. The British effort was not uncontroversial at home; the publicity surrounding the tighter controls generated a surge in license applications and subsequent delays in processing them at the Department of Trade and Industry. At the same time, Customs and Excise increased the number and frequency of searches, further compounding the delays.[22]

The French similarly tightened export restrictions, in part because of U.S. pressure but also as a consequence of an overall reassessment of the link between technology transfer and national security. The Mitterrand government was significantly influenced by the Farewell documents, the contents of which reportedly turned the president overnight into a hawk on technology protection.[23] An immediate consequence was the dramatic expulsion from France of forty-seven Soviet officials accused of seeking the illegal acquisition of civilian technology with military applications. The Mitterrand government also revamped the French export control system, drawing up a list of "core" technologies to be accorded special attention in licensing and enforcement and granting defense specialists greater influence in the control process.[24]

France's new policy affected two key export cases involving the state-owned electronics firm Thomson. In 1981, Thomson prepared to fulfill a $350 million order for computerized monitoring equipment, which was to be used on the Siberian pipeline. Before allowing the sale, the Mitterrand government, judging the computers involved to be more powerful than necessary and fearing their military application, forced Thomson to downgrade the technological sophistication of the equipment. Similarly, Thomson was forced to modify its 1979 deal with the Soviets, which involved $110 million in sophisticated digital equipment and technology and to which the United States had objected in CoCom. Much to the displeasure of the Soviets, French officials allowed Thomson to export the components, but not the technology needed to produce them, believing it would significantly enhance the ability of the Soviets to produce their own microprocessors.[25]

22. *East European Markets*, December 10, 1984, p. 1. See also *New Society*, February 21, 1986, pp. 316–17.

23. Marie-Hélène Labbé, "Controlling East-West Trade in France," in Gary K. Bertsch, ed., *Controlling East-West Trade and Technology Transfer: Power, Politics, and Policies* (Durham: Duke University Press, 1988), p. 201.

24. Ibid., pp. 191–92, and Interviews, Office of the Prime Minister, Secrétariat Général de la Défense Nationale (SGDN), Paris, June 17, 1982; Foreign Ministry, Paris, June 15, 1982; and U.S. embassy, Paris, June 10, 1982. SGDN officials argued that the Farewell papers made U.S. pressure on France more credible and less resistible and helped to resolve an internal debate over whether export controls should became a national priority.

25. On the monitoring equipment, see *Economist*, November 14, 1981, p. 87 and October 31, 1981, p. 82, and *Le Monde*, October 21, 1981, p. 1. On the switching equipment, Interview, Ministry of External Commerce, Office of External Economic Relations (DREE), Paris, June 16, 1982, and *Financial Times*, December 17, 1983, p. 2. Despite modifications, the United States was

Both cases were unprecedented in that they involved the modification of major contracts, ones already signed and with political significance to the Soviets, in the interest of national security. Overall, the new French approach was particularly striking in light of the fact that during the 1970s France had, in the frank words of one high-level French official, "no strategic trade policy, no systematic control of exports."[26] The new policy contributed to a stronger CoCom both by restricting access to French technology and by removing the opportunity for other members to resist tighter controls on the grounds that such controls would be futile considering France's laissez-faire policy.

The West German government expressed support in principle for the need to improve enforcement, yet it proved reluctant in practice to increase significantly the resources it devoted to the task. Both U.S. and West German officials in 1985 characterized the West German enforcement effort as a "resource problem," stressing that the West Germans "do what they can within their limits."[27] The West German government did not devote priority attention and resources to enforcement until 1989, after it became publicly known that West German firms had exported chemical weapons technology to Libya and other controlled destinations.

In the early 1980s the West German government did seek to increase the awareness of the business community by publicizing the technology diversion problem in brochures and seminars. It also provided greater training and technical assistance to customs officials, whose willingness to cooperate fully with U.S. Customs in joint CoCom investigations was rated very high by U.S. officials. U.S. customs officials did emphasize, however, that West German efforts were strong when the United States took the lead, but not on their own. According to one U.S. official, the West Germans had a "nine-to-five mentality" regarding the problem and, given that they conceived of trade as a right rather than as a privilege, were unlikely to pursue and prosecute violators with the zeal of the United States.[28]

---

still unhappy that the sale was completed. The United States and France eventually agreed to disagree and dropped the subject from discussion in CoCom.

26. Interview, Office of the Prime Minister, June 17, 1982. The same French officials later revealed (in informal discussions with the author in Paris, June 1989) that when Carter administration officials sought French support for the no-exceptions policy in 1980, senior French officials were at first mystified. They later queried SGDN staff members: "What is this CoCom the Americans kept referring to?"

27. Interviews, Economics Ministry, Bonn, July 16–17, 1985; Foreign Ministry, Bonn, July 18, 1985; U. S. embassy, Bonn, July 18, 1985; and Federation of German Industry, Cologne, July 18, 1985.

28. Interview, U.S. embassy, Bonn, July 18, 1985. The United States could justifiably be accused of being overzealous in its enforcement effort. In the very same conversation in which West Germany's lack of aggression was discussed, the U.S. customs official complained that someone in the DoD was pressuring him to "turn half of Europe upside down to track down some box of widgets."

Finally, it was clear by the early 1980s that Japan, in view of its remarkable progress in the development and commercialization of dual-use technologies, was critical to the overall Western control effort. U.S. officials considered Japan to be a primary target of Soviet diversion efforts, because of both the richness of Japanese technology and the laxity of its security environment. The Soviets themselves made no secret of their desire to acquire advanced Japanese technology; in 1985 the Soviet government presented a wish list to a Japanese business mission; it included items in CoCom-controlled categories such as microprocessors, robotics, biotechnology, advanced composites, and flexible manufacturing systems.[29]

Before the Toshiba diversion (discussed below), Japan's response to the enforcement problem was, like that of West Germany, fairly modest. Some steps were taken to mollify the United States. In 1983 the government required that importers submit end-use certificates to prevent the reexport of technology to the Soviet bloc. The tightening of enforcement at the urging of the United States led to the uncovering of several diversion schemes, resulting in prosecutions and modest fines. The Ministry of International Trade and Industry (MITI) issued a set of guidelines to industry that provided advice on detecting and deterring industrial espionage efforts.[30] Overall, however, there was little commitment of additional resources and, more important, no basic reorientation in the thinking of Japanese business and government officials or in the export control system that operated on the basis of that thinking. In the postwar era, Japanese public and private officials came to conceive of security almost exclusively in terms of commercial success. The technology diversion problem remained a remote concern during the early 1980s, viewed more as a political problem in dealing with the senior alliance partner than as a national security threat.

In summary, by 1985 the United States made some, albeit uneven, progress in increasing awareness and improving its own enforcement efforts and those of its major allies. The contrast with the previous decade, in which enforcement was not a priority for either the United States or any of its major allies, is significant. Nevertheless, in absolute terms, technology diversion continued to pose a major problem. The Toshiba case was only the most publicized of a series of diversions; these, taken together, suggested that with perhaps greater effort and at greater cost, the Soviets could continue to acquire militarily relevant technology from the West.[31]

29. *Japan Economic Journal*, January 4, 1986, p. 2.

30. See *Japan Economic Journal*, April 19, 1983, p. 2; *Foreign Broadcast Information Service* (*FBIS*) *Daily Report* (Japan), August 30, 1985, p. C1, September 11, 1985, p. C3, and October 21, 1985, p. C1.

31. One case involved the sale to the Soviets by a Scottish subsidiary of a U.S. firm (Consarc) of equipment and technology to produce carbon-carbon, a material used in the United States to

With the third initiative, the Reagan administration sought to address what it perceived as CoCom's inadequate institutional capacity. U.S. officials, primarily those in the DoD, complained frequently about the extraordinary organizational weakness of CoCom—its minuscule budget (approximately $500,000 per year), limited staffing, cramped facilities, lack of secure communications, and, ironically, lack of access to modern data processing equipment. CoCom reportedly relied on a hand-cranked mimeograph machine until the DoD donated a photocopier in 1983.[32] The most pressing DoD concern, echoing one voiced by the Truman administration at the creation of CoCom, was the lack of adequate defense input in the multilateral process. U.S. officials complained that West European CoCom delegations were heavily weighted in favor of trade interests, to the point of having industry representatives seated at the negotiating table. The Reagan administration urged other members to permit greater defense ministry participation in their national export control processes and, more important, to agree to the creation of a permanent military subcommittee in CoCom. U.S. officials hoped such institutional changes would assure that the "defense perspective" carried greater weight in CoCom.

The results, again, were mixed. The British and French governments found the idea of a greater defense ministry role in national processes uncontroversial, because in their cases defense officials already played (or, as in the French case, were coming to play) meaningful roles.[33] In the cases of West Germany and Japan, the Reagan administration again came away less than fully satisfied. The Japanese Defense Agency remained virtually absent in the Japanese export control process, which was traditionally dominated by the close interaction of MITI and private

---

reduce the wobble and increase the accuracy of nuclear warheads. See *New York Times*, November 9, 1987, p. 1. Another concerned the Oregon-based firm Tektronix, whose West German subsidiary unwittingly transferred forty state-of-the-art computer work stations to a West German smuggler, who transferred them to Austria and then the East. See *New York Times*, October 10, 1986, p. 1, and *Business Week*, October 27, 1986, p. 58. Between 1983 and 1985, a French firm shipped to the Soviet Union U.S.-origin equipment and technology to produce bubble memory chips. See *New York Times*, October 17, 1987, p. D1.

32. Richard Perle, "The Eastward Technology Flow: A Plan of Common Action," *Strategic Review* 12 (Spring 1984), 24–32.

33. British Ministry of Defence (MoD) officials reviewed all British cases for exports to the East requiring validated licenses, reviewed all exceptions requests of other CoCom members, and also chaired the interagency Security Export Controls Working Group, which decides the most contentious cases and develops Britain's position on list review proposals. MoD officials hold informal veto power in the export control process and view themselves as a friendly counterweight to the pro-trade inclinations of the Department of Trade and Industry. Interview, Ministry of Defence, London, June 2, 1982. See also Gary Bertsch and Steven Elliott, "Controlling East-West Trade in Britain: Power, Politics, and Policy," in Bertsch, ed., *Controlling East-West Trade and Technology Transfer*, pp. 204–38. In the French case, the reconsideration of export control policy noted earlier elevated the positions of the Defense Ministry and the SGDN at the expense of the Trade Ministry, which traditionally dominated French policy on this issue. Interview, Office of the Prime Minister, June 17, 1982, and Labbé, "Controlling East-West Trade in France," p. 200.

industry.[34] In West Germany, the Defense Ministry continued to play a secondary role in a consensus-oriented process controlled by the Economics Ministry. Defense officials claimed in 1982 that they reviewed only the "most important" cases and contended that although they lacked a veto in the control process, they rarely disagreed in any event with the judgments of the dominant actors.[35]

The United States similarly faced difficulties in its proposal for a CoCom military subcommittee. West European economic and foreign ministry officials claimed that CoCom already did well to balance economic and security interests, and thus it was unnecessary to complicate the process by adding a military subcommittee. Some expressed concern that such a group might achieve undue influence and tip the balance too decisively in favor of security concerns.[36]

U.S. officials maintained pressure, and by 1985 a compromise was reached to create the Security and Technical Experts' Meeting (STEM), which would neither be a formal part of CoCom nor play a day-to-day role in multilateral policy. STEM was conceived to serve as an advisory body to CoCom, gathering information on emerging technologies of military significance and providing long-term assessments of the link between technology transfer and Western security.[37] It is composed of national delegations of scientists, engineers, and defense specialists. STEM's creation earned a rebuke from the Soviet government, which contended that the United States was now adding military censorship to the existing CoCom censorship of trade.[38]

The above survey of CoCom developments suggests that in overall terms, the multilateral mechanism was strengthened during the 1981–1985 period. The awareness among Western governments and firms of the problem of technology diversion was increased. The control list was expanded, extending coverage into previously underemphasized areas of military significance; enforcement became a higher priority, especially among the United States, Great Britain, and France; and a mechanism was created to encourage greater defense input to the control process. The reaction of Soviet officials served as an additional indicator of the revival of the regime. By 1984 and 1985, Soviet complaints about CoCom were becoming more explicit and routine. The Soviets informed West Germany that the extension of the control list was the greatest problem hindering economic cooperation between them and had led to a certain loss of trust. Similar points were made in discussions with Great Britain

34. Gordon Smith, "Controlling East-West Trade in Japan," in Bertsch, ed., *Controlling East-West Trade and Technology Transfer*, p. 148.
35. Interview, Defense Ministry, Bonn, July 13, 1982.
36. Interviews, Economics Ministry, Bonn, July 7, 1982, and Ministry of Defence, London, June 2, 1982.
37. Interview, Foreign Ministry, Bonn, July 18, 1985, and *Economist*, October 19, 1985, p. 49.
38. *FBIS* (USSR International Affairs, United States and Canada), October 31, 1985, p. A6.

and Japan. More generally, Soviet commentators warned that Western Europe and Japan had taken a "risky step" toward complicity in U.S. technological warfare against the Soviet Union. Czechoslovakian officials went further and registered a formal complaint with the GATT, claiming that CoCom was an unfair non-tariff barrier to the free flow of trade.[39]

The experience of 1981–1985 suggests three general points. First, to the extent CoCom became stronger, it did so only within boundaries, established by West European preferences, that had endured since the early postwar period. Agreement to strengthen CoCom did not signify a willingness to conduct economic warfare by broadening the control criteria beyond direct military significance. In addition, it did not signify a willingness to alter CoCom's informal, confidential character fundamentally. The rejection of a military subcommittee as a formal part of CoCom reflected this concern, as did West European wariness of U.S. efforts to publicize CoCom agreements and their resistance to the idea, floated informally by U.S. officials, of converting CoCom to treaty status. Within the established limits, however, other CoCom members were willing to accept stricter controls, even though it necessitated greater economic sacrifices.

Second, CoCom's institutional character posed an obstacle to its ability to increase its effectiveness. Given the requirements of a multilateral embargo, the more uniform the policies and practices across member states, the greater the potential for effectiveness. CoCom's rules and procedures, however, encourage uniformity in some respects, yet they discourage it in others. Although it is multilateral, the implementation and enforcement of the control list has been left to national discretion. Consequently, each member has developed over time its own policies, procedures, and traditions of dealing (or not dealing) with the problem. The intelligence capabilities, laws, administrative procedures, and willingness to commit resources have varied considerably among member states. And, as the United States discovered in its efforts to bolster enforcement and inject defense input, established procedures and practices are difficult to change. The compromise struck in the late 1940s, establishing an informal regime that maximized the national discretion of its members, rendered more difficult the revival of CoCom in the 1980s.

Third, the strengthening that did take place can be attributed primarily to the exercise of U.S. leadership. The United States led by example, by tightening its own controls and enforcement. It took the initiative multilaterally, by bringing proposals to CoCom and establishing their

39. See *FBIS* (Federal Republic of Germany), April 16, 1985, p. J1; *FBIS* (USSR International Affairs, United States and Canada), December 26, 1985, p. A4; and *Financial Times*, November 23, 1984, p. 7.

strategic merit as well as by pressuring others to enhance their enforcement efforts. It took responsibility for establishing control over non-CoCom sources of supply, both in the context of the 1984 list review and, as discussed below, more generally. In the absence of such behavior, it is difficult to imagine that CoCom would have been strengthened to the extent that it was.

The key point, however, is that U.S. leadership was a mixed blessing. Just as the United States provided the impetus behind CoCom's revival, its neglect or abuse of its leadership responsibilities constituted the chief reason why CoCom remained highly controversial among member states during 1981–1985 and why the progress made in CoCom was limited and subject to considerable strain.

The Inadequacy of U.S. Leadership

U.S. policy hampered CoCom in three ways. First, U.S. officials allowed foreign policy considerations to intrude on CoCom, thereby compromising the integrity of the control process and raising suspicions regarding U.S. motives in CoCom. Second, they strained political relations with other member governments by violating the diplomatic norms of CoCom. Third, they proved unwilling to decontrol less sensitive items and failed to sharpen the categories of control, thus exacerbating the administrative and economic burdens of the control process.

An important example of the first problem involved the no-exceptions policy adopted following the Soviet invasion of Afghanistan. As I argued in Chapter 7, the United States adopted that policy as much for foreign policy reasons (to punish the Soviets for their behavior) as for strategic reasons (to weaken Soviet military capabilities). Other members grudgingly acquiesced in light of the modest costs and the knowledge that the United States could enforce the policy with its veto power if necessary.

Discontent in CoCom heightened during the early 1980s as the control list was extended with minimal decontrol of less strategic items. General exceptions traditionally facilitated trade in such items when decontrol efforts lagged. The no-exceptions policy in effect removed this option in trade with the Soviet Union. Moreover, U.S. officials exacerbated alliance discontent by reneging on a previous agreement to exempt from the policy exports related to health and safety and those involving spare parts for previously exported equipment. According to West European officials, CoCom governments that resented the maintenance and manipulation of the policy retaliated by selling controlled items to the Soviets without bringing cases to CoCom.[40] Such sales could

40. Interviews, Economics Ministry, Bonn, July 8, 1982, and Foreign Ministry, Paris, June 15, 1982. The point was confirmed by William Root, at the time director of the U.S. State Department's Office of East-West Trade. Personal correspondence, December 30, 1986.

be undertaken "legally" by adopting a generous interpretation of which exports were allowable under national discretion as administrative exceptions.

U.S. officials made matters worse by unilaterally extending the no-exceptions policy to Poland in 1982, after the imposition of martial law. Again, the United States actually used CoCom to impose a foreign policy preference on its allies. The policy was implemented through the exercise of the U.S. veto, leaving other members once more with the choice of reluctant consent or tacit or open defiance of CoCom's rules.[41]

An equally contentious problem arose over the treatment of China. Beginning in the late 1970s, U.S. policy evolved to allow China to purchase U.S. technology of increasingly higher levels of sophistication.[42] By 1983, China was being treated, for export control purposes, as a "friendly" state, albeit one subject to case-by-case review and a presumption of denial for the most militarily sophisticated technologies. The overall result of this rapid liberalization was a barrage of license requests from U.S. firms to sell controlled items to China. The number of licenses approved jumped from fifteen hundred in 1981 to over eighty-five hundred in 1985 (see Table 10). The Commerce Department sought to minimize delays by establishing different "zones" of China requests, ranging from "green," comprising 75 percent of cases and with a presumption of quick approval, up to "red," comprising items of the highest military significance and with a presumption of denial. Even many "green" cases, however, required multilateral approval as exceptions in CoCom.[43]

Although some CoCom members (e.g., Japan) had reservations regarding the foreign policy wisdom of relaxing technology controls to such a degree for China, in CoCom the more divisive issue was the implementation of the policy.[44] Seeking, perhaps, to maintain flexibility so as to influence future Chinese behavior, U.S. officials sought not to shorten the CoCom list for China significantly but to maintain the list and use the exceptions mechanism. The result was the proliferation of exception requests. The situation reached crisis proportions in 1984 and 1985, with the United States alone requesting 3,399 and 3,790 exceptions, respectively, approximately 90 percent of which were for China (see Table 11). During the peak years of détente, the combined excep-

---

41. Interview, Economics Ministry, Bonn, July 8, 1982.
42. An excellent review is Madelyn C. Ross, "Export Controls: Where China Fits In," *China Business Review*, May–June 1984, pp. 58–62.
43. Ibid., and *Aviation Week and Space Technology*, October 3, 1983, p. 23.
44. Japan feared that the United States might be rapidly arming a large, potentially powerful neighbor, which might not always remain friendly. See *New York Times*, January 29, 1984, p. 1, and *Financial Times*, September 28, 1983, p. 4. Some observers in the United States shared Japan's reservation. See Denis Fred Simon, "Technology for China: Too Much Too Fast?," *Technology Review* 87 (October 1984), 39–50.

Table 10. U.S. license approvals for export to China

Fiscal year	Licenses approved	Value (U.S. $ million)
1981	1508	306
1982	2020	355
1983	3314	775
1984	4587	1986
1985	8593	5730

Source: U.S. Department of Commerce figures, cited in Madelyn C. Ross, "China and the United States' Export Controls System," Columbia Journal of World Business (Spring 1986), 27–33.

tion requests of all members in any given year never reached even one-half of what the United States alone requested for China in 1985.

The massive volume of U.S. requests placed an unmanageable administrative burden on CoCom. As CoCom could decide roughly one hundred cases in an average month, there was no way it could keep up; long delays were inevitable for China requests as well as those for Eastern Europe. West European export control officials complained in 1985 of being buried in paper by the United States and of having time to do little more than process U.S. exception requests for China and respond to U.S. proposals for stricter controls for the Soviet Union.[45] At the same time it was urging other CoCom members to devote more staff and resources to export controls, the United States was swamping their national capitals with exception requests.

Not surprisingly, West European governments and firms were especially resentful of the fact that the United States unilaterally liberalized its technology controls for China, seemingly granting its firms an advantage in that competition, while it pressured CoCom to clamp down on Soviet and East European trade, in which West European firms held a comparative advantage. West German officials claimed that the United States felt it could take such actions because it was the "inventor" of CoCom; if West Germany decided unilaterally to liberalize technology exports to Hungary on the grounds that it was a peaceful state, U.S. officials would shoot down the initiative in CoCom.[46] The fact that the United States, acting in a manner reminiscent of the 1970s in the Soviet context, occasionally denied the China requests of other states only made matters worse. One British firm publicly threatened to violate

45. Interviews, Foreign Ministry, Bonn, July 18, 1985, and Foreign Office, London, July 25, 1985.
46. Interview, Economics Ministry, Bonn, July 17, 1985.

Table 11. U.S. exceptions requests for China

Year	Total number of U.S. cases submitted to CoCom	Total number of U.S. China cases submitted to CoCom	China cases as % of total U.S. cases
1982	1150	626	54%
1983	1882	1502	80%
1984	3399	2931	86%
1985	3790	3612	95%

Source: U.S. Department of Commerce figures, cited in Madelyn C. Ross, "China and the United States' Export Controls System," *Columbia Journal of World Business* (Spring 1986), 27–33.

CoCom, claiming its requests had been held up while U.S. firms sold similar equipment through subsidiaries in Sweden. Another accused U.S. officials of preventing it from fulfilling Chinese contracts for four years. U.S. firms, already subject to delays, complained in turn that their own requests were being stalled by other CoCom governments as a measure of retaliation.[47]

A crucial step in alleviating the China bottleneck was finally taken in late 1985, when CoCom members agreed to raise the general exception threshold for China. Much of what previously required formal CoCom review could now be licensed at national discretion. The decision followed, and was contingent on, the completion of what was in effect a second list review for China, which came on the heels of CoCom's 1982–1984 review.[48]

The above episodes suggest the various ways in which U.S. officials compromised the effectiveness of CoCom by injecting foreign policy considerations into the control process. The United States increased the administrative burdens of member governments, making it impractical to conduct careful reviews of exception requests. It strained the credibility of the embargo and increased the incentives for firms (with or without government acquiescence) to violate export controls. It created resentment and raised suspicions of U.S. motives in CoCom, and it called into question the proposals for list addition made by the United States, even those justifiably guided by strategic considerations.

The tendency to politicize the CoCom process can be attributed to what might be called the "sanctions habit"—the reliance of U.S. officials

47. On the problems of the British firms Plasma and Hadland, see *Financial Times*, November 2, 1984, p. 8 and January 11, 1985, p. 4, and "How U.S. Abuses High-Tech Law," *Computing*, January 26, 1984. The complaints of U.S. firms are reported in *Wall Street Journal*, January 3, 1985, p. 14.
48. Interview, Department of Trade and Industry, London, July 24, 1985. The CoCom deal is reported in *Washington Post*, December 17, 1985, p. E3.

on export controls or promotion as instruments of foreign policy. During the 1980s, U.S. officials continued to view the manipulation of export controls as an attractive alternative or complement to diplomacy or military force.[49] Although U.S. foreign policy controls per se are not necessarily detrimental to CoCom, they become entangled with the multilateral regime all too easily when their targets are CoCom destinations and when the sources of U.S. leverage include CoCom-controlled items. The entanglement may be inadvertent, as when export promotion became a key instrument of U.S. policy toward China; it may also be purposive, as when U.S. officials used CoCom as a handy forum to coordinate negative sanctions against Poland or the Soviet Union.[50] In both instances, U.S. policy worked at cross-purposes with U.S. leadership responsibilities and the strengthening of CoCom, even if, at the same time, it adequately served foreign policy objectives.[51]

Second, America's diplomatic style became a source of great controversy during the 1980s. U.S. officials, primarily those in the DoD, were perceived by their CoCom counterparts as arrogant, condescending, and unwilling to listen to or accommodate the views of others. Attempts by defense officials to dictate policy in a context in which compromise and consensus building had been the historical norm led to incoherence in U.S. negotiating strategy as well as conflict in CoCom.

The most well known incident involved negotiations in the critical area of computer controls. The inability to resolve intra-alliance disputes had plagued CoCom since 1979, but negotiations in 1982 and 1983 eventually narrowed the gap between the more restrictive stance of the United States and the more liberal one of other members. A final compromise was planned for October 1983. On the eve of the meeting, however, Defense officials informed their State Department counterparts that no compromise was acceptable and that the United States would not deviate from its initial 1982 position. Defense added that in any event, CoCom was not an appropriate forum in which to negotiate computer control and that the negotiations should take place in an unspecified forum with a senior defense official representing the United States. The DoD's intervention prompted the resignation of William

49. See, for example, Kenneth Abbott, "Linking Trade to Political Goals: Foreign Policy Export Controls in the 1970s and 1980s," *Minnesota Law Review* 65 (1981), 739–889, and Richard J. Ellings, *Embargoes and World Power: Lessons from American Foreign Policy* (Boulder, Colo.: Westview Press, 1985), esp. chap. 5.

50. It is not surprising that U.S. officials are tempted to use CoCom to coordinate foreign policy controls, given that the Western allies have no existing institutional mechanism for that purpose. See Henry Nau, *The Myth of America's Decline* (New York: Oxford University Press, 1990), p. 323. For further discussion of the problems posed for CoCom, see William Root, "Trade Controls That Work," *Foreign Policy*, no. 56 (Fall 1984), 61–80.

51. To say that foreign policy controls worked to the detriment of CoCom is not to deny the intrinsic utility of such controls as an instrument of statecraft. For the argument that the utility of foreign policy controls has been systematically underestimated, see David Baldwin, *Economic Statecraft* (Princeton: Princeton University Press, 1985).

Root, a State Department official involved with CoCom for many years and then head of the U.S. delegation. In his open letter of resignation, Root accused the DoD of seeking to undermine CoCom and the State Department and claimed that the United States was doing its best to "convey to our Allies that their views do not count, that we know best, and that they had better shape up."[52] The computer stalemate was resolved only in the summer of 1984, when the DoD relented and compromised.

A second incident occurred in July 1985, when CoCom members met to discuss a West German proposal to clarify CoCom's telecommunications guidelines.[53] At the meeting, members of the U.S. delegation broke into open conflict among themselves, and some sought to reintroduce the fundamental issues that had been painstakingly negotiated and settled during the list review of the previous year. The U.S. team eventually reorganized itself and relented, but not before it created considerable diplomatic friction and raised questions regarding its ability to lead.

U.S. officials caused further discontent in CoCom by subjecting the performance of individual members to public criticism. The Defense Department and the CIA frequently singled out the West Germans, citing their administrative weakness and prominent trade relationship with the East. A CIA report leaked to the press in 1985 claimed that West Germany was "the leading target of illegal technology transfer among CoCom allies" and suggested that Bonn was unreliable and would place its economic interests above its commitment to CoCom. The report outraged West German officials, as did a public statement by Defense Assistant Secretary Perle, who indelicately advised the West German government to reduce its economic ties with East Germany and focus instead on upgrading its contribution to NATO.[54]

Taken together, these and other incidents created a pattern of behavior that earned the United States a reputation in CoCom as arbitrary, unpredictable, and prone to unilateral action. The extension of the pipeline sanctions, which did not involve CoCom directly but did negatively affect its deliberations, was only the most dramatic manifestation.[55] Although specific consequences are difficult to document, the

52. Root's letter of resignation, which describes the incident, is reprinted in *Transfer of Technology*, pp. 242–45. See also Kevin Cahill, *Trade Wars: The High Technology Scandal of the 1980s* (London: W. H. Allen, 1986), pp. 52–59.
53. Interview, U.S. CoCom delegate, Paris, August 9, 1985.
54. The CIA report and the West German reaction are discussed in "Bonn Protests U.S. Pressure on East Bloc Trade," *International Herald Tribune*, June 25, 1985. Perle's remarks are discussed in *FBIS* (Federal Republic of Germany), December 15, 1986 p. J3, December 16, 1986, p. J1, and December 17, 1986, p. J1. His comments and the West German reaction prompted the U.S. ambassador to Bonn (Richard Burt) to state publicly that Perle was not competent to speak for the United States on policy toward West Germany.
55. See William Root, "CoCom: An Appraisal of Objectives and Needed Reforms," in Bertsch, ed., *Controlling East-West Trade and Technology Transfer*, p. 424.

ill will created by such behavior probably left other members less receptive to U.S. initiatives and with less incentive to implement CoCom's directives faithfully. Equally important, such behavior drew attention in CoCom away from the inadequate control and enforcement practices of other members, focusing it instead on the controversial aspects of U.S. policy.

The decline of U.S. diplomatic effectiveness in CoCom was in part a function of the growing prominence of the DoD in the export control process. Their concern with the technology transfer problem prompted DoD officials to play a major role not only nationally but also multilaterally. Defense officials lobbied successfully for a permanent representative to the U.S. CoCom delegation, and they also contributed funds to bolster CoCom organizationally.[56] Yet as they increased their influence at CoCom, they were also profoundly skeptical of it. Leading DoD officials proved impatient with diplomacy and were inclined to seek unilateral solutions unless the allies fell easily into line behind the United States. Conflict within the U.S. delegation, ostensibly led by the State Department, and within CoCom were thus inevitable. The State Department shared responsibility for this outcome: it capitulated and allowed the DoD to play a lead role where State was empowered to do so by both law and tradition.[57]

At a deeper level, the problem reflected distinctive features of U.S. export control policy: a lack of consensus over policy substance, a fragmented policy process, and the absence of initiative and direction from the White House. Since the mid-1970s, U.S. policy had been characterized by an ongoing struggle in which the State and Commerce Departments lined up against the DoD or the NSC staff, with each side backed by a coalition in Congress. The White House intervened only infrequently, allowing such conflicts to fester unresolved. As they played themselves out domestically, they tended to spill over into the multilateral context, compromising the ability of the United States to lead consistently and effectively.

Third, and most significant, the United States compromised its leadership role by failing to initiate or encourage a streamlining of the control list. The expansion of the list during the first half of the 1980s took place without a concomitant effort to liberalize less strategic items. U.S. officials were unreceptive to decontrol for a variety of particular reasons: their commitment not only to a stronger strategic embargo but also to economic warfare; the fact that U.S. industrial exports to the Soviet

56. *Multinational Business* (Winter 1986), 28–29, and U.S. House of Representatives, *Export Task Force Weekly Report*, May 13, 1985, pp. 4–5.

57. William Root, "State's Unwelcome Role," *Foreign Service Journal* (May 1984), pp. 26–29. See also National Academy of Sciences, *Balancing the National Interest* (Washington, D.C.: National Academy Press, 1987), p. 131.

Union had already been stifled by the Afghanistan and the Polish sanctions; and the heavy influence of the DoD, which kept domestic advocates of decontrol on the defensive. The U.S. position was more deeply rooted, however, in the fact that even after 1969, American officials never reached a consensus that the economic benefits of exporting to the East were worth defending as a national interest. The idea that economic benefits in East-West trade were of national significance found its way into important legislation, such as the Export Administration Act; in practice, however, strategic and political considerations consistently proved overriding. The one exception, perhaps, was agriculture. In industrial trade, the United States continued to be insensitive to the economic costs of controls nationally, which reduced its incentives to minimize the cost of controls in the multilateral context.

In the first half of the 1980s, other CoCom members did not initiate a large-scale liberalization effort themselves, concentrating instead on responding to and containing U.S. list expansion proposals. In any case, any such attempt would probably have been frustrated by U.S. veto power. Consequently, by 1985 the widespread sentiment in Western Europe (and in the U.S. business community) was that the control list was unnecessarily restrictive, containing items that were technologically obsolete and/or subject to widespread foreign availability. The frustration of other CoCom members was captured by a French official: "If CoCom had existed in 10,000 B.C., the wheel would still be on the list today."[58]

The problem was not simply the length of the list but the lack of specificity in many of the categories controlled. Vague or loose definitions of controlled items hamper the embargo effort by encouraging differences in list interpretation and creating uncertainty for exporters and enforcement officials. As the NAS put it, "Overly broad coverage reduces the credibility of the control system and encourages laxness on the part of public officials and industry."[59] This problem was exacerbated by the 1984 list review, because a number of key additions to the list lacked clear definition, including the all-important computer, software, and telecommunications entries. The concern among industry officials over the broad coverage of the software entry was so great that both the U.S. and the British governments quickly decided that the new controls would apply to only the East rather than to all destinations.[60]

---

58. See *Business Week*, April 4, 1983, p. 95.

59. National Academy of Sciences, *Balancing the National Interest*, p. 139.

60. After receiving some four thousand complaints, in April 1985 the Commerce Department recategorized numerous software items as technical data, which removed the validated license requirement for export to Western destinations. See Dean Overman, "Reauthorization of the Export Administration Act: Balancing Trade Policy with National Security," *Law and Policy in International Business* 17 (1985), 345, and *East European Markets*, May 13, 1985, p. 7. On the British decision, see *Financial Times*, June 14, 1985, p. 3. The British followed the example of the United States in implementing the software controls more narrowly. Interview, London, July 24, 1985.

And, despite the CoCom compromise, the computer entry maintained controls on a range of personal computers and integrated circuits so widely available from non-CoCom sources that any effort to deny them to the East would be futile.[61]

The clarification of control categories was a complex and tedious task, one that CoCom members began to tackle following the 1984 review. Significant liberalization of less strategic items, however, required a political commitment. By 1985 there was still no consensus with the U.S. government on the need for decontrol. Despite growing pressure from industry, some members of Congress, and the Commerce Department, the DoD remained unreceptive. The White House, by failing to intervene and establish a clear policy preference, in effect supported the DoD.

To other CoCom members, the United States was not only unresponsive to decontrol but also appeared to apply relentless pressure to make controls even more restrictive.[62] The concern was compounded by the ongoing U.S. effort to integrate the ideas of the Bucy Report into the control process. CoCom members by now accepted the basic premise, that the control of technology should be granted the highest priority. They were still uncertain about how to enforce such controls but, more important, about how far the United States intended to go. Implementing the directives of the report in the United States had led to the MCTL, which dwarfed the U.S. Commodity Control List and the CoCom list. The DoD relied on the MCTL for export control guidance, and Congress in 1985 reasserted its directive that the MCTL should be integrated into the existing U.S. CCL. The allies feared the end result of that integration would be even more comprehensive U.S. controls and subsequently more pressure to adopt such controls in CoCom.[63] The DoD did little to allay such fears; late in 1985 it announced its intention to maintain control of civilian technologies relevant to the Strategic Defense Initiative "before, during, and after the laboratory."[64] This effort involved classifying emerging technologies still in the development stage and placing them on both the MCTL and, eventually, the CoCom list.

By 1985 an alliance stalemate was evident. To the United States, decontrol was not a priority; the list could stand to be even more extensive

---

61. Industry officials noted that the new regulations were so confusing that neither they nor commerce officials could interpret them with confidence. *Datamation*, September 1, 1984, p. 70, and Mary Jo Foley, "Tying Up High Tech Exports with Red Tape," *Electronic Business*, June 1, 1985, pp. 136–37.

62. As a West German official put it in 1985, "We're concerned that our resources will be dissipated in trying to enforce a controlled items list that, as a result of certain voices in the Pentagon, keeps getting bigger and bigger." See "Bonn Protests U.S. Pressure on East Bloc Trade," *International Herald Tribune*, June 25, 1985.

63. See North Atlantic Assembly, *Interim Report of the Subcommittee on Advanced Technology and Technology Transfer* (Brussels, October 1985), p. 6.

64. *East European Markets*, November 15, 1985, pp. 4–5.

and, most important, enforcement needed to be improved. To America's major allies, the existing list already strained the credibility of the embargo, and enforcement resources were already stretched too thin. British license requests, for example, jumped from thirty-seven thousand in 1983 to over eighty thousand in 1985.[65] To the allies, without relief at the lower end there was no way CoCom could be strengthened further. In fact, the gains already made were placed in jeopardy.

The stalemate had implications beyond CoCom. U.S. dissatisfaction with CoCom increased its incentives to act unilaterally by tightening intra-Western controls and extending U.S. control authority extraterritorially.

THE EXTENSION OF INTRA-WESTERN RESTRICTIONS

Throughout the postwar era, a central thrust in U.S. policy had been to minimize the impact of East-West trade controls on U.S. economic relations with other non-Communist states. This trend began with the creation of CoCom and continued through the 1960s and 1970s, culminating in the Carter administration's 1978 refusal, contrary to the advice of the Bucy Report, to extend controls on technology transfers to allied and neutral states. Despite the fact that applications for export to the non-Communist world typically accounted for 90 percent of all validated license requests, until 1982 the Commerce Department had expedited the great majority of them with little or no review. Indeed, in 1982 the GAO characterized the process as "more of a paper exercise than control" and recommended the elimination of non-Communist license requirements for low-technology items.[66]

In the first half of the 1980s, the Reagan administration reversed the long-standing practice of segregating East-West controls from intra-Western trade by extending and tightening restrictions on U.S. technology transfers to both CoCom members and neutral states. This shift in policy can be traced, first, to the administration's skepticism regarding the commitment of other states to strategic export controls generally and to the protection of U.S. technology in particular. West European defiance on the pipeline deal and the difficulties of quickly obtaining a consensus for stricter controls in CoCom reinforced the conviction among key U.S. officials, mainly in the DoD, that the allies could not be trusted. Neutral states such as Sweden and Austria, which had no official export control commitments and a pattern of illicit reexports during the

65. *Financial Times*, June 14, 1985, p. 3.
66. GAO, *Export Control Regulation Could Be Reduced without Affecting National Security* (Washington, D.C., May 26, 1982).

1970s, posed an even greater risk. U.S. officials were convinced that the most significant pattern of illegal transfers to the East did not involve direct exports from the United States but rather the transshipment of U.S. technology through Western Europe.[67] Ironically, the more the renewed enforcement efforts of U.S. officials turned up evidence of illegal technology transfers, the more it appeared that selected states in Western Europe and the Pacific Rim were weak links in the chain of protection.

Second, U.S. officials, again primarily in the DoD and again reflecting a Bucy Report theme, viewed intra-Western restrictions as a source of leverage. The threat or the practice of technology denial could be used to compel other governments and their firms to comply with U.S. reexport controls and to improve their own control and enforcement efforts.

The more vigilant approach to intra-Western trade manifested itself in the U.S. export control process as well as in policy. Of most importance was the expansion of Defense Department licensing authority. The Jackson Amendment to the Export Administration Act of 1974 had granted the DoD a formal role only in the review of U.S. exports to Eastern destinations, and that role had been confined primarily to providing technical assessments of the military significance of license requests. The Reagan DoD sought to extend its role to intra-Western cases, not only to render technical judgments, but also to monitor points of diversion and the reliability of Western end users. Following a protracted struggle between the Defense and Commerce Departments, in January 1985 the National Security Council provided the DoD with the right to review all license applications in eight critical technology areas—including computers, scientific instruments, and semiconductor production equipment—for export to fifteen nonproscribed destinations deemed to be probable transshipment points for the diversion of U.S. technology. The targets reportedly included Austria, Spain, Finland, Switzerland, and Sweden in Western Europe as well as Hong Kong, Malaysia, and Singapore in the Far East.[68] By enhancing Defense's role over the objections of U.S. industry and the Commerce Department, the Reagan administration signaled its desire to provide maximum protection for U.S. technology and to enlist the support of third countries in the export control effort, even at the risk of disrupting America's intra-Western economic links. Defense was more likely than Commerce to render conservative judgments on intra-Western license applications, and just as the expanded role for

67. See *Transfer of High Technology to the Soviet Union*, pp. 225–27, and GAO, *The Government's Role in East-West Trade: Problems and Issues* (Washington, D.C., February 4, 1976), p. 44.

68. GAO, *Export Licensing: Commerce-Defense Review of Applications to Certain Free World Nations* (Washington, D.C., September 1986), p. 10. The Memorandum of Understanding (MOU) list of target destinations is classified. The license requests subject to review by the DoD comprise an estimated 85 percent of all export applications filed. See *Export Control News* 1 (December 30, 1987), pp. 2–5.

Customs pressured Commerce to be more vigilant in enforcement, pressure from Defense led to the same outcome in licensing.[69]

The newly restrictive U.S. stance was felt most forcefully in reexport cases. During the early 1980s, British and West German firms complained of long delays in the processing of reexport requests and noted that Commerce sometimes returned them eventually "without action," an insidious way of turning them down.[70] Early in 1984, in what became a major controversy, IBM reminded its British customers that the reexport of U.S.-origin equipment or technology to Western Europe or even to other firms *within* Britain required permission from the Commerce Department. A major target of this reminder was ICL, Britain's major computer manufacturer. ICL was fined by the U.S. government in 1982 for exporting a computer containing U.S. components to South Africa and was also warned to obtain U.S. reexport licenses for all of its transactions or face a cutoff of its vital U.S. sources of supply.[71] One of Spain's major electronics firms, Piher Semiconductor SA, did receive a denial order for reexporting U.S. semiconductor manufacturing equipment to Cuba and the Soviet Union. Piher was denied access to U.S. supplies from 1982 to 1986.[72] Major firms in Sweden and Austria also found their access to U.S. sources constrained or jeopardized by their failure to secure reexport permission. The full exercise by the United States of its contested extraterritorial authority caused considerable concern across Western Europe, given the reliance of many firms (particularly those in computers and electronics) on U.S.-origin supplies.

An equally troubling U.S. initiative involved distribution licenses, which had been introduced in 1968 to expedite multiple shipments of controlled items to approved end users. Over seven hundred U.S.-based firms held such licenses, which governed 40 percent of all licensed exports and 15 percent of all U.S.-manufactured exports, worth $20 billion annually.[73] For both U.S. firms and their foreign consignees, distribution licenses were critical in easing the burdens of the control system. In 1984, Commerce proposed a comprehensive review and tightening of the system and suggested that certain licensees might have their privileges revoked or significantly modified. More controversially, Commerce requested that each licensee provide, on a quarterly basis, a list of all customers in non-CoCom countries that its overseas distributors expected to supply.[74]

Strong objections from U.S. firms and West European governments

69. GAO, *Export Licensing*, pp. 3–4.
70. *East European Markets*, December 10, 1984, pp. 1–2, and interview, Bonn, July 16–17, 1985.
71. Interview, ICL House, London, July 26, 1985. See also Cahill, *Trade Wars*, pp. 82–84.
72. *New York Times*, September 6, 1985, p. D1.
73. *Christian Science Monitor*, June 13, 1984, p. 8.
74. The proposed rules are found in *Federal Register* 49 (January 19, 1984), 2264–67.

forced Commerce to make a partial retreat. It eliminated the customer listing requirement in exchange for a self-policing program in which each holder, and its consignees, would be expected to establish internal review mechanisms designed to prevent illegal diversion. Commerce reserved the right to audit license-holders and their consignees randomly. It demonstrated its more vigilant approach by restricting the distribution license privileges of Digital Equipment Corporation (DEC) after it discovered that DEC computers had been transshipped through Western Europe to the Soviet Union. DEC was required through most of 1984 to obtain individual licenses for export to its customers in Great Britain, West Germany, Austria, and Sweden. During this period few DEC computers or spare parts reached these destinations from the United States.[75]

Other examples of tighter intra-Western controls abound. West Germany's Max Planck Institute and Japan's NTT faced restrictions on their access to U.S. supercomputers. ICL officials publicly protested that U.S. officials demanded that the firm obtain an export license for the information carried in the minds of American engineers they hired.[76] U.S. officials also determined that oral presentations at scientific meetings at which foreigners were present were covered by U.S. controls. Foreign nationals were selectively denied access to scientific papers, computer programs, and conferences.[77]

Intra-Western Restrictions and the Third-Country Effort

The expansion of intra-Western restrictions, like U.S. initiatives in this period generally, had mixed consequences for CoCom. The use of such restrictions in relations with *non-CoCom* countries helped to strengthen CoCom by providing U.S. officials with an important source of leverage to extract the cooperation of alternative suppliers of CoCom-controlled items. Intra-Western restrictions similarly provided U.S. officials with some leverage in dealing with *CoCom member states*. The use of what was widely perceived as a unilateral instrument proved counterproductive

75. A good discussion of the DEC case is Janet Lunine, "High Technology Warfare: The EAAA of 1985 and the Problem of Foreign Reexport," *NYU Journal of International Law and Politics* 18 (Winter 1986), 680–86. See also *New York Times*, March 19, 1984, p. D1, and Cahill, *Trade Wars*, pp. 127–28.
76. *Financial Times*, December 18, 1984, p. 6, and *Japan Economic Journal*, January 22, 1985, p. 5. On ICL, see *Economist*, February 25, 1984, pp. 66–67.
77. *Wall Street Journal*, January 25, 1985, p. 1. Late in 1984, the Reagan administration abandoned any systematic effort to limit the dissemination of scientific and academic papers. The decision was influenced by a National Academy of Sciences report that found little evidence of damage to national security as a result of information obtained by the Soviets from academic sources. See National Academy of Sciences, *Scientific Communication and National Security* (Washington, D.C.: National Academy Press, 1982).

in the multilateral context, however, creating considerable resentment among member states and reducing the incentives for effective coopera-tion.

Throughout the 1980s, U.S. officials engaged in what became known as the "third-country effort"—a wide-ranging attempt to bring non-CoCom suppliers of strategic technology into compliance with the multi-lateral export control regime by manipulating their access to U.S. tech-nology and equipment. In Europe, Sweden and Austria were important and particularly vulnerable targets, given the overwhelming dependence of their key firms on the United States as a source of supply. Following the discovery of its unauthorized reexport of U.S.-origin air-traffic-control technology, the Swedish firm Datasaab was fined over $3 million in 1984 by the U.S. government and was blacklisted. The Swedish firms Ericcson and Asea, similarly threatened with U.S. denial orders after their respective reexport violations were uncovered, lobbied the U.S. government for softer penalties while they sought to convince their own government that policing U.S. exports was in the national interest. The U.S. DoD reinforced the message by reminding Sweden that vital U.S. technology for its next-generation fighter aircraft could be with-held. The Swedish government, in exchange for the continued and im-proved access of its firms to U.S. technology, committed itself in 1986 to strengthening enforcement (including cooperating fully with U.S. inves-tigations) and producing new restrictions on transit trade in strategic goods of foreign origin. Government officials clearly conceded to these policies with the greatest reluctance, given Sweden's commitment to neutrality and nonalignment. The dilemma prompted a leading daily, *Dagens Nyheter*, to editorialize that bending to U.S. demands was consis-tent with nonalignment because the policy required the functioning of civil society at a reasonable level and that, in turn, required access to U.S. technology.[78]

The Reagan administration placed similar pressure on Austria, which had the dubious distinction of being the largest foreign source of U.S. reexport diversions.[79] U.S. officials warned Austria in 1984 that it would be left with only "pastries and 1950s machinery" if it did not develop an export control apparatus and tighten its restrictions on the reexport of U.S. technology. The Commerce Department delayed for several years a joint venture between Austria's Voest and America's Microsystems to

78. See *FBIS* (Sweden), March 7, 1986, pp. 6–10. See also *Wall Street Journal*, January 15, 1987, p. 1, and Jan Stankovsky and Hendrik Roodbeen, "Export Controls Outside CoCom," Paper presented at the conference "Forty Years After: CoCom, Alliance Security, and the Future of East-West Economic Relations," Leiden, the Netherlands, November 1989.

79. A U.S. customs official suggested that if all the hardware that entered Austria remained there, each citizen would have the capacity to build his or her own ICBM. Interview, U.S. embassy, Bonn, July 18, 1985.

produce customized ICs, and it threatened to remove Austria from the list of destinations for which U.S. firms could obtain distribution licenses. The Austrian government complied, passing legislation in early 1985 that prohibited unauthorized reexports of U.S. technology and established significant penalties for violators. In 1986, Austria entered into a customs agreement with the United States to facilitate further the policing of U.S. high technology exports. A prominent Austrian official emphasized that the new initiatives were not a "voluntary gesture of good will to the U.S., but rather an important economic necessity."[80]

As its new law failed to cover transit trade, Austria was subject to further U.S. pressure. In exchange for some liberalization of U.S. controls, Austria passed additional legislation in 1987 that gave Customs the power to intercept suspicious items in transit and granted the United States the right to conduct investigations regarding such items. The new legislation also empowered the Austrian government to restrict Austrian exports of strategic goods that incorporated technology of foreign origin.[81]

U.S. officials employed the carrot as well as the stick. Switzerland and Finland were granted special treatment in recognition of their willingness to maintain fully effective export control systems.[82] Though neither had formal links to CoCom, each established "CoCom-comparable" controls over indigenous as well as CoCom-origin technology, made full use of import certificates, and consulted regularly with the United States as a means to monitor and prevent the diversion of controlled technology. In 1987, after strengthening its controls on transit trade, Switzerland became the first country outside CoCom to receive the full range of licensing privileges granted by the United States under Section 5(k) of the Export Administration Act.[83] Shortly thereafter, Finland received the

80. The quote is found in *Wall Street Journal*, November 30, 1984, p. 35. See also *Business Week*, November 26, 1984, p. 66, and *New York Times*, January 2, 1985, p. D1. Other Austrian officials emphasized that Austria's neutrality was military (as opposed to economic) and thus cooperation with the West on technology controls was not inconsistent with neutrality. *Export Control News* 1 (October 30, 1987), 10–12.

81. *Wall Street Journal*, September 16, 1987, p. 31, and November 10, 1987.

82. Given its geographical proximity to and political relationship with the Soviet Union, Finnish cooperation with the technology embargo is a delicate matter that has rarely been publicly discussed. Occasionally, items surface in the press that suggest the vigilance of Finnish enforcement. See, for example, *International Herald Tribune*, March 17, 1989, which reports that a Finnish court sentenced two businessmen to thirty-four months in prison on charges of treason for selling U.S.-controlled technology to the Soviet Union. Informal discussions with American and Finnish officials revealed that the two countries consult regularly on export control matters and that Finland relies on guidance from the United States on the appropriateness of certain Finnish exports to the East.

83. Section 5(k) directs the State Department to lead negotiations with countries outside CoCom aimed at securing their cooperation with multilaterally agreed export controls. It stipulates further that where such negotiations lead to the establishment of "CoCom-comparable" restrictions by a third country, the secretary of commerce shall treat exports to that country in the same manner as exports to CoCom members are treated. See *Export Administration Amendments Act of 1985*, reprinted with an introduction and legislative history in John R. Liebman et al., *Export Controls in the United States* (New York: Law and Business/Harcourt Brace Jovanovich, 1986),

same benefits. Finland's preferential status triggered a debate in neighboring Sweden, which received fewer licensing benefits from the United States because of its failure to establish a fully CoCom-compatible control system. Interestingly, the Swedish business community appears to have directed its discontent at being placed at a disadvantage vis-à-vis its Finnish competitors not at the U.S. government but at its own, for the latter's inability to do what was necessary to placate U.S. officials and obtain the full range of benefits.[84]

America's demand that governments adopt stricter controls in exchange for continued access to U.S. technology informed numerous other bilateral relationships. Spain—the newest NATO member, yet undecided as to whether to join CoCom—was informed in 1984 that AT & T's proposed $200 million microchip plant, the cornerstone of Spain's electronics development program, would not be allowed unless the Spanish government devised a more satisfactory system to protect U.S. technology. The government complied by establishing new regulations for export controls similar to those of CoCom members. Shortly thereafter, Spain formally joined CoCom.[85]

India was subject to similar pressure and in 1985 signed a Memorandum of Understanding (MOU) with the United States that committed the Indian government to develop reexport control regulations, including an import certification system. In exchange, the Reagan administration approved a sale of nineteen computers that had been held in abeyance for three years and subsequently granted India permission to purchase a U.S. supercomputer, the first state outside the Western alliance to be allowed to do so.[86] Singapore, upon learning of its status as a

---

pp. 331–32. U.S officials came to interpret a "CoCom-comparable" system as having to include (a) controls on all items, either imported or produced indigenously, that appear on the CoCom list; (b) an IC/DV system; (c) procedures to prevent the unauthorized reexport of controlled items, including items in transit; (d) an effective enforcement regime, including government-to-government cooperation with other controlling countries; and (e) procedures for pre-license screening and post-shipment checks of exporters and end users. The benefits granted to 5(k) countries include expedited processing of license requests, eligibility for certain general licenses, and the ability to reexport U.S.-origin items to China without prior authorization. See *Export Control News* 1 (August 28, 1987), 9–10, and National Academy of Sciences, *Finding Common Ground: U.S. Export Controls in a Changed Global Environment* (Washington, D.C.: National Academy Press, 1991), pp. 66–68, 122–26.

84. During 1989 the Swedish technical weekly *Ny Teknik* published a series of articles airing the views of the Swedish business community on the export control issue. See "The U.S. Does Not Relieve Export Controls," April 20, 1989, and "Government Plans to Sharpen Swedish Export Controls," May 11, 1989, informal translations provided by the U.S. embassy in Stockholm. One Swedish businessman suggested that if the Swedish government was finding it difficult to negotiate with the Americans, it should ask the Finnish government for assistance. The problem, however, was not negotiating prowess but Sweden's failure as of early 1989 to agree to control indigenous technology and to establish an IC/DV system with the force of Swedish law behind it.

85. *Economist*, February 23, 1985, p. 70, and *Financial Times*, February 28, 1985, p. 30, and September 20, 1985, p. 2.

86. *Financial Times*, November 20, 1984, p. 4, and March 13, 1985, p. 5, and *Washington Post*, November 27, 1984, p. D1, and November 1, 1986, p. C1. In 1987 the DoD began to block certain U.S. technology exports to India because of its concern over how the Indian import certification

destination under DoD review, pledged to cooperate with the United States and CoCom. It announced a new policy of controlling CoCom list items, implementing the IC/DV system, and requiring government approval for the reexport of U.S. technology. In return, Singapore became the first non-CoCom state in Asia to receive some of the 5(k) licensing benefits already enjoyed by Switzerland, Sweden, and Finland.[87]

South Korea signed an MOU with the United States in 1987, agreeing to establish controls on indigenous technology and on that originating in other cooperating countries. Two years later South Korean officials announced their desire to join CoCom. They reasoned that if South Korea, under pressure from the United States, was to abide by CoCom regulations, it would do well to join the organization and thereby participate in shaping those regulations.[88] A similar calculation was made by Australia, which became the seventeenth member of CoCom in April 1989. On joining, Australia became eligible for the full range of licensing benefits available from the United States in trade with CoCom member states.[89]

The above examples suggest that U.S. third-country efforts were ambitious and fruitful. The effort did entail diplomatic costs—governments tend to resent being coerced, and U.S. officials tended not to be subtle—as well as economic ones, as I discuss below. Moreover, the success enjoyed by U.S. officials was not unqualified. Some third-country governments, particularly on the Pacific Rim, have been slow to put export control systems into place as promised or to devote adequate resources to enforcement. As of 1990, the diversion of CoCom-controlled items through third countries continued to be a problem for CoCom.[90]

Despite its costs and limitations, a strong third-country effort was essential to U.S. officials to protect U.S. technology and as a necessary counterpart to the U.S. effort to strengthen CoCom. Furthermore, responsibility for the task fell squarely on the United States, as it had since the 1950s. Although, at U.S. urging, CoCom members formally and collectively adopted the third-country initiative in 1984, in practice most

system was being administered. The Indian government subsequently adopted a stricter IC/DV system. See *Export Control News* 1 (September 28, 1987), 14–16, and 2 (July 11, 1988), 12–14.

87. *Financial Times*, June 19, 1986, p. 6, and *Wall Street Journal*, October 21, 1987.

88. *FBIS* (Northeast Asia), February 1, 1988, p. 16, and December 22, 1988, pp. 15–16; *Journal of Commerce*, June 23, 1987, p. 24, and April 25, 1988; *Wall Street Journal*, December 7, 1989, p. A10. South Korean officials promised to ban the export and reexport of all items on the CoCom list. They claimed that South Korea could indigenously produce six list items in 1988 but would be able to produce twenty within five years. See *FBIS* (Northeast Asia), April 22, 1988, p. 37. As of 1991, South Korea had not joined CoCom.

89. See *Export Control News* 3 (April 30, 1989), 8–9. Australia's already effective export control system required few changes to become fully CoCom-compatible. The initiative to join CoCom came wholly from the Australian side, and it was required to approach and gain the acquiescence of each CoCom member individually before being formally accepted as a member.

90. See National Academy of Sciences, *Finding Common Ground*, pp. 30, 288. The report tends to underestimate U.S. leverage in the third-country context, claiming that U.S. threats to restrict U.S. technology were "hollow" (p. 123). The evidence presented above suggests otherwise, even if the United States was not successful in every case.

of the progress made in extracting compliance was due to the efforts of U.S. officials. No other CoCom member possessed the combination of interests and resources necessary to undertake the effort and to do so with a reasonable degree of effectiveness. A crucial difference between the 1950s and the 1980s, however, was that in the latter period, interdependence and the diffusion of technological capabilities rendered the third-country effort considerably more difficult, because it involved many more states in the context of a more integrated world economy. As of late 1989, U.S. officials were continuing consultations with most of the states noted above and had opened discussions with numerous others, including Brazil, Indonesia, Malaysia, Taiwan, and Pakistan.

Intra-Western Restrictions and CoCom

The willingness of the Reagan administration to expand intra-Western restrictions also provided some leverage in U.S. relations with CoCom member states. The threat of technology denial clearly enhanced the U.S. bargaining position in CoCom's 1982–1984 list review and in bilateral negotiations with CoCom allies over enforcement, although to what extent is not altogether clear.[91] Similarly, by linking the participation of West German firms in the Strategic Defense Initiative to West German export control efforts, the DoD gained a commitment from the West German government to strengthen its relevant laws and enforcement practices.[92]

In overall terms, however, intra-Western restrictions proved counterproductive in the CoCom context because they created profound political resentment and reduced the incentives for effective cooperation. First, to the extent the restrictions were applied extraterritorially (i.e., reexport controls), they kept alive the bitter alliance dispute made public by the pipeline episode. West European governments hoped that confrontation would lead the United States to reconsider its unique assertion of authority, and they undertook an intensive lobbying campaign during 1983 to influence the revision of the Export Administration Act. In a series of strongly worded, formal diplomatic statements, they argued that U.S. reexport controls were contrary to international law and detrimental to friendly relations among states. They urged the United States to relinquish its claim and issued thinly veiled threats of retaliation

91. For strong statements suggesting the utility of intra-Western restrictions in the CoCom context, see Cahill, *Trade Wars*, pp. 94, 128, and Harold Tittman, "The Drawbacks of Having the Pentagon in Charge," *Financial Times*, April 4, 1985.

92. See the "secret" letters of Richard Perle and Lorenz Schomerus in "Federal Republic of Germany-United States: Agreements on the Transfer of Technology and the Strategic Defense Initiative," March 27, 1985, reprinted in *International Legal Materials* 25 (July 1986), 957–77. DoD officials extracted explicit—in the eyes of some West Germans, humiliating—concessions from the West German government regarding technology protection before signing an agreement.

against U.S. subsidiaries operating within their borders.[93] To their dismay, the EAA retained its extraterritorial assertion, which was even strengthened in that the Executive was given the discretion to impose *import* sanctions, along with export denial, on foreign firms that violated U.S. controls.[94]

The sharpness of the diplomatic dispute demonstrates a point made in Chapter 7, that in the aftermath of the pipeline, the exercise of U.S. reexport controls could not remain depoliticized. That is, U.S. officials could no longer count on it as a low-profile, low-cost means to control U.S.-origin technology and equipment worldwide. Other Western governments were compelled to object, both as a matter of national pride and in response to domestic political pressure. The customer listing requirement in the U.S. distribution license proposals prompted an angry response from West European governments, which claimed it forced their firms to act as overseas "agents" of the U.S. national security apparatus. Similarly, the British government publicly denounced the IBM letter (discussed above), and the West German government forced the Reagan administration to back down in its effort to prevent IBM's West German subsidiary from supplying a computer to another West German firm owned by the Soviet Union.[95]

Second, the extension of West-West controls proved costly to CoCom by reducing incentives for effective cooperation. To other member governments, such controls represented a U.S. preference for unilateralism and a vote of no confidence in the multilateral mechanism. French officials raised this point explicitly in a diplomatic exchange, protesting that U.S. extraterritorial controls "reveal an unacceptable lack of confidence in the way we implement and enforce the CoCom embargo."[96] Not surprisingly, it was widely perceived in Western Europe that the extension of intra-Western controls was not motivated by security concerns but by the desire to maintain U.S. technological supremacy. An internal study by ICL warned of growing "technological imperialism" in the

93. "Letters and Aide-Memoires from Foreign Governments concerning the Export Administration Act," reprinted as appendix 34 in U.S. Congress, House, Committee on Foreign Affairs, Subcommittee on International Economic Policy and Trade, *Extension and Revision of the Export Administration Act of 1979*, hearings and markup, 98th Cong., 1st sess., February 24–May 26, 1983, pp. 1854–99.

94. For discussion, see D. Collette Gonzalez, "How to Increase Technology Exports without Risking National Security—An In-Depth Look at the Export Administration Amendments Act of 1985," *Loyola Los Angeles International and Comparative Law Journal* 8 (1986), 498–501.

95. The Thatcher government encountered severe criticism in Parliament for its failure to take on the United States more forcefully over the IBM letter. Liberal member of Parliament Paddy Ashdown argued that Britain's overwhelming dependence on U.S. computers in banking, the stock exchange, and the Ministry of Defence constituted a national security threat, in light of America's technology denial policy. See *Parliamentary Debates* (Hansard), February 17, 1984, pp. 496–504. See also *Aviation Week and Space Technology*, May 7, 1984, pp. 18–19; *Financial Times*, January 24, 1984, p. 1; and *New York Times*, September 29, 1987, p. A1.

96. Letter from France's ambassador to the United States to Representative Don Bonker, March 16, 1984, in "Letters and Aide-Memoires from Foreign Governments," p. 1874.

United States and claimed reexport controls were used to maintain U.S. lead time over Western competitors. Studies conducted by the Institute of German Economy and for the Research and Technology Ministry of the Bonn government reached similar conclusions.[97] The United States, it was argued, used its licensing requirements to pursue a divide-and-conquer strategy in Western Europe, creating barriers to intra-European trade, technological collaboration, and the achievement of economies of scale. These attitudes were reinforced by the proposed customer listing requirements for overseas distributors and by the revelation that U.S. intelligence kept track of the activities of over three hundred West European high-technology firms, ostensibly to anticipate and prevent illegal diversion. A popular book in Great Britain began with the hyperbole that "the United States of America is embarked on an attempt to gain control of all high technology in the free world."[98]

To many West Europeans, the extension of U.S. unilateral controls and the strengthening of CoCom became indistinguishable. CoCom was perceived as simply another instrument of U.S. "technological imperialism." In 1986 the European Parliament passed a resolution attacking both reexport controls and CoCom as violations of the Treaty of Rome, and in 1987 some members proposed that CoCom be dissolved precisely because it was associated with the U.S. assault on West European sovereignty.[99] By encouraging such attitudes with its behavior, the United States kept open the possibility that defiance of its highly contentious unilateral controls would spill over into defiance of CoCom, since both were perceived to spring from the same source.

The above discussion suggests that the targeting of intra-Western restrictions against CoCom members tended to carry higher costs, and fewer benefits, than those directed at third countries, primarily because in the former context, national systems and a multilateral mechanism for coordinating controls were already in place. For the United States to rely on unilateral instruments where a multilateral regime existed—and to do so in a manner that violated the sovereignty of other participants—reduced the incentives for others to participate faithfully and diverted attention from simultaneous U.S. efforts to strengthen the regime. Those consequences are particularly significant in the case of CoCom, given its informal nature and the wide discretion it grants member governments in the implementation of its directives.

---

97. Jeremy Strachan, "U.S. Trade Law: The Triple Threat. The Stifling of High-Technology Business," *ICL Internal Working Paper* (London, 1983). For discussion, see Stuart Macdonald, "Controlling the Flow of High Technology Information from the United States to the Soviet Union: A Labour of Sisyphus?" *Minerva* 24 (Spring 1986), 62–63.

98. Cahill, *Trade Wars*, p. 1. The CIA's targeting of West European firms was revealed in a speech by Director William Casey before the Commonwealth Club of California, April 3, 1984 (untitled, typed transcript), p. 7.

99. *Aviation Week and Space Technology*, March 17, 1986, pp. 66–67.

This is not to suggest that technology sanctions had no utility in the CoCom context. On the contrary, the benefits may well have exceeded the costs when intra-Western restrictions were targeted specifically at CoCom members (e.g., Japan) whose export control infrastructures and enforcement efforts lagged far below the norm. In such cases, the credible threat of technology denial could have provided the impetus for CoCom member governments to upgrade significantly their national control efforts, as many third countries did during the 1980s. Unfortunately, the intra-Western restrictions of the United States were applied rather indiscriminately, falling on those CoCom states whose efforts were exemplary (e.g., Great Britain) as well as on those whose efforts were lagging. Moreover, rather than as a last resort, they were applied in lieu of less costly and contentious means (i.e., streamlining the control list) of strengthening the commitment of other member states to CoCom. Thus, that they created resentment is hardly surprising.

Intra-Western Restrictions and U.S. Economic Costs

Finally, intra-Western restrictions compromised the competitive position of U.S. high-technology industry, a point that applied in the third-country as well as the CoCom context. During the 1970s, U.S. manufacturing firms gained reputations as unreliable suppliers to the East. During the 1980s, they acquired the same reputation *globally*, where there was far more at stake. By segregating East-West and intra-Western controls, U.S. officials had been able to manage the conflict between, and to realize simultaneously, America's economic and narrowly defined national security objectives. With the blurring of that distinction, national security concerns intruded on economic (i.e., competitiveness) objectives in a fundamental way and at a time when the latter were becoming a subject of increasing national concern.

By the mid-1980s, evidence was accumulating of what the National Academy of Sciences termed "de-Americanization"—the movement of foreign firms away from U.S. sources of supply in sectors most affected by U.S. controls. In telecommunications, French firms sought to reduce reliance on U.S. components, and the Hong Kong subsidiary of the British firm Cable and Wireless, citing persistent reexport delays, announced in 1984 that it would turn from the United States to other suppliers. The West European Airbus consortium took steps to design U.S. components out of its products, and a similar pattern appeared in West European and Japanese space systems. Its inability to obtain U.S. licenses prompted the French government to launch a project to develop advanced supercomputers. In energy projects, French firms heeded the post-pipeline advice of the Soviet Union and sought to replace American with West German technology.[100]

The costs of designing out American components can be considerable, and numerous firms found it impractical to do so in the short run, despite the burden of U.S. controls. According to officials at the Federation of West German Industries, this position was that of the majority of West German firms in 1985.[101] Britain's ICL took a similar position at the time.[102] Nevertheless, even if adjustment were to take place slowly, continued reliance on intra-Western controls would logically lead more and more firms to search for alternative sources of supply. The long-term threat to the competitive position of U.S. high-technology firms was significant and demonstrated the paradox of U.S. power. Because its firms have traditionally dominated global markets, the U.S. government has potential leverage over foreign firms and governments. The sustained use of that leverage, however, tends to erode the very foundation on which it is based by driving foreign firms away from reliance on the U.S. private sector.

BREAKING THE STALEMATE:
THE JANUARY 1988 AGREEMENT

The alliance controversy in CoCom and over America's tightened intra-Western restrictions peaked in 1985. By January 1988, the United States and its CoCom allies agreed at high political levels to a deal similar to that struck in 1954. U.S. officials promised a reduction in the control list and a relaxation of intra-Western restrictions in exchange for a greater commitment to enforcement on the part of other CoCom states. The deal is best explained by the confluence of two events: a reassessment of the economic costs of controls at the highest levels of the U.S. government and the revelation of a major CoCom diversion involving Japanese and Norwegian firms.

100. See, for example, *Financial Times*, October 15, 1983, p. 1, May 30, 1984, p. 1, and August 15, 1984, p. 1; *Computer Weekly*, February 3, 1983, p. 1; *Washington Post* (National Weekly Edition), May 21, 1984, p. 21; *Aviation Week and Space Technology*, March 9, 1987, pp. 88–91; and *New York Times*, October 18, 1987, p. E5. The most publicized account of foreign firms designing out U.S. components is found in National Academy of Sciences, *Balancing the National Interest*, p. 247.
101. Interview, Federation of West German Industries, Cologne, July 18, 1985.
102. Interview, London, July 26, 1985. See also Macdonald, "Controlling the Flow of High Technology Information," p. 60. Recognition of the costs of moving away from U.S. sources is also reflected in the policies of Western governments, even those most vehemently opposed to reexport controls. The British, for example, maintained a policy of refusing to recognize the legitimacy of U.S. controls, while leaving the actual decision of whether to comply in the hands of British firms, who were advised to act on the basis of commercial prudence. Interview, London, July 24, 1985, and *Security Export Control*, pp. ii–iii. The British government even acquiesced to the controversial U.S. plan to audit foreign firms, provided that the firms in question consented to the audit. See *Financial Times*, February 23, 1987, p. 6.

Within the United States, intra-Western restrictions had been a subject of contention from the time controls were tightened in the early 1980s. The Commerce Department, the U.S. export community, and a majority in the House of Representatives worked politically to temper the DoD-inspired policy. Until 1985 they managed little success. Commerce and U.S. exporters remained on the defensive in an atmosphere of U.S.-Soviet hostility, while in Congress, supporters and opponents of more liberal export controls were locked in a bitter struggle over the renewal of the Export Administration Act. The intensity of that struggle was reflected by the fact that it persisted long after the EAA formally expired, forcing the Executive, in a legally dubious maneuver, to resort to the International Emergency Economic Powers Act in order to maintain export control authority.[103]

The legislation that finally emerged in 1985 represented a compromise. Advocates of tight controls could point to expanded enforcement authority for the Customs Service and harsher penalties for violators, the renewal and expansion of the Executive's extraterritorial authority, and a mandate for the DoD to integrate its MCTL into the existing CCL administered by Commerce. One of the most contentious proposals, for a formal DoD role in intra-Western licensing, was omitted from the legislation after President Reagan granted the DoD the authority in the above-discussed executive order. U.S. exporters and other proponents of relaxed restrictions were served by provisions directing Commerce to eliminate licensing requirements on administrative exception-level items for CoCom destinations; process intra-Western licenses more quickly; liberalize controls on embedded microprocessors; and create a comprehensive operations license (i.e., a distribution license for technology).[104]

The Commerce Department, prodded by U.S. industry, took initial steps to relax intra-Western controls in 1984 and 1985.[105] The real impetus for significant liberalization, however, came early in 1987, when the White House threw its support in favor of a national export promotion strategy.[106] Until then, the president tended either to remain neutral or to side with the DoD in interagency export control disputes. By early 1987, concern over America's record-breaking and seemingly intractable trade deficits led the White House to launch a competitiveness program to enhance the position of U.S. firms in world trade. At virtually the same time, an NAS study publicized the link between declining

103. Overman, "Reauthorization of the Export Administration Act," p. 369.
104. For detailed discussion, see ibid., and Gonzalez, "How to Increase Technology Exports without Risking National Security."
105. Late in 1984, Commerce introduced a "fast-track" procedure to expedite West-West license requests. In 1985 it carried out the mandate of the new law and removed licensing requirements on administrative exception items. See *Business America*, October 28, 1985, p. 12.
106. *Wall Street Journal*, February 6, 1987, and *Journal of Commerce*, May 18, 1987, p. 1.

U.S. competitiveness and the expansion of intra-Western export controls. It argued that as presently administered, the U.S. export control system overemphasized national security objectives at the expense of national economic objectives. The unnecessarily broad scope of U.S. controls damaged the export performance of U.S. firms in world markets at a time when export competitiveness had become essential to the health of the domestic economy.[107]

The reassessment on the economic side coincided with a relaxation of U.S.-Soviet tensions, symbolized by the resumption of superpower summitry beginning in 1985. Although not leading to a large-scale relaxation of controls, the improved U.S.-Soviet environment made it less precarious politically to support the decontrol of less sensitive items. Early in 1987, U.S. officials removed restrictions on energy technology and equipment, which had initially been imposed on the Soviet Union for foreign policy purposes almost a decade earlier.[108]

Immediately thereafter, Commerce Secretary Baldrige announced a set of measures to liberalize intra-Western controls. Among the most important was the creation of "gold card" privileges—the elimination of validated license requirements for export to Western firms certified as reliable by the Department of Commerce.[109] Commerce also relaxed reexport controls by announcing its intention to remove the licensing requirements for finished goods whose U.S.-origin components did not exceed 25 percent of the goods' total value. Previously, the DoD had forced compliance even when U.S.-origin components were of negligible value.[110]

That the proponents of intra-Western liberalization gained the upper hand did not mean the DoD simply abandoned its opposition. Even without the full support of the White House, the DoD had sufficient bureaucratic power to delay implementation of liberalizing procedures or counter them with new restrictions. During 1987, DoD officials watered down plans to relax reexport controls, controls on trade with China, and restrictions governing personal computers. The department also proposed new controls on ICs and wrested from Commerce the right to review exports to Communist-controlled enterprises in CoCom member states. Representative Don Bonker, one of the strongest con-

107. National Academy of Sciences, *Balancing the National Interest*, p. 152.
108. *Washington Post*, January 16, 1987, p. A1.
109. Firms possessing gold card status could receive controlled commodities from the United States within one or two days. Gold card holders could also ship controlled goods among themselves without reexport authorization. The privilege was initially granted to firms owned or controlled by CoCom member governments; eventually it would become available to firms based in CoCom countries and then to those in reliable neutral states. See *Christian Science Monitor*, February 11, 1987, p. 6, and *American Metal Market*, February 16, 1987, p. 1.
110. Commerce Department officials hoped the new standard would eliminate 60 percent of reexport authorizations. *Wall Street Journal*, March 25, 1987, and *American Metal Market*, July 6, 1987, p. 1.

gressional advocates of relaxed controls, expressed his concern early in 1987 that the Commerce Department would "get down to the one yard line and never cross the goal line."[111]

Yet, even if Defense opposition could be overcome, the victory for proponents of decontrol might be short-lived unless the root cause of the problem—U.S. dissatisfaction with the enforcement efforts of its allies—was addressed. Without improved enforcement, unilateral U.S. decontrol initiatives in the intra-Western context would remain vulnerable to conservative opposition. For the allies to improve enforcement, however, would require clear evidence that the United States was indeed willing to carry out what it had long resisted—meaningful liberalization of the control list.

The potential for a trade-off was thus evident but required political will to be realized. Ironically, the catalyst for an alliance compromise was provided by the revelation of a major diversion of CoCom-controlled technology to the Soviet Union.

The Significance of Toshiba-Kongsberg

Early in 1987 it became known in the United States that between 1982 and 1984, Toshiba Machine, an independent subsidiary of the Toshiba Corporation, and Kongsberg Vaapenfabrik, a Norwegian state-owned firm, sold sophisticated milling machines and computer controllers to the Soviet Union. The equipment was installed at the Leningrad shipyard and was used to produce propellers for Soviet submarines. Military assessments of the sale vary. To some analysts, it did massive damage to U.S. and Western security by affording the Soviets a revolutionary technological advance that enabled a breakthrough in efforts to quiet their submarines. To others, Toshiba's machines made at most a modest contribution to an ongoing, well-advanced Soviet effort to quiet its fleet, an effort enhanced most significantly by information obtained via espionage involving an American naval officer.[112] As was the case following the Bryant Grinder sale, this dispute will probably remain unresolved. What is not in dispute is that the Toshiba sale represented a blatant violation of CoCom regulations. At the time machine tools with more than two axes, or independent cutting units, were multilaterally controlled. The Toshiba machines possessed nine independent axes.[113]

The Toshiba-Kongsberg diversion had its roots in the weakness of

111. *Wall Street Journal*, March 13, 1987, p. 52. See also *Business Week*, December 7, 1987, pp. 114–15.

112. DoD officials vacillated between the more and less alarmist positions. See *New York Times*, March 14, 1988, p. 1. A strong statement of the view that Toshiba's sale mattered less than did the Walker spying incident is found in the *Japan Economic Journal*, August 8, 1987, p. 7.

113. A detailed account is in *New York Times*, June 12, 1987, p. 1.

CoCom during the 1970s. According to investigations conducted by the Japanese and Norwegian governments, Toshiba Machine was first approached by the Soviets to sell multiaxis machine tools in 1974. The Japanese firm refused, citing CoCom restrictions. The Soviets then turned to a French firm, with more success. Forest Line (then known as Ratier-Forest) shipped nine machines, in violation of CoCom, to the Soviets between 1976 and 1979. French government officials offered a retrospective justification of the sales in 1987, claiming in effect that the violations were understandable because CoCom lacked credibility during the 1970s.[114] The French violation ultimately did double damage: Toshiba Machine officials, bitter over having lost the initial sales, proved more receptive in 1979 when the Soviets approached again, this time requesting even more sophisticated machines. Kongsberg was willing to provide the computer controllers; it had done so on a regular basis since 1974, again clearly in violation of CoCom rules.[115]

The reaction in Congress was immediate outrage, directed far more at Toshiba and Japan than at Kongsberg and Norway. Revelations of the diversion compounded an already high level of discontent over what were perceived as unfair Japanese trade practices and an unwillingness on the part of Japan to shoulder a defense burden commensurate with its economic power. Four members of Congress deemed it appropriate to vent their anger by smashing a Toshiba radio on the steps of the Capitol. The more substantive response involved the introduction of punitive legislation; the Senate, by a vote of 92–5, proposed a ban on Toshiba's exports to the United States of from two to five years. Toshiba's parent firm and all its subsidiaries were targets of the legislation, despite their claim to having had no knowledge of the illegal activity of Toshiba Machine.[116]

The Senate response was precedent-setting in two ways. It called for U.S. sanctions against a foreign firm that had violated CoCom regulations, as opposed to U.S. regulations (no U.S.-origin items were involved), and its penalty involved the denial of the U.S. market, as opposed to U.S. exports. Given Toshiba's successful penetration of the U.S. market, this latter feature made the measure all the more attractive to anti-Japanese protectionists, who were happy to join anti-Soviet conservatives in its support.

Though similarly angered by the sale, executive officials sought from the outset less to punish Japan than to use the occasion as an opportunity to convince the Japanese government to strengthen its export control system. The Department of Commerce did suspend Toshiba America's

114. *Wall Street Journal*, October 21, 1987, and *Washington Post*, September 14, 1987, p. 24.
115. *New York Times*, October 22, 1987, p. D1, and *Washington Post*, October 22, 1987, p. A1.
116. *Washington Post*, July 1, 1987, p. A1. The proposal did allow an exemption for military items in order to enable the navy to continue its purchases of Kongsberg's Penguin anti-ship missiles, if it so desired.

distribution license, and the Pentagon did opt for a U.S. supplier over the previously favored Toshiba to fulfill a $100 million contract for laptop computers.[117] The usually combative executive agencies agreed, however, that the imposition of strict import sanctions would be counterproductive and might trigger an attitude of defiance in Japan. Even hard-liners in the DoD warned that the United States should avoid "shooting itself in the foot." The Executive was not above using the *threat* of congressional legislation, however; administration officials made it clear they would do little to blunt the Senate's initiative until the Japanese tightened their controls.[118]

On learning in 1986 of the diversion, the Nakasone government reacted slowly and did not view the incident as a matter of grave concern. Pressure from U.S. defense secretary Weinberger prompted it to complete an investigation and impose penalties in May 1987. Toshiba Machine was banned from exporting to Communist states for one year, and C. Itoh, the trading company that facilitated the deal, for three months. Although mild by U.S. standards, the penalty against Toshiba Machine was the most stringent allowable under existing Japanese law and one that until then had never been imposed on a Japanese firm.

U.S. officials, however, demanded not just a strict application of existing regulations but also a revision of those regulations as well as a revamping of the Japanese control system. It was clear by the early 1980s that Japan was simultaneously among the technologically most advanced CoCom members and among those with the weakest control systems. A high MITI official frankly conceded that Japanese firms paid little attention to security considerations in exporting to the East.[119] MITI's own policy can best be described as having been one of benign neglect. Hopelessly understaffed, with thirty officials attempting to process over one hundred thousand annual license requests, the agency was forced to rely almost exclusively on the good faith of Japanese firms to assure export control compliance. The problem was psychological as well as administrative; Japanese public officials and corporate officials lacked sensitivity to the strategic risks of illegal technology transfer. As one Japanese observer put it, "Even if the Japanese understand the Russians are coming, they may tend to think that is something the Americans will look after."[120]

In July 1987, Nakasone signaled his willingness to acquiesce to U.S. demands by publicly accusing Toshiba Machine of having betrayed Japan. Shortly thereafter the Japanese Cabinet proposed, and the Diet approved, an increase from three to five years in the maximum prison

117. *Wall Street Journal*, July 8, 1987, and *New York Times*, August 12, 1987, p. D1.
118. *Business Week*, July 6, 1987, pp. 46–47, and *Washington Post*, July 17, 1987, p. F1.
119. *Japan Economic Journal*, May 30, 1987.
120. Japanese professor Masashi Nishihara, quoted in *New York Times*, July 19, 1987.

sentence for export control violators and an increase in the maximum administrative sanction from a one- to a three-year ban on exports to Communist states. For the first time, Japan's Foreign Exchange and Foreign Trade Control Law was revised to refer explicitly to the maintenance of "security" as grounds for denying export licenses. MITI's export control staff was to be more than doubled, and the agency would no longer enjoy essentially exclusive responsibility for export controls. The Foreign Ministry was given a role in the process, and a high-level interagency committee was created to monitor and review Japanese policy. Finally, Japan agreed to contribute more to CoCom and assist the United States in gaining the cooperation of third countries.[121]

The Toshiba diversion also affected the procedures of Japanese firms. MITI ordered all Japanese firms that deal in controlled commodities to develop plans to assure compliance with CoCom regulations. By December 1987, MITI claimed to have received responses from 2,178 of 2,327 firms.[122] Not surprisingly, Toshiba responded with a far-reaching plan— the creation of a division, staffed by over one-hundred individuals, whose assignment was to screen the sale of controlled commodities to both domestic and foreign customers.[123]

Although the long-term impact of Japan's export control revisions remains to be seen, the Toshiba incident does appear to have marked a turning point. It led the Japanese government to tighten controls, take its commitment to CoCom more seriously, and increase the awareness of Japanese firms of consequences of diversion.[124] The Japanese government appears to have been driven more by a concern over the force of the U.S. reaction (which it clearly did not anticipate) than by a fundamental reappraisal of the military risks of technology transfer. Nevertheless, at least in the short run, the Toshiba situation affected Japanese attitudes and behavior considerably.

The importance of this case went beyond Japan's role and affected CoCom as a whole by focusing the attention of member states squarely on the enforcement problem. Investigations into the incident brought numerous other cases to light, exposing an underside of illegal trade in CoCom-controlled items during the 1970s and early 1980s. The most

121. *Wall Street Journal*, May 18, 1987, p. 22; *New York Times*, August 1, 1987, p. D1; and *Japan Economic Journal*, August 8, 1987, p. 4.
122. *Japan Economic Journal*, January 16, 1988.
123. *Japan Economic Journal*, November 21, 1987, p. 32.
124. In a survey of CoCom member enforcement systems conducted in late 1988, Japan was rated an "up and comer" whose system and commitment had improved dramatically since 1987. See *Export Control News* 2 (December 22, 1988), 5. In informal discussions, U.S. officials confirmed this finding, noting that the number of diversions via Japan decreased significantly after 1987 and that Japan participated more actively in CoCom. One U.S. official observed approvingly that the Japanese were behaving like "reformed alcoholics." In 1989 the Japanese government publicized a self-initiated crackdown on illegal technology transfer as tangible evidence of its new commitment. See *Wall Street Journal*, May 25, 1989, p. A14.

compelling evidence was that Kongsberg had illegally shipped over eighty computer controllers to the Soviet Union since 1974, in conjunction with British, French, West German, and Italian machine tool firms.[125] As more and more cases came to light, it was evident that not only Japan and Norway but also in a political sense all the major CoCom allies were implicated. U.S. officials, armed with the enhanced credibility this evidence provided and pointing to the uneven application of CoCom regulations by members, redoubled their campaign for improved enforcement.

In the wake of the Toshiba incident, West European CoCom members could ignore neither the evidence nor the proposed remedy. They did cling to their previous position, however, that enforcement could be improved significantly only if the CoCom list were reduced in scope. While U.S. officials stressed, once again, the urgency to inject political vigor into CoCom, West European officials emphasized the need for "higher fences around less technology."[126]

The deal was struck formally in January 1988, at a special CoCom meeting at which member states were represented at senior political levels.[127] For their part, America's allies agreed to devote priority attention to enforcement. Members committed to augment inspection staffs, impose stiffer penalties for violations, share intelligence more completely, and join the United States more fully in the third-country effort. CoCom members also agreed to inform each other when they turned down a domestic request to export a controlled item, as a way to alert all members to the possibility of a diversion. Finally, it was reaffirmed that each CoCom member would retain the final responsibility for implementing and enforcing controls in accordance with its own laws and institutional structures. CoCom's fundamental task remained to achieve uniformity in outcome through diversity in the application of its directives.

For its part, the United States finally committed at a high level to a significant liberalization of the control list. Although no official would elaborate on what or how much would be decontrolled, categories cited in news reports as probable candidates included computers, civilian aircraft, and telecommunications. As an indication of the liberalizing trend, CoCom approved an exception in June for Western Europe's Airbus consortium to sell aircraft to East Germany.[128]

U.S. officials also indicated that the CoCom agreement would permit a further relaxation of intra-Western trade. They speculated that the list

125. *Wall Street Journal,* July 8, 1987, and *New York Times,* August 12, 1987, p. D1.
126. *New York Times,* October 18, 1987, p. E5.
127. The U.S. delegation was led by Deputy Secretary of State John Whitehead. *New York Times,* January 21, 1988, p. D1; *Wall Street Journal,* January 29, 1988, p. 12; and *Export Control News* 2 (February 3, 1988), 14–16.
128. *New York Times,* February 15, 1988, p. 1, and *Wall Street Journal,* June 27, 1988, p. 9.

reductions might eventually eliminate thirty-six thousand licenses to CoCom destinations annually and another ten thousand to reliable non-CoCom states. Of more immediate significance was that the Defense Department announced at the conclusion of the CoCom meeting its intention to facilitate the transfer of highly sophisticated U.S. military technology to Western Europe. Texas Instruments would be allowed to transfer radar technology to Thomson CSF for use in France's advanced combat aircraft program, and Hughes Aircraft could participate on similar terms in the next generation of the European fighter aircraft program.[129]

Beyond January 1988

The political agreement reached in January 1988 left CoCom members with the difficult task of implementation. The initial effort, from January 1988 through the fall of 1989, illustrates the overall argument of this chapter regarding the inability of the United States to lead fully and effectively because of its insensitivity to the economic costs of controls and to the fragmentation in the U.S. control process.

First, in the view of virtually all members, little progress was made in 1988 and 1989 on the streamlining of the CoCom list. This failure was a source of considerable dissatisfaction among America's allies, particularly since the January 1988 meeting had raised expectations that U.S. officials would quickly adopt a more accommodating posture. Dramatic political developments in Eastern Europe during 1989, which lent an increased urgency to demands for export control liberalization, exacerbated the discontent. Press accounts, reflecting the strong views of West European CoCom members, reported a "16 to 1" division in discussion of streamlining at the high level meeting of October 1989.[130]

While it was committed in principle to streamlining and had pledged to carry it out, in practice the United States maintained its traditional approach to the list review process, adopting a conservative definition of "military significance" in item-by-item reviews and stressing that further progress in enforcement must accompany any relaxation of controls.[131]

---

129. The decisions seemed to have been timed to coincide with the CoCom agreement and might have been intended as a U.S. signal of its willingness to facilitate intra-Western trade even at the highest levels of technology, provided its major allies kept to their part of the bargain. See *Wall Street Journal*, January 29, 1988, p. 12, and *Financial Times*, January 29, 1988, pp. 1, 2.

130. See *New York Times*, October 27, 1989, p. D3.

131. Some CoCom members (e.g., West Germany, Japan, and Norway) argued that enough progress had already been made in enforcement to warrant prompt U.S. action on streamlining. Norway, like Japan, bolstered its enforcement in the wake of the Toshiba-Kongsberg diversion. See *Export Control News* 2 (December 22, 1988), 5. West German improvements came in 1989 and included increases in staffing and in the maximum fines and prison sentences allowable under the Foreign Trade Law. The changes were prompted not by CoCom issues but by evidence that West German firms had assisted the Libyan and other developing country regimes in the produc-

The government of the United States, in contrast to those of Western Europe, viewed streamlining not as *the* most pressing CoCom priority, one requiring immediate attention, but as one of several issues (e.g., enforcement, the third-country effort, and intra-CoCom trade) that required attention simultaneously. The alliance conflict resulting from these divergent perspectives was most evident in the review of machine tool controls, which many CoCom members publicly proclaimed to be a litmus test of the organization's commitment to streamline. Despite complaints from other governments (and firms, including American ones) that controls in this area were fifteen to twenty years out of date, U.S. officials proved unwilling to endorse a large-scale liberalization, pointing to evidence that the Soviet military continued to make wide use of Western machine tools.[132]

The U.S. government did continue to move fairly rapidly in 1988 and 1989 to liberalize *intra-Western* controls, in which the stake of U.S. firms was established and relatively significant.[133] In *East-West* controls, in which other CoCom members had significant stakes economically (and, increasingly, politically), U.S. officials proved far more lethargic, even at the cost of further straining the CoCom consensus.

Second, disagreements over policy among the principal U.S. export control agencies persisted and continued to compromise the ability of the United States to lead effectively in CoCom. Late in 1988, the Soviet decision to withdraw from Afghanistan and concomitant pressure from other CoCom members forced the United States to reconsider the no-exceptions policy. Prior to the date of final Soviet withdrawal on February 15, 1989, Secretary of State Shultz sought to lay the groundwork for relaxing the restriction by arguing in letters to Defense Secretary Carlucci that a foreign policy sanction imposed when the Soviets invaded should be removed when they withdrew. Carlucci disagreed, contending that over time the ban had taken on national security significance and should be retained indefinitely. Impatience in CoCom turned to indignation following the Soviet withdrawal because U.S. officials gave no indication the sanctions would be lifted and either vetoed exception requests for the Soviet Union or requested extensions in considering them so as not to have to make any decisions. Three months after the Soviets withdrew, and only when the issue threatened to become a major

---

tion of chemical weapons. The new regulations were to be applied across the board, however, which suggested that West German control efforts in the CoCom context could potentially benefit. See *Financial Times*, February 15, 1989, p. 24, and *Journal of Commerce*, June 23, 1989, p. 5A.

132. *New York Times*, October 9, 1989, p. D1.

133. In July 1989 the Commerce Department introduced its G-COCOM license, which allows previously controlled U.S. exports (up to the China Green Line) to be exported to CoCom members, Switzerland, and Finland without the need for individual validated licenses. Commerce also announced that virtually all U.S. controlled items could now be reexported to the same destinations without prior U.S. approval. See *Journal of Commerce*, July 13, 1989, p. 3A.

diplomatic dispute with close allies, President Bush finally intervened to resolve the bureaucratic impasse and announce the revocation of the policy.[134]

Finally, the United States maintained its penchant for unilateralism. The Toshiba diversion prompted Congress not only to impose mandatory sanctions against the Japanese firm but also to include a provision in the Omnibus Trade and Competitiveness Act of 1988 that granted the president the authority to impose sanctions against *future* violators of CoCom controls.[135] Thus, while it supported in CoCom the principle that export control enforcement was the exclusive preserve of national governments, the United States reserved in its own national legislation the right to enforce CoCom controls *for* member governments, if it saw fit to do so. The legislation even directed the president to calculate damages and seek compensation from foreign violators proportionate to the costs incurred by the United States in devising countermeasures to Soviet weapons enhanced by the illegal transfer of technology.

The Japanese protested the (watered-down) Toshiba sanctions as a violation of their sovereignty and an expression of great-power arrogance. West European governments similarly reacted predictably, contending that the new authority claimed by the United States was "unacceptable as a matter of law and policy" and warning that they would "move to protect all their legitimate GATT rights," if necessary.[136]

CoCom thus entered its fifth decade mired in controversy. It also faced its most significant structural challenge of the postwar era—the end of the cold war and the dissolution of Communist power in Eastern Europe. Chapter 9 discusses the prospects for CoCom in a radically altered security environment.

---

134. *New York Times*, January 25, 1989, p. A1; *Washington Post*, January 26, 1989, p. A22; and *Export Control News* 3 (April 30, 1989), 6–8, and 3 (June 30, 1989), 4–6.
135. *Omnibus Trade and Competitiveness Act of 1988*, PL 100-418, August 23, 1988, secs. 2443 and 2444. Senate and House conferences agreed on a three-year ban on sales in the United States of Toshiba Machine and a one-year ban on federal government purchases from the parent firm.
136. *New York Times*, January 30, 1988, p. D2.

CHAPTER NINE

World without Cold War:
Is There a Role for CoCom?

In August 1989, a West German official, commenting on possible obstacles to a proposal by Siemens to manufacture digital switching systems in the Soviet Union, noted that the United States was still "master of the CoCom game."[1] Although U.S. officials subsequently vetoed the proposal in CoCom, the timing renders that depiction somewhat ironic. The year 1989 was a watershed, marking the transition from the cold war to a post–cold war era, an era in which it is uncertain whether the United States will remain master of the CoCom game or, indeed, whether that game itself will continue to be played. That the reference was made by a West German official renders it doubly ironic, as a unified Germany, situated geopolitically and economically in the heart of Europe, can play a major role in determining whether CoCom persists, and for how long, in the post–cold war era.

This concluding chapter explores the prospects for CoCom and Western export controls in the radically altered security environment of the 1990s. In the next section, I review the experience and contemplate the future of postwar Western cooperation and conflict in pursuit of economic warfare and linkage strategies. The following three sections focus on CoCom's strategic embargo. First, I reconstruct the logic of the argument regarding the maintenance of effective cooperation in CoCom and consider its theoretical implications. I also assess the extent to which CoCom mattered in the conduct and resolution of the cold war. Second, I analyze the effort by CoCom members to adjust the strategic embargo to the new realities of the 1990s. Third, I take up the question of whether even a reformed CoCom can survive in a post–cold war environment. A final section briefly examines the lessons of the CoCom experience

---

1. The official, unnamed, is quoted in *Financial Times*, August 15, 1989, p. 6.

and their application to Western governments struggling to create new regimes to prevent the spread of weapons of mass destruction to the developing world.

COOPERATION AND WESTERN ECONOMIC STATECRAFT, 1949–1989

Chapter 2 presented four strategies for trade with a potential military adversary—economic warfare, strategic embargo, tactical linkage, and structural linkage. The evidence put forth in Chapters 3 through 8 suggests that there has been sustained Western cooperation in only one, the strategic embargo.

Cooperation in pursuit of economic warfare was maintained only from 1950 to 1953, during the Korean war. West European governments shifted their export control preferences to complement those of the United States primarily because they anticipated an imminent military conflict with the Soviet Union. In light of that threat, which proved extraordinary in the postwar era, they were prepared to allow the demands of rearmament to take precedence over those of economic recovery, to view the Soviet economy as a "war economy" devoted mainly to the mobilization of Soviet military power, and to accept the economic sacrifices and political risks of economic warfare. To be sure, West European officials were unlikely to initiate economic warfare on their own. U.S. leadership was needed to translate congruent preferences into a strategy of multilateral cooperation, and as I discussed in Chapter 3, U.S. officials effectively completed the task.

When the threat of war receded, however, so too did the willingness of West European officials to maintain economic warfare. After 1953, export control preferences conflicted. The failure of the United States to sustain support for its preferred strategy cannot be adequately explained by focusing on America's material power resources. U.S. officials retained an ample reserve of coercive instruments, even after the utilization of Marshall Plan aid. More important explanatory factors included the fundamental divergence of interests and the unwillingness of the United States to compromise its broader alliance objectives in pursuit of a victory in the more narrow East-West trade issue area.[2] The "successful operation/dead patient" analogy loomed large in the minds of top U.S.

---

2. In the energy issue-area, Ethan Kapstein finds that U.S. hegemonic resources were sufficient to determine alliance outcomes, even in the face of conflicting interests. See *The Insecure Alliance: Energy Crises and Western Politics since 1944* (New York: Oxford University Press, 1990), pp. 10–11. In those cases, however, U.S. officials possessed within the issue-area a key source of leverage whose functional equivalent was lacking in the export control area—control over emergency supplies needed by the allies during oil shortages.

policy-makers. Consequently, America's weaker partners managed to determine the course of alliance strategy.

This result, coming at the time of U.S. hegemonic ascendance, suggests that it is useful to reexamine the sources and limits of American influence during the early postwar era and to focus on, in addition to power resources, such factors as the degree of convergence or divergence of state interests and the extent to which U.S. objectives complemented or contradicted one another. Recently, several such reappraisals have been made.[3] The result also suggests a more cautious approach to the common linear generalization that the United States dominated alliance outcomes during the 1950s and 1960s yet became less capable of doing so as its power resources declined during the 1970s and 1980s.[4]

The United States continued to pursue economic warfare from 1954 to 1969 in the absence of alliance support, driven more by political and ideological than by strategic imperatives. It revived the strategy and its efforts to achieve multilateral coordination during the early 1980s, for reasons both political—the return of a confrontational foreign policy vis-à-vis the Soviet Union—and strategic. The Reagan administration's decision to accelerate the arms race and to force the Soviets to bear more fully the economic burden of their overseas military commitments enhanced the appeal of economic warfare, as did the shift in military strategy that called for the United States to prepare for a protracted conventional conflict with its global adversary. As military competition intensified and absorbed more of the resources of both the U.S. and the Soviet economies, economic warfare acquired the status of a strategic imperative. U.S. officials might have pursued it even more forcefully had they possessed information that later came to light; these data suggested that the Soviet economy was significantly smaller than had been conventionally assumed in the West and thus that the defense and empire burden on the economy was significantly greater.[5]

3. For example, G. John Ikenberry finds that the United States "got both less than it wanted and more than it bargained for in the early postwar period." See "Rethinking the Origins of American Hegemony," *Political Science Quarterly* 104 (Fall 1989), 376. Susan Strange makes a similar point, noting the inability of the United States to dominate outcomes across a range of issues, including air transport, GATT rules, and the structure of Western Europe's custom union. See "The Persistent Myth of Lost Hegemony," *International Organization* 41 (Autumn 1987), 561–62. The most systematic reappraisal of the sources and limits of U.S. influence in the early postwar period is found in Joseph Nye, *Bound to Lead* (New York: Basic Books, 1990).

4. That generalization informs much of the work in the East-West trade issue-area. See, for example, Bruce Jentleson, *Pipeline Politics* (Ithaca: Cornell University Press, 1986); Beverly Crawford and Stephanie Lenway, "Decision Modes and International Regime Change: Western Collaboration and East-West Trade," *World Politics* 37 (April 1985), 375–402; and Gary Bertsch, *East-West Strategic Trade, COCOM, and the Atlantic Alliance* (Paris: Atlantic Institute for International Affairs, 1983). The latter two rely on Adler-Karlsson's interpretation of the early postwar period. Jentleson underemphasizes the 1954 CoCom list review, presumably because his focus is on energy cases. He finds that the United States dominated in the early 1950s, got much of what it wanted in the early 1960s, and completely failed in the 1980s. An examination of national security controls obviously leads to a different assessment.

5. Prominent Soviet economists argued in 1990 that the CIA had consistently overestimated

For America's CoCom allies, however, neither the political nor the strategic rationale proved compelling, and economic interests dictated the avoidance of the extensive export and credit restrictions called for by economic warfare. Unlike in the 1954 conflict, U.S. officials in this instance did resort to coercion as part of their effort to obtain compliance. The overriding commitment of the Eisenhower administration to alliance harmony was absent by the time of the Reagan administration, whose top officials divided into unilateralist and multilateralist camps. The latter ultimately prevailed, as the coercive attempt proved extremely costly to both the United States and its allies. As in 1954, U.S. officials compromised to preserve their broader alliance objectives. The Reagan administration learned by hard experience what the Truman and Eisenhower administrations seemed to intuit, and once again West European governments managed to prevail by demonstrating resolve in the face of a fundamental conflict of policy preferences.

Given the changes that have occurred in Soviet domestic and foreign policy, and indeed in the Soviet Union itself, the emergence of a consensus for the revival of economic warfare within the United States, much less multilaterally, is highly unlikely. The political rationale disappeared as U.S.-Soviet relations moved from the competition of the 1970s and the confrontation of 1980–1985 to the widespread cooperation of the late 1980s and early 1990s. Similarly, any strategic rationale was vitiated by the decline of U.S.-Soviet military competition, which was reflected by the two sides' ambitious efforts to reverse the conventional and nuclear arms races. Ironically, the two principal outcomes that the United States sought to achieve, without success, through multilateral economic warfare in the early 1980s—a reduction in Soviet military spending and the scaling back of Soviet overseas commitments—were undertaken by the Soviets themselves by the late 1980s, in recognition of the severity of their domestic economic predicament.

Postwar Western cooperation in pursuit of a tactical linkage strategy—the systematic use of economic rewards and punishments to influence Soviet behavior—similarly proved elusive. U.S. interest in tactical linkage emerged during the Kennedy administration, and after nearly a decade of congressional obstruction, the strategy was ultimately adopted during the Nixon administration. Nixon and Kissinger made no attempt to coordinate tactical linkage multilaterally; they instead sought to exploit

---

the size of the Soviet economy and thus underestimated Soviet military spending. See *New York Times*, April 24, 1990, p. 14, and April 25, 1990, p. D2. See also Henry Rowen, "Gorbachev's Best Defense for Less Defense," *Wall Street Journal*, September 29, 1988. Rowen cites an estimate of Soviet defense and empire spending of 25 percent of GNP, much greater than the 11–14 percent commonly assumed in the West and more characteristic of nations on a near wartime footing, such as Israel or North Korea.

America's unique source of leverage as a "monopoly supplier" of the ability to confer great power status on the Soviet Union by normalizing trade relations. After registering some initial successes, they found their strategy and leverage undermined by the passage of the Jackson-Vanik Amendment. The Soviets responded to Congress's own version of tactical linkage by abrogating the 1972 U.S.-Soviet Trade Agreement and turning to Western Europe and Japan as preferred sources of industrial equipment and technology.

When tactical linkage reemerged during the Carter administration, its effectiveness depended on the substantive, as opposed to the symbolic, need of the Soviets for Western technology and trade. Because the United States was not a unique supplier, multilateral coordination was essential. Other Western states resisted, however, leaving U.S. officials frustrated in their efforts to respond with economic diplomacy to Soviet human rights violations and intervention in Africa in 1977–1978, to the invasion of Afghanistan in 1979–1980, and to the imposition of martial law in Poland in 1981–1982. U.S. attempts to coordinate sanctions created diplomatic friction within the West and hindered the ability of the United States to lead effectively in CoCom.

Western states proved incapable of sustaining cooperation in pursuit of tactical linkage during the 1970s and the 1980s for two reasons. First, although the U.S. government was willing to impose on the private sector the economic costs of routine disruptions in trade relations with the Soviet Union, other Western governments were not. "Lightswitch diplomacy" was not an attractive option to the more trade-dependent states of Western Europe and Japan, particularly in a market that rewarded sustained effort and long-term commitment.[6]

Second, even if all had been willing to bear the economic burden, Western states could probably not have agreed on the specific political objectives to be pursued via economic diplomacy. The United States and other CoCom governments clearly differed with regard to whether Soviet human rights abuses and Soviet intervention in the developing world were appropriate targets for Western economic intervention. The one case in which there was agreement serves as the exception that proves the rule. Seeking to avoid the alliance disarray that followed the invasion of Afghanistan, during 1980 and 1981, NATO members prepared a contingency plan, including concerted economic sanctions, in the event

6. Even within the United States, officials came to question whether tactical linkage was worth the cost. Revisions to the Export Administration Act in 1979 and 1985 sought to circumscribe the Executive's use of foreign policy controls. Late in 1984, Secretary of State Shultz publicly criticized the use of tactical linkage against the Soviet Union, stating that the United States "should not sacrifice long-term interests to express immediate outrage." *New York Times*, November 7, 1984, p. D1.

the Soviets invaded Poland.[7] When the Soviets failed to invade outright and relied instead on an internal Polish crackdown, the West was thrust once again into collective indecision.

Cases of tactical linkage shared a similarity with those of economic warfare: the lack of congruent interests among CoCom member states. Thus, effective cooperation similarly required a U.S. effort to extract the compliance of its allies. Again, U.S. officials were constrained by their broader alliance objectives. Tactical linkage cases typically involved political confrontations with the Soviet Union; as Secretary of State Haig recognized during the Polish crisis, sustained U.S. efforts to force other states to coordinate sanctions threatened to distract attention from objectionable Soviet behavior and focus it instead on intra-alliance conflict.

Will Western governments be more likely to pursue a concerted tactical linkage strategy during the 1990s? By 1991, the central issue had clearly emerged, namely, whether Western governments should provide large-scale economic assistance to facilitate economic reform in the former Soviet Union. That year, a well-connected U.S.-Soviet working group floated the idea of a "grand bargain"—a sustained and substantial Western aid package that was *conditional* on continued political pluralization in the Soviet Union and on a credible program for rapid transition to a market economy.[8] Architects of the grand bargain proposed that the United States, the European Community, and Japan share the costs, for three years, of an annual aid package of $15–20 billion.

Before the attempted Soviet coup of August 1991, the prospects for a collective Western strategy appeared bleak. Internal political struggles created uncertainty in the Soviet commitment to reform, and Western governments pursued divergent strategies. The adoption of a coordinated approach was a top priority for the Houston Summit of July 1990, but Western leaders failed to reach agreement. The United States was unwilling to grant aid, arguing publicly that it should be contingent on clear Soviet progress in restructuring the economy on the basis of free market principles, and privately that aid was infeasible as long as the Soviets continued to prop up client states such as Cuba. The West Germans were more receptive, calculating that economic aid might be critical in maintaining Soviet acquiescence in German unification plans. Indeed, the West German government had already offered the Soviets a $3 million bank credit but did not coordinate it with other Western

7. A good discussion is Denis Lacorne, "From Detente I to Detente II: A Comparison of Giscard's and Mitterrand's East-West Policies," in David Baldwin and Helen Milner, eds., *East-West Trade and the Atlantic Alliance* (New York: St. Martin's Press, 1990), pp. 118–44.

8. See Graham Allison and Robert Blackwill, "America's Stake in the Soviet Future," *Foreign Affairs* 70 (Summer 1991), 77–97, and Graham Allison and Grigiry Yavlinsky, "The Grand Bargain to Bring the Soviets to Market," *Washington Post Weekly*, July 15, 1991, p. 24.

governments. Japan resisted, linking the consideration of aid to *its* major bilateral concern, the willingness of the Soviets to relinquish their claim to Japanese territory seized at the end of World War II. France and Italy lined up behind West Germany in supporting aid, whereas Britain and Canada endorsed the U.S. and Japanese position.[9]

The stalemate persisted through the London Summit of July 1991. Although Soviet president Gorbachev managed a historic invitation to attend and plead the case for Western assistance, he came away empty-handed. Following the attempted coup in August, however, Western governments were more receptive to a coordinated strategy. The failed coup consolidated the position of Soviet reformers and also helped to expose more clearly the desperate nature of the Soviet economic predicament. Western governments subsequently accepted an unprecedented invitation from the Soviet leadership to become deeply involved in the reform process. The link between aid and reform was implicit, yet clear.[10]

With Western governments committed in principle to tactical linkage, the execution of a multilateral strategy required leadership. As in the case of the strategic embargo, the existence of common interests did not result automatically in effective coordination. As of late 1991, it was uncertain whether, and from where, leadership would emerge.

The United States, traditionally dominant in the coordination of economic denial strategies, was less well positioned to lead in the coordination of economic assistance. U.S. officials were constrained by persistent and sizable budget deficits and by concern over the domestic political repercussions of providing substantial aid to the Soviets while the U.S. economy experienced recession. Japan was more favorably positioned economically, yet it would need to resolve its territorial dispute with the Soviet Union and overcome its traditional reluctance to assert itself diplomatically. Although the German government proved eager to extend credits with unification pending, it became more cautious afterward, as it approached the daunting problem of reconstructing the former East German economy. Collective leadership was possible but would require Western governments to resolve the problem of burden sharing. Initial efforts were discouraging; late in 1991, a U.S. proposal for a delay in Soviet debt repayment created a conflict with Germany, whose banks were most heavily exposed in the Soviet market.[11]

Nevertheless, as of late 1991 there remained good reasons to anticipate the emergence of a coordinated Western strategy, assuming the

9. *New York Times*, July 9,1990, p. A1, and July 11, 1990, pp. A1, A5. The German offer reportedly involved a twelve-year credit, with a six-year grace period on repayment and interest below commercial rates. See *Wall Street Journal*, June 22, 1990, p. A5.
10. *New York Times*, Octobert 14, 1991, p. A1.
11. *Washington Post Weekly*, October 21, 1991, p. 7.

Soviet economic and political situation continued to deteriorate. The more desperate the Soviet predicament, the greater the incentives for Western governments to cooperate to avoid potential losses. The prospect of nuclear weapons dispersing to warring factions, civil wars spreading into central Europe, and refugees spilling across borders into Western Europe would probably concentrate the minds and resources of Western governments on a strategy that attempted to link economic assistance with progress in assuring political and economic reform.[12]

The potential for Western governments to maintain some commitment to structural linkage, although it was not pursued collectively in the past, similarly appears promising. Structural linkage calls on Western governments to promote, unconditionally, positive trade, financial, and investment relations with the former Soviet Union. It posits that such relations will foster and reinforce positive political and strategic developments, such as democratization and the commitment to transfer resources from the military to the civilian sectors of the economy. West European governments, the French and West German in particular, pursued structural linkage during the 1970s, for example, by subsidizing export credits. The United States endorsed structural linkage in rhetoric, although its actual policy, by mixing positive and negative sanctions as Soviet behavior changed, more closely reflected the logic of tactical linkage.

The current and future appeal of structural linkage rests on the fact that, whatever their particular national concerns, major Western governments have all committed to supporting and furthering ongoing reform efforts in the former Soviet Union and have recognized the intensification of economic relations as a meaningful instrument to accomplish that end. The fact that structural linkage complements, rather than conflicts with, the economic interests of Western states reinforces its appeal. Structural linkage is also less demanding, with regard to coordination, than are strategies that rely wholly or in part on negative sanctions. It does not require Western states to act in lockstep fashion in order to prevent the trade of one or more states from frustrating the sanctioning efforts of others.

A collective Western commitment to structural linkage—albeit loosely defined and coordinated—was articulated at the 1990 Houston Summit. In their final communiqué, Western governments emphasized that "political and economic freedoms are closely linked and mutually reinforcing" and that "each of us stands ready to help in practical ways those

12. During the spring of 1992, as this book went to press, the Bush administration announced that major Western countries had agreed on a $24 billion aid package for the former Soviet Union. Conflicts over sharing the burden, however, persisted. Japan accused the United States of making the plan public before an agreement actually had been reached on the size of each country's contribution. Germany later accused Japan of failing to meet its share of the obligation to assist. *New York Times*, May 6, 1992, p. A1.

countries that choose freedom, through the provision of constitutional, legal and economic know-how and through economic assistance, as appropriate."[13] As suggested above, the provision of large-scale financial assistance would probably emerge as part of a *tactical* linkage strategy, conditional on explicitly articulated steps to be taken by Russia and the other former republics. At the same time, Western governments have agreed to provide "technical" assistance without explicit conditions, on the assumption that such assistance, by its very nature, prods the Soviet Union in the direction of market reform and political democracy.

For example, the United States has pledged to provide teams of experts to help rebuild rail networks, food distribution systems, and other essential services. The British have sponsored privatization projects, and they assisted in setting up a Moscow commodity exchange. The European Community has focused on nuclear safety and business training.[14] The willingness of Western governments to encourage the participation of the former republics in the GATT and in the IMF similarly conforms to the logic of structural linkage by demonstrating a Western commitment to integrate the former Soviet economy into the world market economy.

The ability of Western governments to maintain and strengthen structural linkage depends on two factors. First, a significant political setback (e.g., domestic repression, a successful coup, or civil war among former republics) would likely induce the United States and perhaps other Western governments to terminate positive incentives and resort to negative sanctions. Second, the long-term effectiveness of structural linkage will depend on continued progress in relaxing CoCom controls and encouraging technology transfer. For example, by making available advanced telecommunications networks, the West could help to develop the infrastructure necessary for an advanced market economy and to attract foreign investment to the former Soviet Union. By transferring advanced manufacturing and quality control techniques, the West could help to ensure the success of conversion from military to civilian industry. If reform efforts take hold, Western governments will face a choice between maintaining national security export controls as an insurance policy or liberalizing in order to reinforce reform efforts. Put differently, the logic of structural linkage will directly confront that of the strategic embargo.

CoCom's Strategic Embargo, 1949–1989

In contrast to the fundamental conflict of interests among Western governments over the questions of whether export control policy should

13. *New York Times*, July 11, 1990, p. A4.
14. *Washington Post Weekly*, October 21, 1991, pp. 7–8.

be used to influence Soviet behavior or to weaken—or strengthen—
Soviet economic capabilities, this book has shown that there has long
been a Western consensus on the issue of whether exports with direct
military significance should be restricted in trade with the Soviet Union
and Eastern Europe. Western governments have agreed on the necessity
of a multilateral strategic embargo and have proved willing to participate
in an export control regime to facilitate the coordination of their national policies. To be sure, there have been differences in approach to, and
enthusiasm for, CoCom's strategic embargo. Among the most important
has been America's willingness to err on the side of caution and tolerate
more comprehensive controls versus the preference of other CoCom
members to accept greater strategic risk and to devise and implement a
control list that interferes minimally with the pursuit of economic interests. Nevertheless, it is critical to recognize that this difference and others have been played out in the context of an underlying consensus that
the strategic embargo, coordinated internationally, is a legitimate and
necessary means to protect national security. What might be called the
"hard core" of a common interest in restricting strategic trade with the
Soviet Union sustained cooperation in CoCom for four decades, from
1949 through 1989.

I have further argued that the existence of common interests does not,
by itself, assure effective cooperation. An effective strategic embargo
does not emerge spontaneously but requires conscious policy coordination. The key variable accounting for the effectiveness of that coordination has been the exercise of U.S. leadership.

The combination of America's preponderant and other CoCom members' contingent interest in strategic export controls and the informal,
discretionary nature of the CoCom regime have made it imperative that
the United States undertake a specific, definable set of leadership responsibilities. The United States has had to act as the guardian or conscience of CoCom, by preserving the integrity of the control process,
setting a domestic example, and maintaining the compliance of non-
CoCom suppliers of strategic items. At the same time, in deference to
the concerns of its trade-sensitive partners, it has had to minimize the
economic and administrative burdens of the control system. The extent
to which the United States has performed such tasks adequately has
been the key factor in determining both the effectiveness of participation in CoCom and the extent to which the regime has been a source of
controversy among its member states.[15]

15. In stressing the importance of leadership, I do not mean to imply that the CoCom regime
itself has been insignificant in sustaining and strengthening effective cooperation. The multilateral regime has mattered; it has led to more effective cooperation than would have been the
case if governments had relied on a series of bilateral arrangements. It reduced transactions
costs, conveyed information, and fostered a shared sense of expectations among its participants.
In general, see Robert Keohane, *After Hegemony: Cooperation and Discord in the World Political*

The chapters above suggest, however, that if the United States by necessity has been the "master of the CoCom game," it nevertheless has failed to master that game. In overall terms, America's postwar performance has been inconsistent and inadequate. It proved most inadequate during the 1970s, when the United States failed both to act consistently as the conscience of CoCom and to minimize the burdens of the control system. U.S. performance improved somewhat during the 1980s, as officials launched high-priority efforts to expand the control list, compel other CoCom members to strengthen enforcement, and extract the cooperation of non-CoCom suppliers. At the same time, however, U.S. officials badly neglected their responsibility to minimize the burdens of the system and succumbed to the temptation to politicize CoCom controls, thereby compromising the integrity of the control process.

The inadequacy of U.S. leadership was not confined to the 1970s and 1980s. In the years immediately following CoCom's shift in 1954 to a strategic embargo, U.S. officials similarly resisted the decontrol of less strategic items. They also compromised the integrity of the control process by insisting on the maintenance of a China differential, which made little sense strategically but was intended to placate conservatives at home and punish China for its role in the Korean War.

In relative terms, U.S. leadership was most effective in the years 1958–1968. The United States set a domestic example, preserved the coverage of the control list, and worked to maintain the cooperation of the key non-CoCom suppliers, Sweden and Switzerland. Moreover, by acquiescing in the far-reaching CoCom decontrol of 1958, despite retaining more extensive controls nationally, U.S. officials contributed significantly to relieving the economic and administrative burdens of the control system. Even in this period, however, U.S. performance was not unqualified. U.S. officials injected foreign policy considerations into the control process by encouraging a "Polish differential" for CoCom exceptions, and they increased burdens in CoCom by imposing reexport licensing requirements on items that could be exported by member states at national discretion. And, by 1967–1968, the relatively sharp increase in CoCom exceptions suggested that the decontrol of less strategic items was failing to keep pace with technological developments.

Inconsistent and inadequate U.S. leadership diminished the effectiveness of Western cooperation and created controversy in CoCom. This fact was apparent between 1954 and 1958, when the failure to decontrol

---

Economy (Princeton: Princeton University Press, 1984), and Kenneth A. Oye, ed., *Cooperation under Anarchy* (Princeton: Princeton University Press, 1986). An explanation stressing the functionality of regimes, however, would have difficulty accounting for variations in CoCom's effectiveness, in particular the weakening of effective cooperation in the 1970s and 1980s. All else being equal, a "regime" explanation would expect cooperation to remain stable or perhaps to strengthen over time as predictable channels of information became established and participants internalized norms and expectations.

less strategic items led to a loose interpretation of CoCom controls by member states, and the persistence of the China differential inspired a barrage of exceptions, and even defiance of controls, until it was abandoned in 1957. During the 1970s, the lack of U.S. vigilance influenced the approaches of other members, contributing to lax enforcement, demands for exceptions, and nationally convenient interpretations of CoCom's control parameters. These behavior patterns were reinforced and exacerbated by the U.S. failure to decontrol less strategic items and by suspicions that U.S. officials were manipulating CoCom to national commercial advantage.

During the 1980s, the partial recovery of U.S. leadership was responsible for the strengthening of CoCom. U.S. officials orchestrated the expansion of the control list and struggled to bring increasing numbers of non-CoCom suppliers into conformity with the regime. At the same time, by failing to decontrol less strategic items and by injecting foreign policy considerations into the control process, U.S. officials assured that CoCom remained controversial and that progress in strengthening controls was more difficult to realize. As long as the United States was asserting extraterritorial control, restricting widely available commodities, and keeping alive suspicions of its neomercantilist conspiracy, other governments could credibly resist measures to strengthen CoCom. U.S. behavior allowed attention in CoCom to be diverted from the inadequacy of member states' control and enforcement efforts and focused instead on the contentious and unilateral aspects of U.S. policy.

Clearly, U.S. officials have believed that the effective denial of strategic technology is a critical element in protecting U.S. national security. What, then, explains the failure of the United States to lead fully and consistently in CoCom?

The decline in U.S. power, defined in terms of material resources, provides a partial answer. Economic and technological capabilities clearly matter; as the experience of the 1960s demonstrated, the dominance enjoyed by the United States facilitated its task of leadership by making it easier to gain acceptance for control proposals in CoCom and to prevent non-CoCom suppliers from frustrating the intent of the embargo. These tasks became more challenging during the 1970s and 1980s as U.S. mastery diminished and the ability to develop and disseminate dual-use technologies spread to CoCom and non-CoCom states.

It would be a mistake to conclude, however, that the relative decline in power resources rendered the United States *incapable* of leading effectively in CoCom. During the 1970s and 1980s, the United States continued to possess sufficient resources in absolute terms to lead effectively. The Reagan administration's controversial third-country effort clearly demonstrates this point; numerous governments, including those of Sweden, Austria, India, South Korea, Singapore, Spain, and Australia, were forced to cooperate informally with the Western embargo or face the cutoff of U.S. sources of technology and equipment. Officials who absorbed the full force of U.S. pressure, such as those in Sweden, understandably tended to view the "decline of U.S. power" as an academic construct devoid of practical significance.[16]

U.S. officials retained ample sources of leverage in the CoCom context as well. The well-established dependence of many British, West German, and other West European firms on U.S. firms as sources of supply in computers and electronics and of Japanese firms on the vast U.S. market are but two of the most salient examples. The Reagan administration's military buildup provided additional sources of leverage, because military contracts, including those associated with the Strategic Defense Initiative, were awarded or withheld from U.S. allies in line with their performance in CoCom. In the words of Susan Strange: "There is little question about the combined structural power the United States derives from the security structure, the production structure, the credit (or financial) structure, and the knowledge structure. Neither Europe nor Japan can equal the Americans' performance across all four structures."[17] U.S. officials were not averse to utilizing such power during the 1980s, which did help to prod West European members of CoCom to accept the expansion of controls and Japan to tighten enforcement in the wake of the Toshiba incident.[18]

Moreover, a preponderance of economic power resources was not imperative for the execution of at least some leadership tasks. To minimize the burdens of the control system hardly required it; on the contrary, the relative decline of U.S. economic power should have facilitated this task by enhancing the sensitivity of executive officials to the economic costs of controls. Some evidence that officials increased their sensitivity was apparent by the late 1980s, when the Reagan administration scaled back intra-Western controls in the face of a substantial trade deficit and concern over the competitive decline of U.S. industry. Similarly, to maintain list coverage by pinpointing strategic technologies and making a credible case for their control in CoCom required defense intelligence resources—which the United States maintained in ample supply

16. The point was consistently made by Swedish officials at a conference in Stockholm on "economic security," which was sponsored by the Swedish National Board of Civilian Emergency Preparedness, January 1990. See also Gunnar Sjostedt, "Economic Security Policies of France and Sweden since 1970: Dilemmas of Cooperation," in Baldwin and Milner, eds., *East-West Trade and the Atlantic Alliance*, pp. 181–84.

17. Strange, "The Persistent Myth of Lost Hegemony," p. 571.

18. As I argued above, interests matter. The exercise of U.S. power helped to convince CoCom members to expand or strengthen controls they considered to be legitimate in principle. Where the complementarity of "core interests" was lacking, U.S. power resources were not sufficient to compel other members to accept U.S. preferences—in either the 1950s or the 1980s. The explanatory importance of the interplay of power and interests is recognized explicitly by Ethan Kapstein in his analysis of Western energy policy. See *The Insecure Alliance*, chap. 1.

during the 1970s and 1980s—more so than a preponderance of economic resources. According to West European officials, acceptance by CoCom members of U.S. proposals for list addition in the early 1980s was made easier by detailed U.S. intelligence presentations, which established a clear link between particular Western technologies and particular Soviet military systems.[19]

A more thorough explanation of America's leadership inadequacy has forced us to look beyond relative economic and technological capabilities and to consider the nature of the American state and the relationship between state and society in the United States.[20] As Henry Nau and others have recently emphasized, the U.S. political system and America's choices of purpose and policy have been critical determinants of its postwar influence and its ability to sustain cooperation in international regimes.[21] In the export control issue-area, the above chapters have documented three structural features that have set the United States apart from other CoCom members and have facilitated patterns of behavior which have constrained U.S. leadership in CoCom.

First, the U.S. Executive has possessed far-reaching, highly elastic authority to control U.S. exports for reasons of national security. That authority was granted at a time when Congress and the Executive anticipated an imminent war with the Soviet Union; it remained in place, and was creatively expanded by executive officials, long after the threat of war faded away. Controls were stretched with regard to destination, from East-West to intra-Western targets, and with regard to purpose, from weakening capabilities to influencing behavior. They were applied extraterritorially and, by the late 1980s, were expanded to cover imports into the U.S. market as well as exports. As the cold war has ended, this cluster of powers has been retrained on new targets, for new purposes, in the developing world.

Although powerful, the U.S. Executive has been divided and decentralized. Many agencies have been deeply involved in export control policy, yet no one agency has been in charge. At times the president has

19. Interviews, Paris, Bonn, London, June–July 1985.
20. This line of argument is consistent with Timothy J. McKeown's observation that international structural theories are marred unnecessarily when they neglect to incorporate domestic process variables. See "The Limitations of 'Structural' Theories of Commercial Policy," *International Organization* 40 (Winter 1986), 43–64. Stephan Haggard and Beth Simmons register a similar complaint about the regime literature. See "Theories of International Regimes," *International Organization* 41 (Summer 1987), 491–517.
21. Henry R. Nau, in a wide-ranging study of the Bretton Woods order, finds that U.S. hegemonic power and interest group politics "explain the ups and downs of American influence in the postwar period less persuasively than America's choices of national purpose and economic policy." See *The Myth of America's Decline: Leading the World Economy into the 1990s* (New York: Oxford University Press, 1990), p. 4. Similarly, Susan Strange argues that a "far more plausible explanation for the erosion of so-called international regimes than the decline in American hegemonic power lies within the American political system. . . . Stability in these regimes requires, above all, some consistency on the part of the leading participant." See "The Persistent Myth of Lost Hegemony," pp. 571–72.

been actively engaged; most often, he has not. The result has been a bureaucratic free-for-all and, as the National Academy of Sciences put it in 1991, "a disproportionate amount of bureaucratic resources are expended in resolving disputes, rather than administering and enforcing the export control system."[22] These conflicts were less intense during the first two decades of the cold war, yet they became more so and burst into public view during the 1970s and 1980s, as the cold war consensus collapsed.

Third, the book reveals that U.S. industry has been strikingly ineffective at influencing the essential character of export control policy. Corporate officials have failed to prevent the U.S. government from maintaining a more extensive control list and a more burdensome export licensing process than does any other CoCom member, or from resorting routinely and unilaterally to foreign policy controls. U.S. industry began at a disadvantage, given its own disinterest in Eastern markets and the viscerally negative public perception of East-West trade. It never really caught up. By the 1970s and 1980s, U.S. firms were less tolerant of export controls but were unable to reverse the course of an entrenched system of laws, practices, and institutions that continued to operate according to the cold war logic of the 1950s. The system, in effect, had become too powerful for industry to transform. The most they could manage was to temper its more destructive aspects, for example, by working to circumscribe foreign policy controls or to scale back licensing requirements for intra-Western trade.

Taken together, these features have facilitated patterns of policy behavior at cross-purposes with the ability of the United States to lead effectively in CoCom. For example, fragmentation led to inconsistent and unpredictable policies in the 1970s and 1980s, making it difficult for U.S. officials to maintain the integrity of the control process and to set a domestic example. It also facilitated the rise of the Defense Department, with the attendant problems for U.S. leadership and CoCom described in Chapter 8. The power of the Executive, coupled with the deference of industry, bred the sanctions habit and with it the failure of the United States to set a domestic example. That same combination helped to render the United States insensitive to the economic costs of controls, which cut against its ability to minimize the burdens of the control system. Finally, the extraterritorial application of controls—a direct extension of the Executive's delegated authority—proved to be a constant source of irritation in CoCom and furthered the image of the United States as a coercive rather than a benevolent hegemonic state.[23]

22. National Academy of Sciences, *Finding Common Ground: U.S. Export Controls in a Changed Global Environment* (Washington, D.C.: National Academy Press, 1991), p. 87. The NAS was unable to locate an analogous policy area with a comparable number of involved bureaucracies, regulatory statutes, and overlapping jurisdiction. See ibid.
23. Duncan Snidal emphasizes the distinction between coercive and benevolent leadership in

At times, the United States has managed to override its dominant behavioral tendency and lead effectively. As I suggest in the book, to overcome the system's inertia and break bureaucratic stalemates has required high-level intervention. In 1956, President Eisenhower became personally involved in resolving the destructive CoCom conflict over the China differential, at one point arguing in the NSC that Japan's choice was to trade with Communist China or "pass a tin cup around in San Francisco."[24] At the president's urging, the NSC subsequently decided to acquiesce in the major CoCom list reduction of 1958, which ushered in a period of stability for CoCom. After the Soviet invasion of Afghanistan, high-level intervention was needed to suspend interagency struggles and launch an effort to revive CoCom. When such attention lapsed, the system returned to business as usual. As discussed below, President Bush had to intervene personally in 1990 to shock an immobilized bureaucracy to the realization that CoCom would collapse in the absence of dramatic U.S. initiatives.[25]

CoCom and the East-West Military Competition

In Chapter 1, I distinguished the effectiveness of CoCom, defined in terms of the nature and extent of cooperation among its participants, from the impact of CoCom controls on East-West military competition. As the cold war ends and we look back over the regime's forty-year history, how and to what extent did CoCom matter?

The distinction between economic warfare and the strategic embargo is useful in placing the contribution of CoCom to the West's victory in the cold war in perspective. With the possible exception of the early 1950s, CoCom was ineffective at retarding the economic growth of the Soviet Union or the amount of resources it devoted to military pursuits. As U.S. officials conceded in the early 1960s, after 1954, unilateral U.S. controls could do little to impede Soviet economic growth and development.

During the early 1980s, multilateral economic warfare—the denial of credits, grain, and turnkey factories—would probably have hastened the already precipitous Soviet economic decline. Western states, however, declined to carry out the strategy. To be sure, the strengthening of CoCom's dual-use controls during that period, though directed at the Soviet military, probably had some impact on the Soviet economy, partic-

---

"The Limits of Hegemonic Stability Theory," *International Organization* 39 (Autumn 1985), 579–614.

24. See memorandum at the 282d meeting of the National Security Council, April 26, 1956, reprinted in *Foreign Relations of the United States, 1955–57*, vol. 10, p. 350.

25. The National Academy of Sciences has recognized the importance of presidential leadership, calling for the president to provide an explicit national security directive to guide all future national security export controls and their implementation. See *Finding Common Ground*, pp. 139–40.

ularly in electronics and communications. Nonetheless, given the relative autarky of the Soviet economy and its inability to absorb and diffuse new technology, it is hard to dispute that the collapse of the Soviet economy had far more to do with internal factors—the exhaustion of an inefficient extensive growth strategy—than with Western export controls.[26]

CoCom controls played a far more important role where the strategic embargo was concerned: they helped in maintaining U.S. lead time and in preventing the Soviet Union from attaining technological parity in military systems. It is difficult to be precise; information on the impact of Western technology on the Soviet defense sector traditionally has been limited, and some of what was made available during the early 1980s was probably exaggerated as part of a U.S.-led campaign to mobilize Western publics for more restrictive controls.[27] Moreover, a host of other factors, including the efficiency of the U.S. and Soviet military procurement systems and the effectiveness of espionage efforts, has affected the ability of the United States to maintain technological lead time in weapons systems. Given limited and incomplete information, it is difficult to untangle the impact of export controls from that of other factors.[28]

With these caveats in mind, available evidence does indicate that CoCom mattered. U.S. intelligence assessments of the 1960s, cited in Chapter 4, documented lagging areas of Soviet military development that had been directly affected by CoCom controls. Alternatively, U.S. intelligence studies, the Farewell documents, and the investigations following the Toshiba-Kongsberg diversion suggest that the Soviets, at least in some areas, made significant progress in reducing U.S. lead time through illegal acquisitions during the period of CoCom weakness in the 1970s and early 1980s. In microelectronics, for example, U.S. sources

26. Two of the best sources on the impact of Western technology on the Soviet economy are Philip Hanson, *Trade and Technology in Soviet-Western Relations* (New York: Columbia University Press, 1982), and Bruce Parrott, ed., *Trade, Technology, and Soviet-American Relations* (Bloomington: Indiana University Press, 1985).

27. An excellent effort to unravel these problems is Julian Cooper, "Western Technology and the Soviet Defense Industry," in Parrott, ed., *Trade, Technology, and Soviet-American Relations*. Cooper finds that although U.S. defense and intelligence officials did tend to overstate the military significance of Soviet acquisitions during the early 1980s, on a selective basis the Soviets have successfully exploited Western technology to accelerate their development of new weapons.

28. According to the U.S. Department of Defense, the Soviet Union administered two acquisition programs during the 1970s and 1980s. One involved espionage and overt collection by bloc officials of one-of-a-kind military hardware, blueprints, and product samples for reverse-engineering purposes. The second was a trade diversion program to acquire large numbers of dual-use manufacturing equipment for direct use in Soviet military production lines. The first is a counterintelligence problem; the second is an export control problem. As the characteristics and collection channels of the two programs have overlapped, the DoD noted the difficulty of untangling the precise impact of each or of devising adequate countermeasures. See U.S. Department of Defense, "Soviet Acquisition of Militarily Significant Western Technology: An Update" (Washington, D.C., September 1985).

and the Farewell documents found that the Soviets acquired more than twenty-five hundred pieces of major fabrication and test equipment during that period, for direct use in Soviet military production lines and in some cases to develop and fully equip new production lines. According to the DoD, "the equipment acquired through these efforts is largely responsible for the significant advances the Soviet microelectronics industry has made thus far, advances that have reduced the overall lead in microelectronics from 10 to 12 years in the mid-1970s to four to six years [in the mid-1980s]."[29]

By the early 1990s, there was evidence indicating that the U.S. lead over the Soviets in militarily significant technologies had once again widened, particularly in areas subject to tighter controls, such as computers, software, and microelectronics manufacturing and test equipment.[30] Given the collapse of the Soviet economy and the disruptions introduced by *perestroika*, however, one cannot easily infer the extent to which the strengthening of CoCom controls explains that outcome.

Indirect evidence lends further support to the argument that CoCom mattered. The simple fact that the Soviets organized, across their own government and those of Eastern Europe, a systematic, well-funded campaign to acquire militarily significant Western technologies through illicit trade diversion testifies to the importance of the multilateral control program. The same applies to the increase in Soviet complaints and diplomatic pressure on both CoCom members and non-CoCom states as the strategic embargo was strengthened during the 1980s.

It is important to recognize that CoCom did not, and could not, prevent the Soviet Union from being a great military power. Yet, as the arms race became more technology-intensive and as sophisticated weaponry came to rely more on the combined efforts of military and civilian industry, CoCom controls accentuated the natural advantage of the West over the Soviet Union by preventing the Soviet military from reaping the benefits that the Western, but not the Soviet, civilian sector could provide. The cold war would have begun, endured, and ended without CoCom. Nevertheless, CoCom was a useful weapon in its conduct. And, as I emphasized in the chapters above, the existence of the regime helped to insulate national security export controls from the liberal trading order that provided the foundation for Western military strength.

29. Ibid., pp. 24–25. An annex to the report lists several hundred examples of Soviet military equipment that benefited from Western technology, acquired either through espionage or trade diversion (pp. 31–34).

30. National Academy of Sciences, *Finding Common Ground*, pp. 27–32. In 1986, the United States was superior to the Soviet Union in thirteen of twenty militarily significant technology areas, and the two were equal in the remaining seven. By mid-1990, the United States was superior in fifteen areas, the Soviets were superior in two (conventional warheads and armor materials), and the two sides were equal in three.

The Strategic Embargo and the 1990s

CoCom entered its fifth decade in uncertainty. The international environment that had provided the support and justification for its existence since 1949 was shaken to its foundations by the revolutionary events of 1989–1990. To appreciate fully the impact of those events, it is useful to review the fundamental tenets upon which CoCom's strategic embargo has been based, and how those were challenged or altered at the end of the 1980s. A set of six interlocking assumptions regarding trade, Soviet military power, and East-West relations provided the framework and the rationale for the West's forty-year strategic embargo effort.

The first assumption was that the Soviet Union posed a serious military threat because it possessed the capabilities and motives to engage the Western allies in costly conflicts in Europe, the Pacific, and (of more significance to the United States than to other CoCom members) in parts of the developing world. Second, given the East-West military competition and the technological inferiority of the Soviet Union vis-à-vis the West, the likelihood was strong that advanced dual-use technologies and equipment, if exported to the Soviets by Western suppliers, would find their way into Soviet military systems and make an important contribution to Soviet military capabilities. Third, the military threat was not one of imminent war but involved an ongoing competition, of varying intensity and indefinite duration, in the development and deployment of military weapons. Fourth, the key to war avoidance was deterrence, which required the United States and the West to maintain a qualitative lead in the application of advanced technology to military systems in order to offset the quantitative advantages of the Soviet Union and Warsaw Pact. Fifth, the very existence of the Warsaw Pact, a military alliance reinforced by ideological and economic ties, demanded that the strategic embargo be extended to cover Eastern Europe as well as the Soviet Union. Exports of controlled technology might find their way from the civilian to the military sectors of East European states and, given the close links between military forces and intelligence agencies, to the Soviet military as well. Finally, the East-West military competition was reinforced and complemented by an ideological competition, the existence of which made it easier politically to justify economic discrimination against the East and the economic sacrifices thus entailed for the West.

By the end of the 1980s, each of these assumptions had been seriously challenged, if not completely discredited. First, although the Soviet Union continued to possess sufficient capabilities to threaten the West militarily, evidence abounded of Soviet willingness to carry out significant reductions in those capabilities. By 1991, Soviet defense spending was at least flat and possibly in decline. In conventional arms talks, Soviet negotiators accepted deep, asymmetrical reductions, which if carried out would eliminate the quantitative superiority that has been NATO's obsession since the 1950s. Soviet officials also made important concessions to complete the START treaty in 1991, and shortly thereafter they matched the United States in announcing unilateral reductions in tactical nuclear weapons. The Soviets carried out a phased withdrawal from Afghanistan and adopted plans to reduce forces in preparation for a complete withdrawal from Eastern Europe.

In addition to affecting Soviet capabilities, these and other initiatives obviously influenced Western assessments of Soviet intentions. At no other time in the postwar era were Western governments seemingly *less* concerned by the prospect of Soviet military aggression in Europe, or adventurism globally, than they were by 1990. If more evidence was needed, the Soviets dutifully fell in line behind the United States in the Persian Gulf crisis of 1990, cutting off weapons to its longtime ally Iraq, joining the West in supporting comprehensive economic sanctions, and voting for every United Nations resolution condemning Iraq and demanding its withdrawal from Kuwait.

Second, reductions in military spending and Gorbachev's ongoing effort to liberalize the Soviet economy called into question the assumption that the West's dual-use technology would significantly aid the Soviet military sector. To be sure, the available evidence indicated that the Soviet Union continued its acquisition efforts in 1990 and 1991, although without the assistance of their former East European allies.[31] The importance of such evidence was inevitably viewed, however, in light of the broader changes taking place, in particular the Soviet's desperate need for Western technology to facilitate economic reform and Gorbachev's commitment to defense conversion, or the transformation of military to civilian production. The latter was especially disorienting to Western export controllers, because underlying the strategic embargo was the notion that military production enjoyed a priority claim on the material, technological, and human resources generated by the Soviet Union, indigenously or from abroad. Evidence that some Soviet assembly lines were converting from tank to refrigerator production, or from missile parts to sauce pans, strengthened the hand of those in the West who believed that the risks of diversion were decreasing and worth taking in an effort to lock in desirable changes in Soviet domestic and foreign policy through the transfer of Western technology.[32]

Third, it was clear by 1990 that for domestic political and economic

31. Ibid., pp. 31–32.
32. On Soviet conversion, see Alexei Izyumov, "The Other Side of Disarmament," *International Affairs* (Moscow), no. 5 (1988), and "Conversion of the Soviet Military Industry in an Era of Arms Reductions," unpublished paper, July 1991. See also Ethan Kapstein, "From Guns to Butter in the U.S.S.R.," *Challenge* (September-October 1989), 11–15.

reasons, the Soviet leadership wished to move (in the language of Chapter 2) from stable to declining military competition between East and West and perhaps to eliminate that competition altogether. NATO governments, meeting in July 1990, issued a "London Declaration," asserting that the Warsaw Pact and NATO were no longer adversaries; they invited Gorbachev and other Warsaw Pact leaders to visit NATO headquarters and to establish regular diplomatic ties with NATO.[33] The implications for CoCom were obvious: should a strategic embargo be maintained if Eastern and Western alliance members declared an end to the cold war and no longer considered themselves military adversaries?

Fourth, the chain of reasoning linking war avoidance with deterrence, deterrence with the maintenance of the West's qualitative advantage, and qualitative advantage with an effective strategic embargo was similarly called into question. For the first time since the onset of the cold war, the fundamental issue of whether it was necessary to devote significant Western resources to deter Soviet military aggression could be raised credibly by responsible authorities. And even if the response was "yes, for now," one could further question the overriding importance of maintaining *qualitative* lead time, in the face of evidence that the Soviet government was prepared to negotiate away the *quantitative* advantages it had long enjoyed in the European theater. Could the marginal contribution of the strategic embargo to the West's qualitative lead time—at a time when the latter itself was arguably receding in importance as a component of Western security—justify the maintenance of export controls in light of the economic costs, intra-Western diplomatic friction, and foregone opportunities to assist Eastern reform that such controls entailed?

Fifth, the notion that Eastern Europe should be viewed as an appendage of the Soviet Union for export control purposes was rendered obsolete by the dramatic collapse of Communist power across the region in 1989 and the subsequent determination of the new regimes to distance themselves from Moscow politically, militarily, economically, and ideologically. Even before the collapse, Hungarian and other East European officials were lobbying U.S. officials for the right to receive controlled technology on liberal terms in exchange for official commitments to safeguard that technology from diversion to the Soviet Union.[34] During 1989, U.S. officials considered such an arrangement implausible; during 1990, they accepted it as part of far-reaching plan to reform CoCom. And, in 1991, CoCom members were forced to go even further and contemplate the export control implications of trade with the newly independent Baltic states, whose claims to sovereignty were recognized in the aftermath of the failed Soviet coup.[35]

33. See Peter Hayes, ed., "Chronology 1990," *Foreign Affairs* 70 (1990–1991), 224.
34. See "Hungarians Press U.S. for CoCom Rules Waiver," *Financial Times*, November 30, 1989.
35. See *Export Control News* 5 (September 25, 1991), 2–3.

Finally, the ideological struggle between East and West, which had reinforced the need for both military deterrence and the strategic embargo, virtually disappeared by 1990. As was apparent not only in Eastern Europe and much of the developing world but also within even the Soviet Union, Communism had lost whatever appeal it previously enjoyed as an ideological force. The Soviet Communist party struggled to retain both a political and an economic role in the face of pressure to move toward a multiparty polity and a market-oriented economy and under the even more ominous prospect, realized shortly thereafter, of the complete disintegration of the Soviet Union.

Taken together, these changes raise the obvious question of whether CoCom can survive. The next section examines the attempt by member governments to assure the regime's survival in the short run; the subsequent section explores CoCom's longer-term prospects.

Preserving CoCom: The High Level Meeting of 1990

As I noted in Chapter 8, during 1989 the United States took a business-as-usual approach to CoCom, despite the changes taking place in the Soviet Union and Eastern Europe. U.S. officials remained cautious with regard to streamlining, warding off proposals by other members to expedite that process either through the automatic downgrading of controlled items over time or by relaxing CoCom's unanimity principle and thereby removing America's veto. On machine tool liberalization, which many members cited publicly as a litmus test of America's willingness to streamline, U.S. officials stalled and argued that liberalization should be driven not only by foreign availability but also by the commitment of members to strengthen compliance with existing controls. The U.S. delegation stressed at the October 1989 high level meeting that despite Gorbachev's "new thinking," the Soviet military continued to target Western machine tools for illegal acquisition and was utilizing them in the production of a variety of military systems.[36] Similar caution was exhibited on the question of liberalizing controls differentially for Hungary and Poland; although each had made dramatic progress in political and economic reforms, U.S. officials contended that liberalization would be risky in light of their continued membership in the Warsaw Pact and their cooperation with the Soviets in the illicit acquisition of controlled technology.

Other CoCom members grew increasingly impatient during 1989 and were placated only temporarily by America's lifting of the no-exceptions policy in May. Most were disappointed by the slow progress made on

36. To demonstrate the depth of U.S. concern, President Bush sent personal letters to the leaders of other CoCom governments in September, urging them to redouble enforcement efforts. See *New York Times*, October 9, 1989, p. D1.

streamlining and attributed it to the combination of U.S. intransigence and the tedious incrementalism of CoCom's item-by-item review process. The failure to liberalize machine tool controls was especially troubling, particularly because West Europeans perceived this area as one in which they enjoyed a comparative advantage and therefore resented having their market access obstructed by the U.S. government. Although the frustration of CoCom members did not translate into open defiance of multilateral rules, signs of discontent were apparent. The Belgian government pushed the Alcatel deal and the British the Simon-Carves deal, despite strong U.S. objections at the highest levels.[37] According to U.S. officials, one CoCom member floated in defiance the idea of reserving the right not to implement the results of a list review segment if the review did not produce significant streamlining. The British government adopted its own scheme for more liberally licensing intra-CoCom trade, rather than await the emergence of a multilateral consensus. Finally, in December 1989 the West German Social Democrats introduced a resolution calling for West Germany to pull out of CoCom. The resolution was defeated by the majority coalition but clearly reflected the depth of anti-CoCom sentiment in a key member state.[38]

The pressure for relaxed controls was not new, and the United States may have been able to finesse it once again were it not for the dramatic transformation in Eastern Europe and of East-West relations. In these circumstances, a major streamlining of CoCom controls was required, at a minimum, to maintain the credibility of the regime.

A turning point was reached in February 1990. At a meeting of CoCom's Executive Committee, the United States sought to appease its partners by proposing that CoCom approve license requests up to the China Green Line for Poland, Hungary, and Czechoslovakia. West European members rejected this plan as insufficient; led by West Germany, they argued for the complete decontrol of items up to the Green Line, for the Soviet Union as well as for Eastern Europe.[39] The meeting ended without a resolution, and officials on both sides of the Atlantic began to voice concerns publicly that in the absence of a major breakthrough,

37. See "Allies Defy U.S. on High-Tech Sales to the Soviets," *Wall Street Journal*, May 3, 1989, p. A18. The Simon-Carves case involved a $450 million deal to build a plant in Soviet Armenia that would produce microcomputers and factory automation equipment. The British government refused to submit a general exception request to CoCom, arguing that the technology involved could be licensed at national discretion. The Alcatel deal is described in Chap. 1.

38. *Export Control News* 3 (May 27, 1989), 9, and the interview with Paul Freedenberg in ibid. 4 (March 31, 1990), 7. See also Wolfgang H. Reinecke, "Political and Economic Changes in the Eastern Bloc and Their Implications for CoCom: West German and European Community Perspectives," study prepared for the National Academy of Sciences' Panel on the Future Design and Implementation of National Security Export Controls, May 29, 1990, pp. 68–69. Reinecke quotes one SPD member's sentiment that the United States wished "to replace the Iron curtain with a CoCom curtain" (p. 69).

39. See *Export Control News* 4 (February 28, 1990), 2.

cooperation in CoCom might unravel altogether. Former Commerce undersecretary Paul Freedenberg openly warned that members might begin to ignore CoCom directives and devise and administer their own national controls as an alternative. West European officials echoed this concern, stressing that it was up to the United States to take the initiative and demonstrate to an increasingly skeptical Europe that CoCom was capable of adjusting itself to a new strategic environment.[40] The skepticism was greatest in West Germany, where CoCom was perceived as an obstacle to German security, one that stood in the way of urgently needed efforts to reduce pollution and enhance nuclear safety in neighboring Eastern Europe and the former East Germany.[41]

Given the pace of events in Central Europe, the longer the United States delayed, the more it would take to satisfy other members and restore the credibility of CoCom. Unfortunately, the interagency struggle between Commerce and Defense persisted and, in the absence of high-level intervention, would probably have rendered the United States incapable of an adequate response. The gravity of the situation was recognized, however, and three meetings of the NSC, two with President Bush himself, were devoted to developing a U.S. position in anticipation of the June high level meeting.[42] Officials engaged in the process remarked that they had never seen the export control bureaucracy—under direct orders from the president—move as quickly as it did to resolve disputes. The U.S. position was publicly announced in May and provided the framework for the drastic revisions in CoCom's strategic embargo adopted in June.

The June 1990 high level meeting was widely regarded as "the most significant in the history of export controls."[43] Although the accuracy of that judgment remains to be seen, the decisions taken were clearly profound and, at the very least, of a similar magnitude to the CoCom revisions of 1954 and 1958. Member states undertook four major initiatives. First, they agreed to override the list review process and immediately delete from CoCom's industrial list 30 of its 116 categories. This change was intended to have both a symbolic and a substantive impact; by taking decisive action at the political level, CoCom members sought to demonstrate that they were not bound to a tedious streamlining process which seemed to produce two backward steps for every forward one.

40. Freedenberg interview, and interview with Kurt Steves of the Federation of West German Industries in *Export Control News* 4 (April 30, 1990), 6–7.
41. Reinecke, "Political and Economic Changes in the Eastern Bloc," pp. 36–39. He reports that German residents were outraged as public debate revealed that CoCom was restricting technologies which could help to minimize nuclear accidents across the border in Eastern Europe.
42. Interview with U.S. ambassador Alan Wendt, in *Export Control News* 4 (June 30, 1990), 4–8.
43. See ibid., pp. 2–9, and Lionel Olmer, "CoCom Adopts Strategic Trade Rules," *Bulletin*, Atlantic Council of the United States 1 (July 19, 1990).

Second, CoCom members liberalized controls in the three contentious sectors of telecommunications, computers, and machine tools. In the latter two, U.S. officials finally acceded to the long-standing demands of its allies for far-reaching liberalization.[44] The new machine tool parameters substantially increased the level of accuracy and general sophistication of machines that could be exported under general license. According to the Commerce Department, that category included 75 percent of the machines produced in the United States.[45] The machine tool concession meant that the United States finally passed the litmus test; it also helped to disarm the accusation that U.S. officials were prepared to liberalize only in areas in which U.S. firms enjoyed a competitive edge.

U.S. officials proved less accommodating in telecommunications. Although agreeing to some liberalization, they refused to decontrol advanced switching devices and high-speed fiber optics. The Bush administration signaled its resolve by vetoing a proposal by US West and its West European partners to lay a high-speed fiber-optic cable across the Soviet Union, the grounds being that such a cable would considerably enhance Soviet military capabilities and would frustrate the ability of U.S. intelligence to monitor Soviet communications. U.S. officials indicated that they would seek to prevent other governments, explicitly mentioning that of South Korea, from allowing their firms to replace the sale. The British government supported the U.S. position and announced that it would deny a similar license request from a British firm.[46]

Third, CoCom members accepted a U.S. and British proposal to develop a highly selective "core list" to replace CoCom's existing Industrial List. Members agreed to begin from scratch, placing the burden of proof on those seeking to justify candidates for inclusion on a new list rather than on those seeking to remove items from the existing list. The very idea of controlling a short list of only the most sensitive items was not new. It had been the "ideal" preference of West European governments in CoCom, particularly the French, for at least a decade. It was also endorsed in the Bucy Report, although it obviously never found its way into U.S. or multilateral policy. By the middle of 1990, the United States was championing the idea as a necessary, drastic step to salvage CoCom.

---

44. In computers, systems with a PDR (processing data rate) of 275 megabits/second (Mbits./sec.) were decontrolled, up significantly from the prior threshold of 69 Mbits./sec. For the first time, advanced personal computers (e.g., the IBM PS/2) and computer systems or other items based on microprocessors standard in the West (e.g., Intel's 80386) could be shipped to the East without a validated license. Members also agreed to license, at national discretion, computers with a PDR of up to 550 Mbits./sec. (the current China Green Line level) and to grant favorable consideration to systems up to 1,000 Mbits./sec. The latter was more than twelve times greater than the existing level, which meant that powerful mainframes, previously denied without question, now enjoyed a presumption of license approval. See *Export Control News* 4 (June 30, 1990), 2–3.

45. Ibid., 4.

46. Ibid., 8–9.

U.S. officials did indicate that they expected improved, uniformly effective enforcement to accompany the core list and at the June meeting convinced CoCom members to advance the date for implementation of the "common standard of effective protection" from December to April 1991.[47]

The core list was ratified by CoCom members in May 1991. U.S. officials proclaimed it as an additional 50 percent reduction in controls, following the liberalization of June 1990. Industry officials and analysts were nonetheless cautious, in part because the U.S. government had insisted on the retention of strict controls on telecommunications and computer networking in exchange for its acquiescence.[48]

Fourth, CoCom members adopted plans to transform the countries of Eastern Europe from targets of the embargo to collaborators in its enforcement. The West German government agreed to establish, upon unification, one German export control system, with controls enforced at the former East German border. The new Federal Republic of Germany would remain a member of CoCom.[49] Three other East European countries—Poland, Hungary, and Czechoslovakia—were offered a special relationship with CoCom. Their newly elected governments were encouraged to adopt export control systems, including IC/DV procedures, end-use checks, and on-site verification, to safeguard Western technology from diversion to the Soviet Union. In exchange, they would receive favorable treatment from CoCom—a presumption of approval for most items on the core list. Teams of American export control advisers visited the East European capitals immediately after the June meeting. By early 1991, the three countries had each established CoCom-compatible systems and were granted the promised trade benefits.[50]

The steps taken by CoCom in 1990 were similar to those taken almost thirty-six years earlier, in 1954. In both cases, members sought to adapt the embargo to what was perceived as a new strategic environment. In 1954, they sought to adjust from an atmosphere of intensified military competition and the threat of war to one of a more stable competition of indefinite duration, and, in 1990, from an environment of stable to one of declining military competition and to the end of the cold war. In both cases the control list was significantly reduced to render it more credible and enforceable. And in both, the United States was forced to com-

47. Ibid., 4. The deadline for each member to adopt uniform enforcement standards was subsequently pushed back to January 1992.
48. See Bill Root, "The Core List: Expectations vs. Reality," *Export Control News* 5 (July 29, 1991), 2–5.
49. See Reinecke, "Political and Economic Changes in the Eastern Bloc and Their Implications for CoCom," pp. 25–26. The two Germanies were officially unified on October 3, 1990.
50. *Export Control News* 5 (March 26, 1991), 9–10.

promise as a result of considerable pressure from its principal CoCom allies.

The key difference is that in 1954, the effectiveness of CoCom, not its existence, was in doubt. In 1990, both the effectiveness and, more important, the very survival of CoCom were called into question. It required a major crisis—the genuine prospect of regime collapse—to prompt the United States to resolve its internal conflicts and finally accept the necessity of a significant streamlining of the CoCom list. Crisis inspired effective leadership.

Does CoCom Have a Future?

The adoption of the core list assured CoCom's short-term survival. It alleviated the immediate crisis and helped to divert high-level attention in Western Europe away from CoCom and to other pressing matters, such as the reconstruction of the former East Germany and the economic collapse of the Soviet Union. Streamlining also made it easier for CoCom members to take the domestic steps necessary to meet their enforcement objective, the adoption of a common standard of effective protection. That standard was, in turn, a requirement for the full liberalization of controls on intra-CoCom trade, including U.S. reexport controls. As of late 1991, it appeared that the common standard would be adopted and that, in response, the United States would phase out its reexport controls and other impediments to intra-CoCom trade.[51]

Looking beyond the immediate future, however, is it plausible to expect CoCom to endure in a post–cold war environment? CoCom's demise could take several forms: its complete collapse as an export control institution; the maintenance of CoCom controls over munitions and atomic energy items, but the elimination of controls on dual-use items; or, the preservation of CoCom, but with a shift in mission and perhaps membership (for example, Russia joins CoCom, and the embargo is reoriented to target the flow of technologies of mass destruction to developing countries). Any of the above would constitute a regime change, because the purpose of CoCom cooperation in the postwar era has been the denial of munitions, atomic energy items, and dual-use technologies and equipment, with the Soviet Union as the principal target.

Social science prediction is a notoriously precarious endeavor and one worthy of caution, especially in the context of a rapidly changing international environment.[52] Short of prediction, it may be useful to isolate

51. Ibid. 5 (June 28, 1991), 6, and 5 (July 29, 1991), 21. All CoCom members, with the possible exception of Italy, appeared ready to adopt the common standard by January 1, 1992. Italy's proposed legislation stalled during 1991.

52. I was bolder in a previous stab at this question, arguing that CoCom would probably

and examine the factors likely to bear significantly on the question of CoCom's survival. Three factors—the interests of the United States, the interests of other CoCom members, and U.S. leadership—have emerged in this book as critical and are apt to remain so in the foreseeable future.

Assuming that the United States maintains an interest in the strategic embargo, the preservation of cooperation in CoCom will depend on the preferences of other CoCom members and U.S. leadership, with the former taking on greater importance than the latter. If West European governments no longer prefer to participate, the United States will be unable to sustain cooperation, even if it attempts to do so coercively. At the same time, if those governments remain committed in principle to the strategic embargo, then the survival of CoCom will depend on the deftness of U.S. leadership. In the past, U.S. leadership was required to ensure the effectiveness of cooperation; in the future, it will be needed to sustain cooperation itself, and only if other members remain in principle willing to participate.

Will the United States itself prefer to maintain the strategic embargo in a post–cold war world? The answer is yes, but only as long as U.S. officials perceive the former Soviet Union and in particular Russia to be a principal military competitor, one with the capability to pose a serious threat to U.S. territorial integrity or global interests. For U.S. officials, postwar export controls have been intended to enhance U.S. security, not just that of NATO; even if NATO collapses, as long as some level of military threat from Russia is perceived, the United States will be reluctant to abandon what it has considered a valuable strategic asset in its relations with a technologically inferior, yet militarily formidable great power.

At the very least, U.S. officials will probably prefer as an "insurance policy" the retention of a streamlined strategic embargo to protect against a possible reversal of Russian foreign policy or a resumption of the arms race. Even the deep, mutual reductions in nuclear and conventional arms agreed to in 1991 may not eliminate that preference; on the contrary, at lower force levels, U.S. officials may be even more inclined to protect technological lead time, because qualitative advantages will take on greater marginal significance as the quantitative resources devoted to Russian and U.S. defense are diminished.

Institutional factors may also play a role. During the 1980s, the U.S. defense and intelligence community developed a considerable stake in the export control issue, and it is likely to behave as would most bureaucracies and defend that stake, even as the substantive rationale for controls is undermined. Export controls are sometimes described as the

---

survive the 1990s but would come under increasing strain. See "Technologische Revolution und die Ost-West Beziehungen," *Europaische Rundschau* 2 (1989), 49–58.

"abortion issue" of the U.S. defense community, suggesting that officials hold deep-seated convictions and defend them tenaciously, seemingly impervious to countervailing evidence. To be sure, high-level political intervention can impose change on even the most resistant officials and institutions, as was demonstrated by the U.S. reversal on streamlining in 1990.[53] Not to be counted out is the possibility that U.S. leaders will go further and in the near future abandon the strategic embargo altogether in exchange for Russian cooperation on issues of vital national security concern. In the absence of such high-level initiative, however, it is likely that defenders of the "reformed" strategic embargo within the U.S. government will be able to hold their own against the traditionally weaker bureaucratic supporters of U.S. economic interests in trade with the former Soviet Union.

It is less probable that other CoCom members will similarly prefer to maintain the strategic embargo. For West European states, the main strategic justification for CoCom has been the threat of war and, specifically, a direct Soviet attack on the European continent. That threat virtually disappeared by 1991, with the unification of Germany on Western terms, the collapse of the Warsaw Pact and subsequently of the Soviet Union, the scheduled removal of Soviet forces from Eastern Europe, and the agreement to reduce conventional forces in Europe significantly. In the absence of a perceived, direct military threat, there is little to sustain Western Europe's interest in the strategic embargo. Unlike the United States, other CoCom members lack the institutional imperatives or the sensitivity to great power competition that might serve as substitute justifications for the maintenance of controls.

Equally important, to abandon the strategic embargo would be politically attractive in Western Europe as an opportunity to take a step of symbolic and substantive importance in the integration of the former Soviet Union into the European community of nations. As long as the profound economic crisis in the former Soviet Union persists, West European leaders will probably focus less on residual CoCom controls and more on the "grand strategy" of large-scale economic assistance. As the former Soviet economy improves and if market institutions begin to take hold, however, residual controls will be increasingly viewed as an obstacle to continued Soviet progress. This is particularly true for items relevant to the development of a modern economy and an open polity, such as computer networks and telecommunications. West European officials view reform in the former Soviet Union as a long-term process and the

53. An official with the Pentagon's Defense Technology Security Administration (DTSA), when reminded in an interagency struggle that President Bush personally ordered the core list compromise, responded that the president "doesn't speak for DTSA." See Robert Kuttner, "How 'National Security' Hurts National Competitiveness," *Harvard Business Review,* January–February 1991, p. 148.

liberalization of controls as a means to facilitate that process rather than as a "reward" for its successful completion.[54] Patience with CoCom will wear even thinner if West Europeans continue to perceive controls as a hindrance to their ability to assist their East European neighbors in the tasks of pollution control and nuclear safety.

Indications that West European governments planned to abandon the strategic embargo "prematurely" would leave U.S. officials with two undesirable options. In the interest of political harmony, they could acquiesce and thereby undertake (as they did in 1954, 1958, 1982, and 1990) yet another strategic retreat, in this case the ultimate one. The U.S. government could subsequently seek, as a fallback with obvious problems, to coordinate bilaterally with other governments as a way to protect the most sensitive items. For example, given Japan's formidable capabilities, one could easily imagine a U.S. effort to negotiate a bilateral arrangement to restrict critical technologies. One could just as easily imagine the deal breaking down, as an increasingly assertive Japanese government refuses to accept the handcuffing of its firms while some in Europe forge ahead in the former Soviet market.

Alternatively, U.S. officials could resist and seek to preserve multilateral cooperation by employing the coercive tactics adopted by the Reagan administration in the early 1980s. Those tactics, recall, were somewhat successful in the context of CoCom's strategic embargo, but they were clearly unsuccessful in the case of economic warfare and the pipeline. A replay in the 1990s would probably resemble the outcome in the latter case—a costly confrontation, with the United States ultimately defied.[55] The reasons are numerous: the newfound prestige and assertiveness of the European Community, by virtue of its 1992 program; the growing perception in Europe that the United States is no longer crucial to the solution of European problems; and the lack of a military threat to generate even minimal sympathy for the U.S. position. In addition, the United States would lack an important source of leverage it enjoyed in the early 1980s, one that was employed effectively by the Defense Department: the ability to offer or deny lucrative military contracts in exchange for export control cooperation.

To be sure, West European and Japanese firms would remain dependent on the United States as a high-technology market and source of

54. See Reinecke, "Political and Economic Changes in the Eastern Bloc and Their Implications for CoCom," pp. 43, 55.

55. In 1991, U.S. officials floated the idea of a two-tier CoCom: those members with adequate enforcement would receive U.S. technology freely, whereas those without would face licensing requirements. Greece and Turkey, ostensible targets, reacted furiously and threatened to leave CoCom if such a practice were adopted. See *Export Control News* 5 (April 29, 1991), 5. After 1992, that stance would be backed by the entire European Community, because all barriers to trade within the EC, including differential licensing requirements, will be eliminated. Indeed, member governments may be compelled to adopt a common European export control policy. See ibid., pp. 80–101.

supply, though to a lesser degree than a decade earlier. Severing those links would be costly to other CoCom members but obviously to the United States as well. Indeed, it is questionable whether U.S. officials would even embark on such a course in light of the foreseeable repercussions.[56] In addition to sparking a political conflict undiluted by the existence of a common security threat, the United States would risk encouraging the fragmentation of the global economy into three "techno-nationalist" blocs, an outcome its economic diplomacy assiduously sought to avoid during the 1980s.

Whether, and for how long, to maintain the strategic embargo will probably be decided collectively by members of the European Community. In that context, the new Germany is positioned to be the key player. This role may be less apparent in the short term, as German leaders remain preoccupied with the absorption of the former East German economy and as they maintain a low profile on export control policy because of revelations that German firms exported chemical weapons technology to controlled destinations in the Middle East. Over time, however, unification will provide a boost to Germany's political prestige and already formidable economic resources. Indeed, by 1990, Germany emerged as the strongest member of the European Community and as a European power situated geopolitically in the center of the continent, with a stake in developments in both the West and the East.

Germany's economic interest in the East has deep historical roots and was nurtured to the extent possible during the cold war. As the construction of a new Europe is undertaken over the next several years and Russia and Germany seek to redefine their roles and relationships, either may wish to introduce the issue of the strategic embargo into what promises to be a complex equation. It is unlikely that the United States would be able to prevent, or decisively influence, a German-Russian arrangement that included the removal or severe attenuation of residual national security controls.

The trade relationship between Russia and the former East Germany may push events in that direction. As of 1990, 20 percent of the industrial work force in the former East Germany was employed producing goods and equipment for export to the Soviet Union. The technological sophistication of those exports will increase as unification progresses, and some of the items (e.g., computers and precision instruments) will fall under CoCom control. To remain faithful to CoCom, Germany will be required either to prohibit such exports, thereby violating German-Soviet trade treaties that both sides have pledged to observe, or to sell less sophisticated products, thereby requiring separate production lines

56. See National Academy of Sciences, *Finding Common Ground*, p. 123.

for the former Soviet Union and the West.[57] Neither option is likely to garner much support among German government or industry officials, particularly while Soviet troops remain in the former East Germany.[58]

Given the core list compromise and the unpredictability surrounding politics and economics in the former Soviet Union, West European governments may be prepared to tolerate residual controls for quite some time. For as long as other members are willing to maintain the strategic embargo in principle, whether it persists in practice will depend on U.S. leadership. In a post–cold war environment, even governments willing to cooperate will probably have little enthusiasm for controls and little patience for the economic sacrifices and political compromises that accompany the effort. Consequently, the United States will have little margin for error. It must remain on its "best behavior" to sustain what is likely to be the grudging acquiescence of others for the continuation of even a drastically streamlined embargo.

For example, U.S. officials must abide by their promise to update the control parameters of the core list. The initial core list was regarded by other CoCom governments, and U.S. firms, as a necessary first step, not a panacea.[59] If the United States reverts to a business-as-usual approach to streamlining the remaining controls, cooperation in CoCom will be threatened.

Similarly, it will be crucial to set a domestic example. Suspicions that U.S. firms have been granted favorable treatment to export controlled items to the former Soviet Union may overwhelm the ability of other CoCom governments to restrain their firms. Those governments can also be expected to have even less patience than they did during the 1970s and 1980s for the injection of foreign policy considerations into the control process.

Finally, the United States must follow through on its plan to eliminate reexport controls in the CoCom context. West European governments will not only continue to object in principle but, more important, will also refuse to tolerate them as a practical matter. The removal of internal barriers to trade in accordance with the 1992 program will leave little

57. See Reinecke, "Political and Economic Changes in the Eastern Bloc and Their Implications for CoCom," pp. 61–65. A compromise would be for Germany to sell controlled items to the former Soviet Union under strict verification procedures, perhaps including on-site inspections. The National Academy of Sciences advocates this general approach to CoCom in the 1990s. See *Finding Common Ground*, pp. 45–46, 111.

58. Before unification, West German economics minister Helmut Haussman noted that "East Germany can't immediately end its military cooperation with the Soviet Union; . . . if the East Germans have to deliver spare parts, they will." *Wall Street Journal*, June 22, 1990, p. A5. Soviet troops are scheduled to depart completely by 1994; until they do, Soviet officials retain a source of leverage in relations with Germany.

59. See Dan Hoydysh, "Computer Networking—The Core List's Achilles Heel?" in *Export Control News* 5 (April 29, 1991), 5–7. At a workshop sponsored by the National Academy of Sciences in September 1991, one industry official noted that the initial core list "just removed the stuff no one could agree to keep on with a straight face."

room for European firms to observe already contentious U.S. restrictions on the reexport of U.S.-origin items across the national boundaries of EC members.[60]

In short, the post–cold war survival of CoCom, not just its effectiveness, requires the United States to do what it evidently could not during the cold war era—undertake, fully and consistently, its traditional leadership responsibilities. Moreover, even flawless leadership will not guarantee CoCom's survival; that will ultimately depend on West European preferences.

New Threats, Old Patterns?

Finally, regardless of how long CoCom lasts, export controls will remain an important public policy issue during the 1990s. The early part of that decade witnessed a major shift: as Western governments pared down East-West controls, they mobilized to obstruct the flow of nuclear, chemical, and ballistic missile technologies to belligerent governments in the developing world. The experience of CoCom and the arguments of this book should be instructive in understanding the emerging effort to strengthen nascent regimes to meet new threats in the North-South arena.

Consider, for example, the key role of common interests and shared perceptions of threat. The Persian Gulf War of 1990–1991 was a major catalyst, prompting Western governments to strengthen existing informal regimes such as the Nuclear Suppliers Group, the Australia Group, and the Missile Technology Control Regime.[61] The ability of Saddam Hussein to launch ballistic missiles, and threaten the use of chemical and, ultimately, even nuclear weapons, conveyed a sense of urgency to Western governments regarding the potential for proliferation to create national security risks. The fact that Western countries each contributed in some way to the development of Iraq's capabilities reinforced the belief that more effective export controls were necessary.

The experience of CoCom suggests, however, that the existence of

60. See *Europe 1992*, Report of the Advisory Committee for Trade Policy and Negotiations, November 1989, pp. 33–41, and Reinecke, "Political and Economic Changes in the Eastern Bloc," pp. 80–101.

61. The Nuclear Suppliers Group was formed in 1974 and currently includes twenty-six countries. The Australia Group is an informal organization of twenty Western countries formed in 1984 to control chemical warfare agents. The Missile Technology Control Regime was formed in 1987 and as of 1991 had fifteen members. For background, see Zachary S. Davis, "Non-Proliferation Regimes: A Comparative Analysis of Policies to Control the Spread of Nuclear, Chemical, and Biological Weapons and Missiles," *CRS Report for Congress* (Washington, D.C.: Congressional Research Service, April 1, 1991). Some argue that a "super regime" is needed to coordinate controls across these areas. See Paul Freedenberg, "Export Controls in the 1990s: The Proliferation Dilemma," *Siemens Review*, April 1991, 4–9. CoCom is an unlikely candidate for such a role, given its association with the cold war and the fact that the cooperation of its two principal targets, the Soviet Union and China, will be needed for the new regimes to be effective.

common interests and shared threat perceptions will not automatically lead to enhanced cooperation. Leadership is required to coordinate controls of appropriate scope and coverage and to assure participating countries that cooperation will not leave them economically disadvantaged. As in the East-West context, the United States has attempted to provide leadership in fashioning North-South controls. The Bush administration, seeking to capitalize on its victory in the war and lead by example in the chemical weapons area, launched the Enhanced Proliferation Control Initiative (EPCI) in March 1991. The EPCI established strict unilateral controls on chemical weapons precursors and production equipment and technology. U.S. officials subsequently used the new U.S. controls as a benchmark in an effort to strengthen the controls of Australia Group members. Similarly, the administration revised and expanded the U.S. Nuclear Referral List, and it used that as the basis for an attempt to extend the coverage of multilateral controls over nuclear proliferation items.[62]

The CoCom experience also reveals that one cannot take for granted that the United States will lead consistently and effectively. Even at an early stage, it was apparent that the problems which plagued U.S. leadership in the East-West context might recur. By 1991, the Commerce, State, and Defense Departments were engaged in a bitter dispute over which agency was best suited to formulate and administer proliferation controls. The debate was a replay of that which took place following the Soviet invasion of Afghanistan, to the point of commerce officials once again being accused of selling out U.S. security and building the military power of an adversary, in this case Iraq, in order to advance U.S. economic interests by relaxing export controls. For their part, some members of Congress revived the post-Afghanistan idea of consolidating all export control authority in a single agency, so as to overcome the destructive and seemingly interminable interagency squabbling.[63]

More important, the tendency of the United States to seek restrictions on items widely available in international trade was similarly obvious. U.S. officials approached the proliferation problem with the same zeal for overcontrol that had characterized their early efforts in the East-West arena. Proposed controls included validated licensing requirements for *any* item that an exporter "knows" will be used in the design, development, production, stockpiling, or use of chemical or nuclear weapons. This plan would force exporters to obtain the knowledge of ultimate end use that was previously the responsibility of government, and it would indiscriminately cover the export to many destinations of such seemingly innocuous items as slide rules, thumbtacks, and typewriters. "Use" would

62. By the summer of 1991, U.S. officials had made progress in both areas. See *Export Control News* 5 (June 28, 1991), 16–19.
63. Ibid., 6–9, and (July 29, 1991), 7–8.

supersede "significance" as the governing export control criteria. In addition to exporters, freight forwarders, banks, and cargo insurers would be covered by the regulations.[64]

In multilateral negotiations, it was quickly evident that the United States was seeking to go well beyond what other governments were prepared to accept in the way of additional nuclear and chemical controls; it proposed, for example, restrictions on widely available computers and machine tools recently removed from the CoCom list.[65] The potential for multilateral cooperation was probably at its strongest in the immediate aftermath of the Persian Gulf War. As the memory fades and threats become more diffuse, economic interests are likely to reassert themselves in other trade-dependent Western states. This possibility would leave the United States, as it was so often in the East-West context, with the problem of whether to maintain national controls beyond the multilateral consensus.

If history is a guide, the struggle in the United States between business and government that has taken place since the 1960s over export control policy will continue. As of 1991, executive officials were in the process of transferring items (including television descramblers) decontrolled by CoCom to the Munitions and Nuclear Referral Lists; extending controls on items on the latter list to all destinations; and seeking new authority to notify exporters, even orally, that a previously uncontrolled item now required a license because it had been discovered to have a chemical or nuclear weapons use.[66] U.S. industry, for its part, was seeking to figure out how to avoid gaining the reputation of being an unreliable supplier in the North-South context and how to dispute the further expansion of executive authority without appearing to be soft on proliferation. Industry could take some comfort in the fact that the new export control debate was taking place not in an era of uncontested U.S. economic hegemony but at a time in which globalization trends made it imperative that the United States export successfully in order to remain competitive. Whether that reality leads the United States to adjust the balance between economic and national security interests in its export control policy remains to be seen.

64. Ibid. (August 25, 1991), 8–14.
65. Ibid. (June 28, 1991), 16–19.
66. Ibid. (August 25, 1991), 10, and (September 25, 1991), 25.

Index

Abbott, Kenneth W., 32n, 256n, 282n
Acheson, Dean, 71n, 75, 91
Adler-Karlsson, Gunnar, 5n, 6n, 8n, 21n, 69n, 72n, 82–84, 94, 101n, 109n, 110, 114n, 118–19, 120n, 129–30n, 197n, 312n
Administrative exceptions, 111, 116, 122n; and no exceptions policy, 230
AEG-Kanis, 248–49, 251
Afghanistan invasion: and export credit restrictions, 229; and grain embargo, 223, 228–29, 234; interagency review of technology controls, 224–28; and Kama trucks, 221; U.S. response, 220, 223–28; West European response, 228–33
Airbus, 298, 306
Alcatel, 1–3, 332
Alcoa, 232–33
Allison, Graham, 315n
Alsthom Atlantique, 251
Amdahl, 208
American Motors Corporation, 135
Anglo-French list, 75, 79
Arab oil embargo, 53, 248
Armco, 232–33
Asea, 291
Ashdown, Paddy, 296n
AT & T, 1n, 177, 293
Australia: decision to join CoCom, 294
Australia Group, 342–43
Austria: and Bucy Report, 197; cooperation with United States, 185n; export control legislation, 291–92; target of intra-Western restrictions, 288–92

Baer, George W., 53n
Baldwin, David, 7n, 8n, 21n, 27n, 39n, 46n, 49n, 53n, 60n, 119n, 156n, 261n, 282n
Ball, Desmond, 2n
Ballistic missile proliferation, 11
Baranson, Jack, 208
Battle Act, 83–89, 210
Belgium, 3
Berman, Harold J., 73n, 102n, 116n, 118n, 124n
Bertsch, Gary, 8n, 15n, 27n, 83, 119n, 145n, 174n, 275n
Block, Fred, 64n
Boeing Corporation, 157, 186
Bonker, Don, 301–2
Bottleneck effect, 45, 50–51, 56, 128
Bracken, Paul, 43n
Brady, Lawrence, 235
Brodie, Bernard, 41n, 90n
Brown, Harold, 199n, 211, 221
Bryant Grinder case, 171–73
Brzezinski, Zbigniew, 153, 204
Bucy, J. Fred, 186, 191, 195–96, 203–6, 209, 215
Bucy Report, 35, 45, 174n, 226–27, 286; and alliance coordination, 216–19; and Defense Department role, 198–206, 211–12; implementation of, 199–216; interpretation of, 187–99; and non-CoCom suppliers, 197–98; and reexport controls, 195–98, 218; support for economic warfare, 192–94, 203–6; support for intra-Western restrictions, 195–98; support for strategic embargo, 188–

Bucy Report (*cont.*)
  92, 199–206; West European views,
  217–18

Cable and Wireless, 298
Cahill, Kevin, 295n, 297n
Cannon Amendment, 83, 86–87
Carlucci, Frank, 308
Carr, E. H., 39n
Carrick, R. J., 8n, 16n, 22n, 118–19,
  209n, 218
Carter, Jimmy, 153–56, 203–6, 220
Carter Doctrine, 220–21
Casey, William, 240, 297n
Caterpillar Tractor, 240, 245, 250
Central Intelligence Agency, 72, 127–28,
  183–84, 267–68; tracking of West
  European firms, 297
Chemical weapons proliferation, 11
Cheysson, Claude, 260, 262n
China Committee, 98–99
China differential, 98–100, 110, 114
Chincom. *See* China Committee
Churchill, Winston, 93, 96
Chuthasmit, Suchati, 72n, 101n
C. Itoh, 304
Clancy, Tom, 55n
CoCom. *See* Coordinating Committee
CoCom-comparable controls, 292–93
Cohen, Benjamin J., 252n
Commodity Control List, 141, 145, 147;
  and MCTL, 214
Compartmentalization, 241–42
Comprehensive operations license, 300
Computers, 117, 118n, 192n; and Bucy
  Report, 194; Defense Department role,
  282–83; and 1978 list review, 181; and
  1982–84 list review, 269; and 1990
  liberalization, 334; post-Afghanistan
  proposals, 231; Processing Data Rate,
  269n, 334n; and safeguard program,
  189
Consultative Group, 80, 109n, 114
Continental System, 40–41
Contraband rules, 47–48
Cooper, Julian, 326n
Coordinating Committee (CoCom):
  administrative burden, 99–100, 115,
  280–81, 286–87; authority of delegates
  in Paris, 178; budget, 5, 275; China
  Green Line, 279–81, 332; common
  standard of effective protection, 336;
  comparison of 1954 and 1990 revisions,
  335–36; confidentiality, 6, 80–81;
  conflicts among members, 2–4, 28–33,

79, 93–100, 102, 109, 116, 175–84,
  231–32, 244, 278–87, 307–9; control
  list, 12, 80, 91–92, 109–10, 218–19,
  230–31, 306–7; cooperation, 12, 15–
  33; core list, 13n, 334–35; criteria, 69–
  70, 92–93, 114–15, 192, 307; defense
  participation, 81–82, 275–76; during
  detente, 170–85; effectiveness, 12, 15–
  17, 99–100, 170–85, 196, 281, 295–
  98, 325–27; exceptions, 16–17, 88,
  99, 110–11, 174–79, 279–80; formation
  of, 78–82; future of, 336–42; as
  "gentlemen's agreement," 19, 81, 105,
  277; high level meeting of 1982, 243–
  44, 268; high level meeting of 1990,
  331–36; impact on cold war victory of
  West, 325–27; institutional weakness,
  275, 277; as international regime, 5–6,
  15–24, 78–82, 297, 319–20n, 336; and
  intra-Western trade, 5, 11, 33–36, 101–
  5, 121–24, 195–99; January 1988
  agreement, 299–307; key role of
  Germany, 340–41; list coverage and
  interpretation, 16, 99, 109, 179–82;
  membership, 5n; military subcommittee,
  276; and nuclear safety, 333, 339;
  proscribed destinations, 5n; relationship
  to NATO, 6n, 81; reverse China
  differential, 279–81; role of Europe
  1992 project, 339, 341–42; scholarly
  attention, 7–8; special relationship with
  Poland, Hungary, and Czechoslovakia,
  335; strategic assumptions, 328–30;
  strategy, 12; streamlining, 16–17,
  22, 180–82, 284–87, 307–8, 331;
  unanimity principle, 22, 331; and U.S.
  leadership, 10, 18–33, 89, 92, 106, 113–
  18, 170–85, 277–87, 307–9, 320–25,
  341–42; and U.S. power, 9–10, 23–26,
  296n; and U.S. sanctions against other
  members, 196–97, 210, 218, 339; and
  U.S. trade policy, 10–11. *See also*
  Economic warfare; Enforcement;
  Export controls; List reviews; No
  exceptions policy; Non-CoCom
  suppliers; Strategic embargo; Tactical
  linkage; Third country effort; Toshiba
  Machine-Kongsberg case
Crawford, Beverly, 8n, 83n, 126n, 254n,
  261n, 312n
Creusot-Loire, 166, 232, 248, 258–59
Critical technologies, 117–18, 200, 204–6,
  218–19
Critical technologies approach, 200. *See
  also* Military Critical Technologies List

Grosser, Alfred, 122n
Gufstafson, Thane, 45n, 46n, 52n, 172n, 191, 215

Haggard, Stephan, 323n
Haig, Alexander, 240, 243, 245, 255, 315
Hanson, Philip, 8n, 30n, 40n, 45n, 268n, 326n
Hardt, John P., 15n, 56n, 146n, 262n
Harriman, Averell, 67
Heckscher, Eli, 40n
Hegemony, 9–10, 18
Hewett, Edward A., 237n
Hirschman, Albert, 9n, 54, 58n
Holtzman, Franklin, 56n
Hong Kong, 288
Horizontal escalation, 239–40
Houston Summit, 315, 317
Hufbauer, Gary C., 8n, 29n
Hughes Aircraft, 307
Hughes Tool Company, 263
Hull, Cordell, 66
Hungary, 331–32
Huntington, Samuel, 54, 153

IBM letter, 289, 296
International Business Machines (IBM), 135, 208, 289, 296
International Computers Limited (ICL), 145n, 299; study of "technological imperialism," 296–97; target of intra-Western restrictions, 289–90
International Harvester case, 239
Ikenberry, John, 10n, 12n, 312n
Import Certificate/Delivery Verification System (IC/DV), 103–4, 111, 182–83
India, 5, 293
Indonesia, 295
Integrated circuits, 51, 110, 115, 179–80, 183–84, 191, 196
International regimes, 9–10. *See also* Coordinating Committee
Intra-Western trade, 5, 11, 36; and Bucy Report, 195–99, 208–13; Carter administration report, 212–13; Defense Department views and authority, 211–12, 288; during the 1960s, 121–24; and H.R. 3216, 209; impact on CoCom, 295–98; and January 1988 agreement, 306–7; Reagan administration restrictions, 287–99; and scientific meetings, 290; and Siberian pipeline, 264–65; as source of leverage, 288; and third country effort, 290–95, 306, 308, 321; and U.S. economic interests, 208,

298–99; U.S. industry views, 104, 210–11; and West European enforcement, 101–5. *See also* Coordinating Committee; Non-Cocom suppliers; Reexport controls; Technology transfer
Izyumov, Alexei, 329n

Jackson, Henry, 203, 210, 215
Jackson-Vanik Amendment, 53, 149–53
Jacobsen, Hans-Dieter, 8n
Japan: assistance to Soviet Union, 316; and China differential, 98; dry dock case, 180, 268; Foreign Exchange and Foreign Trade Control Law, 305; and Siberian pipeline, 256; supercomputer restrictions, 290. *See also* Toshiba Machine-Kongsberg case
Jentleson, Bruce, 8n, 31n, 54n, 55n, 73n, 83n, 89n, 128–30n, 152n, 236n, 245n, 255n, 261–62n, 312n
Jervis, Robert, 19n, 23n
Jewish emigration. *See* Jackson-Vanik Amendment
John Brown, 248, 251, 259

Kama River truck plant, 145, 166, 171–72, 221, 227–28
Kapstein, Ethan, 7n, 311n, 322n, 329n
Katzenstein, Peter, 10n, 26n
Kem Amendment, 83, 86–87
Kendall, Donald, 159n
Kenen, Peter, 231, 232n
Kennan, George, 67
Keohane, Robert, 9n, 12n, 17n, 56n, 78n, 319–20n
Keynes, John Maynard, 55
Keystone equipment, 190
Kindleberger, Charles, 9n, 18n
Kissinger, Henry A., 144–45, 148, 153, 168
Klitgaard, Robert, 49n, 161n, 192n
Knorr, Klaus, 41n, 43n, 83
Kongsberg Vaapenfabrik. *See* Toshiba Machine-Kongsberg case
Korean War, 14, 89–90
Krasner, Stephen, 5n, 9n, 17n, 26n, 44n, 64n
Kurth, James, 64n

Labbé, Marie-Hélène, 272n, 275n
Lacorne, Denis, 315n
Lake, David A., 9n, 10n, 12n
League of Nations sanctions, 53
Legvold, Robert, 56n, 220n
Lenway, Stephanie, 8n, 83n, 261n, 312n

Cornell Studies in Political Economy

EDITED BY PETER J. KATZENSTEIN

Library of Congress Cataloging-in-Publication Data

Mastanduno, Michael.
   Economic containment : CoCom and the politics of East-West trade /
Michael Mastanduno.
      p. cm. — (Cornell studies in political economy)
   Includes bibliographical references and index.
   ISBN 0-8014-2709-6. — ISBN 0-8014-9996-8 (pbk. : alk. paper)
   1. North Atlantic Treaty Organization. Coordinating Committee on
Export Controls.   2. Export Controls—International cooperation.
3. National security—International cooperation.   4. East-West trade
(1945–)   5. Technology transfer—Communist countries.   I. Title.
II. Series.
Hf1414.5M27   1992
382'.09171'301717—dc20   92-52766